Inspiring air

A history of air-related science

Pere Grapí
CEHIC, Universitat Autònoma de Barcelona, Spain

Series on the History of Science
VERNON PRESS

www.vernonpress.com

In the Americas:
Vernon Press
1000 N West Street,
Suite 1200, Wilmington,
Delaware 19801
United States

In the rest of the world:
Vernon Press
C/Sancti Espiritu 17,
Malaga, 29006
Spain

Series on the History of Science

Library of Congress Control Number: 2019934451

ISBN: 978-1-62273-738-3

Also available:

978-1-62273-614-0 [Hardback]; 978-1-62273-655-3 [PDF, E-Book]

To Avel·lina

For rapidity of working and delicacy of measurement eudiometers leave nothing to be desired; indeed, as regards delicacy, it may be doubted whether amongst all the apparatus for measurement in this exhibition, there is one which can, like some of these instruments, give a distinct value, in weight or volume, to the one fourteen-millionth part of a gram of matter. Their drawback is their fragility, and any improvements tending to diminish this would doubtless be welcomed by chemists.

Edward Frankland's address to the Section of Chemistry in the Conferences held in connection with the Special Loan Collection of Scientific Apparatus, 1876.

Table of Contents

List of Figures

List of Tables

Acknowledgments

The gestation of this book began in the 2003-2004 academic year when I was the advisor of a research paper written by three of my final year high school students. That research was about making a replica of Volta's pile and reproducing some of the accompanying historical experiments. At that time, I was involved in the approach of the history of science to science education for secondary school students and science school teachers, and the replication of historical instruments and experiments had proved to have didactical value.

From that time onwards, I became increasingly interested in exploring the emergence of electricity, in the form of sparks, in the practice of chemistry. I therefore felt obliged to familiarize myself with the contributions of prominent figures such as Beccaria, Volta, Monge, Berthollet, Lavoisier, Priestley, Cavendish or van Marum. At the end of the eighteenth century, Volta's eudiometer was the most emblematic representation of the electrification of chemistry. The examination of Priestley's contribution in this field led me to a deeper knowledge of another important non-electrical eudiometrical device, the nitrous air eudiometer. My retirement from science teaching in 2012 provided me with the time required to complete this book. I hardly need to say that my first thanks go to Maria Rosa, Olga and Càndida, those three former students who were able to remake Volta's pile. Quite unwittingly, they gave me the opportunity to focus my attention on eudiometers instead of other instruments.

I also wish to express my appreciation to many others who have contributed to this project. Bruno Cavalchi, Francis Gires, Jørgen From Andersen, Antoni Quintana, Marta Quintana and Encarna Aihcart gave me complete freedom to photograph, or have provided me with photographs, of the eudiometers under their care. The librarians of the *Fons Històric de Ciència i Tecnologia* at the Universitat Politècnica de Catalunya, Barcelona, and of the *Biblioteca Panizzi*, Regio Emilia, who facilitated access to their documentary fund. Of course, I am also indebted to the published work of so many scholars quoted in this book. Special thanks are due to Agustí Nieto-Galan for his comments and suggestions to improve the final version of this work. Last but not least, I thank my wife Avel·lina for her support and understanding during the composition of this work, which detracted from the time we could otherwise have shared together.

Introduction.
Instruments and procedures
in practical chemistry

Sources and objects in the practice of chemistry

This is a work about eudiometers, a family of instruments originally devised to check the goodness (i.e. salubrity) of common air. The presentation of each eudiometrical device includes descriptions that are as close as possible to the textual sources (original publications, laboratory notebooks and monographs) and accompanied by the corresponding illustrations. Ideally, one should pay close attention to the following points when describing the functioning of any experimental device: material equipment, reagents, and the series of experimental procedures involved in that functioning. These procedures can be classified as manipulative (experiment setup, laboratory operations, step-by-step procedure, the practical skills required, the gathering of qualitative and quantitative data); cognitive (processing and interpretation of data, detection of errors and the setting of results) and communicative (experimenters' comments, the recording of incidents and the drawing of conclusions).

Unfortunately, verbal and non-verbal sources usually lack a complete description of the abovementioned particulars, largely due to partial or missing accounts of experimental procedures that were neglected or considered too complex to be recorded by writings or drawings. It has been claimed that the experience, understanding and skill provided by work on the replication of historical experiments can enrich the understanding of the textual sources. Replication may become unavoidable when an instrument no longer exists. Furthermore, replication may provide gestural knowledge that constitutes a resource in its own right, complementary to the textual and material sources.[1] In order to compensate for these

---

[1] Gooding, 1989, pp. 63-67; Sibum, 1995, pp. 27-28; Höttecke, 2000, pp. 344-347, 358; Usselman *et al.*, 2005, pp. 42-43. The replication of historical experiments and instruments is a field that has mainly been pursued from the perspective of the history of physics rather than that of the history of chemistry. Nevertheless, the following case studies deserve to be mentioned: the replication of Kirchhoff's and Bunsen's flame spectroscope (Henning, 2003); the replication of the Herme's tree, an alchemical recurrent image (Principe, 2000); the replication of Liebig's *kaliapparat*

unavoidable limitations of verbal and non-verbal sources, the descriptions of each eudiometrical device have been developed from original documents in the most understandable way for the reader.

Scientific practice and its related apparatus received little attention from historians of science during the twentieth century. The history of science was mainly regarded as the history of theory, according to which instruments were considered as materialized theories that might help to quantify concepts. However, in the 1990s, instruments ceased to be passive in the eyes of historians of science. For instance, in her paper on the integration of Lavoisier's calorimeter in his chemical system, Lissa Roberts was already adopting this approach of including scientific instruments as an essential factor in scientific practice for providing insights that remained unperceived if the history of science was regarded as an enterprise organized solely around the development of theory.[2] The landmark publication in 1994 of a volume of *Osiris* on *Instruments*, edited by Albert van Helden and Thomas L. Hankins, marked a pragmatic turning point in the interest of historians of science for the material culture of science. Instruments ceased to be perceived as unproblematic or uninteresting and were brought into focus within the history of science.[3] In 1985, Robert G. W. Anderson wrote a chapter devoted to *Instruments and Apparatus* for the book *Recent Developments in the History of Chemistry*, edited by Collin A. Russell. According to Anderson, things had gone particularly badly for the chemical apparatus in the field of the history of chemistry. The few existing uncritical accounts of chemical devices had not been sufficient for understanding the experimental practice of chemistry.[4] Anderson placed the works of the previous twenty years into various categories: general, instruments of particulars chemists, instruments with a particular function, catalogues of collections and instrument making. The fact was that by the end of the twentieth century very few historical studies of chemical instruments were regarded as

(Usselman *et al.*, 2005) and the replication of Lavoisier's and Laplace's calorimeter (Heering, 2005).

[2] Roberts, 1991, p. 200.

[3] van Helden & Hankins, 1994, pp. 2-3; Bensaude-Vincent, 2000, p. 189; Taub, 2009, pp. 339-340.

[4] Anderson, 1985, p. 217. Fifteen years later he was not much more optimistic yet, Anderson, 2000, p. 5.

recommendable for potential researchers in this area.[5] Nevertheless, the publication at the turn of the century of the collective work *Instruments and Experimentation in the History of Chemistry*, edited by Frederic L. Holmes and Trevor H. Levere, reflected to some degree the turning point in the 1990s and brought chemical instruments and experiments to the fore. Subsequent to these contributions, it is worth mentioning the publication in 2002 of the collective work *From Classical to Modern Chemistry. The Instrumental Revolution*, edited by Peter J.T. Morris, which dealt with the replacement of traditional methods in chemistry by automatic machinery in the 1950s and 1960s. With regard to eudiometry, there are very few indications or references to primary or secondary sources. Apart from the general *A History of Chemistry*,[6] the following works have been indispensable for me in the writing of this book: *An Historical Account of the Development of Methods for Determining Oxygen*,[7] *Eudiometrie, 1772-180*,[8] *Measuring Gases and Measuring Goodness*[9] and *Priestley's Quest for Airs and Ideas*.[10]

Chemical knowledge depended on a permanent merger of hand and mind; it was practice-laden as well theory-laden. Broadly speaking, experimental devices with their procedures are inexorably linked to the contemporary conceptual frameworks. Instruments can determine theory because they determine what is possible, and what is possible can condition what can be thought, i.e. theory.[11] Thus, instruments cannot be separated from changes in the conceptual frameworks or from the context in which they evolved. This central idea has inspired and guided the development of this book.

---

[5] The most relevant include Maurice Daumas' *Lavoisier, Théoricien et Expérimentateur*, Paris, Presses Universitaires de France, 1955; Ferenc Szabadváry's *History of Analytical Chemistry*, Oxford, Pergamon, 1966 (reprinted by Gordon and Breach Science Publishers, 1992); Jon Eklund's *The Incompleat Chymist*, Washington, Smithsonian Institution Press, 1975 and Frederic L. Holmes', *Eighteenth-Century Chemistry as an Investigative Enterprise*, Berkeley, Office for the History and Philosophy of Science and Technology, 1989.

[6] Partington, 1961-1970, Vol.3

[7] Benedict, 1912, pp. 3-68.

[8] Watermann, 1968.

[9] Levere, 2000.

[10] Boantza, 2013a, pp. 145-170.

[11] van Helden & Hankins, 1994, p. 4; Beretta, 2002, p. 23.

Each chapter of the book seeks to go beyond a mere inventory of eudiometers presented in chronological order. The aim is firstly to explore and comprehend how eudiometers work, the materials used in making them and the reagents employed in each eudiometrical test, all with especial attention paid to the experimental procedures involved over the course of the test. Secondly, as previously stated, eudiometers took on a life of their own in many different contexts; human and animal health, quantification, gas analysis, chemical theory, medical therapeutics, plant and animal physiology, atmospheric composition, chemical compound composition, gas lighting, chemical revolution, experimental demonstration and the chemical industry. Thus, in order to understand eudiometers, it is essential to stress the interplay between the instruments themselves and their contextual environment.

The first chapter is devoted to establishing the foundations of eudiometry and to presenting the origins of the nitrous gas (nitrogen monoxide) test; in particular, to the figure of Priestley, who in 1772 designed that first chemical eudiometrical test. Two years later, in Italy, Landriani and Fontana repeated Priestley's test and provided it with an instrumentalist format. These first steps in eudiometry were promptly followed by the contributions of Magellan and Gérardin with their own instruments. This not only set in motion a competition in the production of nitrous gas eudiometers but also led to the emergence of a number of problems that the test would have to surmount in the near future.

Some of these drawbacks could be overcome using hydrogen instead of nitrous gas. Chapter Two deals with the spark eudiometer conceived by Volta in 1778 and based on the detonation of a mixture of common air with hydrogen. Actually, the development of this eudiometer was a joint venture involving many actors and subsequently engendered diverse versions of the instrument, the latest ones being characterized by their modular structure. Volta and his eudiometer were involved in a core issue of the chemical revolution, the composition of the product (water) arising from the ignition that took place inside the instrument, one of the phenomena that revealed the potential of Volta's eudiometer as a gas mixture analyser.

The nitrous air eudiometer experienced its most impressive rise during the 1780s with the contributions made by Priestley, Fontana, Cavallo, Magellan, Cavendish, Lavoisier and, above all, Ingenhousz. The developments of the instrument, as well as the growing criticism it received during that decade, form the content of Chapter Three. The Dutch physician Ingenhousz emerged as the leading eudiometrist thanks to his efforts to standardize the instrument both materially and procedurally, as well as to make it profitable for research beyond that of determining the respirability of common air. This process of standardization involved the overcoming of many obstacles and uncovered problems inherent to the transmission of tacit practical knowledge between expert and novice practitioners of eudiometry.

Volta's first spark eudiometer was a portable instrument. The pressing need to determine the respirability of atmospheric air *in situ*, without having to transport the collected air samples from distant locations to the laboratories, gave momentum to the rise of portable eudiometers. Chapter Four is devoted to the early portable nitrous air and phosphorous eudiometers. The development of the latter was significant first of all for revealing the influence of theoretical constructs on the design of instruments, and secondly for the behaviour of eudiometers in the shift between competing chemical theories.

Portability restored simplicity to eudiometry. The complex nitrous gas and spark eudiometers existed alongside very simple phosphorous and alkaline sulphide eudiometers. The phosphorus eudiometer, designed by the Piemontesian chemist Giobert, and the calcium sulphide eudiometer, devised by the Catalan naturalist and chemist Martí-Franquès, are two exemplary cases of instrument simplicity. The conception and progress of instruments based on the alkaline sulphide eudiometrical tests constitute the subject of Chapter Five and makes explicit a case of transference of experimental designs between different kinds of eudiometers.

Furthermore, while eudiometers existed in the midst of competing theories, disputes between theories also led to the emergence of new eudiometers, a situation that is analysed in the sixth chapter, which addresses the eudiometer based on the slow combustion of phosphorous, ideated by Berthollet, and with a new fast combustion phosphorous eudiometer invented by Spallanzani. This latter type of phosphorus eudiometer underwent a remarkable development with the researches carried out on animal respiration by Séguin, who took advantage of Lavoisier's experiments on the combustion of phosphorous, which constituted a key factor in the chemical revolution.

In spite of the many of criticisms levelled at the nitrous gas test, it still continued to survive at the end of the eighteenth century with the adoption of different variants. Humboldt was a pioneer in this new mutational phase of the nitrous gas test. Chapter Seven begins by presenting Humboldt's iron sulphate variant of the nitrous gas test, and also deals with the disputes surrounding this test with the fast and slow combustion of phosphorus tests. Humboldt's contribution to eudiometry constitutes an important example of his frequently forgotten facet as a chemist involved in the working conditions in mining as well as in plant physiology. Humboldt's foray into eudiometry opened up a new path in the reinterpretation of the nitrous gas test. In the same direction, Davy's researches into the oxides of nitrogen provided the appropriate framework for a new variant of that test. Indeed, from 1800 onwards, the increasing involvement in eudiometry of British chemists, such us Davy, Hope, Pepys and Henry, was highly noteworthy.

Humboldt's investigations revealed the effect of the size of eudiometrical vessels on the results of the nitrous gas test, a fact that did not go unnoticed by Dalton in his reassessment of that test in regard to his deduction of the law of the multiple proportions of combination. Dalton's work also contributed in some way to restoring the original instrumentalist simplicity to the test. This provides the point of departure for Chapter Eight, which is devoted to examining the reappraisal of the nitrous gas and the Volta eudiometers. Gay-Lussac's research work in 1809 on the formation of the oxides of nitrogen thus constituted a further recreation of the nitrous gas eudiometer. Gay-Lussac himself, once again with Humboldt, definitively confirmed the concentration of oxygen in atmospheric air by means of Volta's eudiometer, a development that was to enhance the prestige of the instrument as a powerful gas mixture analyser.

The analysis of combustible gas mixtures for industrial purposes, such as gas lightening, coal distillation and cast-iron production, and for the research on animal respiration in the first half of the nineteenth century, placed Volta's explosion tube at the core of new eudiometrical apparatus. Regnault in France, and above all Bunsen in Germany, were the main developers of these new eudiometrical devices in which the control of the sources of error was of capital importance. The ninth chapter of this book addresses the shift undergone by eudiometry towards gasometry in the nineteenth century, thereby gaining greater respectability in analytical chemistry. Nevertheless, Volta's eudiometer continued to be used in other fields of chemistry, such as the determination of the chemical composition of organic compounds.

Presented in Chapter Ten, the conclusions consist of an analysis of the content of the previous chapters by focusing on the following questions: the experimental procedures and the material equipment of the different kinds of eudiometers; the adaptation of eudiometrical designs to the changes in chemical theories; the diverse contexts in which eudiometry was involved above and beyond air respirability, and the role played by eudiometers in the chemical revolution. This final chapter also provides the opportunity for addressing a matter that is common to a number of scientific instruments: the enjoyment of a new life as a didactic instrument.[12]

As explained at the outset, detailed descriptions of the functioning of the different eudiometers are provided in each chapter. For the sake of the readership, it has been considered appropriate that only the descriptions of those eudiometers that, in my view, are necessary for a more comprehensive understanding of the historical development of eudiometry should be included in each chapter. However, in order not to exclude completely those eudiometers whose descriptions were not regarded as entirely essential for the final work, they have found their place in the appendix of the book.

A further modest objective of the book is to contribute towards redressing the imbalance that exists in the history of chemistry regarding the attention given to the theoretical aspects of chemistry to the detriment of chemical practice and apparatus. This is quite a different situation to that found in other fields such as the history of physics or astronomy.[13] The first eudiometrical device was conceived in 1772, and from that year onwards different kinds of eudiometers begin to emerge and spread throughout Europe. Each device followed its own course of development; some were modified to ensure their continuation, while others disappeared almost completely in the nineteenth century, very few surviving in fields except research. In any event, by the early nineteenth century eudiometers had attained the status of standard laboratory instruments, in keeping with the growing use in chemistry laboratories of instruments traditionally employed in experimental physics, together with the evolution in the configuration of these laboratories. Even so, this is a story whose origins are found in the relocation of the apparatus that had already been used by artisans since the late seventeenth century.

The migration of instruments from artisanal to academic chemistry

Up to the middle of the eighteenth century, the equipment used in chemical laboratories resembled that found in the workshops of metalworkers, smelters, apothecaries, soap boilers or distillers. Indeed, academic chemists of the early eighteenth century made use of apparatus, utensils, techniques and materials employed by apothecaries and

---

[12] Scientific instruments were used not only for research and/or educational purposes but also for enteretainement, which often went hand in hand with education. In this way, amusement also provided an inducement to knowledge (Bensaude-Vincent & Blondel, 2008).

[13] Levere & Holmes, 2000, p. viii.

chemistry related artisans. The same kind of evaporation vessels, calcination dishes, mortars, pestles, filters, glass tubes, vials, retorts, alembics, pelicans and receivers used by academic chemists in the eighteenth century had also been employed by distillers of acids, spirit of wine and fragrant oils, as well as by apothecaries for preparing remedies.[14]

François-Gabriel Venel (1723-1775) was a distinguished example of this generation of academic chemists in France. He published more than seven hundred articles in Diderot and D'Alambert's *Encyclopédie*, among which the influential article *Chymie*, published in 1751, is particularly worthy of mention. Despite Venel's opposition to the attempt of experimental philosophers to introduce quantification into chemistry in order to make it a more rigorous discipline, he did not underestimate a chemical approach to the most instrumental dimension of physics, that is, experimental physics. The purpose was to make the practice of chemistry more respectable for experimental philosophers, and thereby gain the same recognition that some elegant and optical and electrical instruments already enjoyed. It was Venel's intention to link a part of the future of chemistry to its instrumental dimension, which is why, in addition to his seminal article *Chymie*, his contribution to the article *Instruments* in the same *Encyclopédie* must also be taken into account.[15]

At this point, however, some clarifications of terminology are required. In order to simplify matters, experimental philosophy may be regarded as a means of inquiry into nature within the domain of natural philosophy, as fueled by Newtonianism. Moreover, during the eighteenth century, natural philosophy became an alternative domain to natural history, the practice of which was grounded on the observation, description and classification of entities and phenomena in the natural kingdom. Unlike naturalists, experimental philosophers performed experiments that went beyond mere observation by either distorting nature to reveal what was imperceptible to the naked eye with the use of optical instruments (e.g. lenses, mirrors, prisms, telescopes and microscopes), or by intervening into nature in order to subject it to unnatural conditions (e.g. vacuum or electrification) and thereby disentangle its secrets. The air pump and the electrical machine would be two significant instruments of this latter type.[16] Both types of instruments came to be known as philosophical instruments as opposed to

---

[14] Holmes, 1789, pp. 17-32; Klein & Lefèvre, 2007, pp. 33-34.

[15] Venel, 1751, Vol. 3, p. 410.

[16] Hackmann (1989, pp. 39-40) distinguishes these types of instruments as "active" in opposition to the other "passive" instruments.

mathematical instruments.[17] In France, experimental philosophy was well represented by experimental physics (*physique expérimentale*), which should be understood in terms of miscellaneous topics belonging to physics (statics, hydrostatics, optics, electricity, pneumatics, or magnetism), exemplified by demonstration lectures using a diversity of instruments.[18] It has been argued that eighteenth-century chemistry belonged neither to experimental philosophy nor to natural history, but was rather a type of experimental history that bridged natural history, technology and experimental philosophy.[19]

The term "instrument" had a broad meaning in the practice chemistry of the eighteenth century. On the one hand, it included laboratory wares or equipment (*attirail*); that is, the assortment of furniture, laboratory utensils and tools (*supellex chimica*) used for the preparation of chemical and mechanical operations. On the other hand, the term "instrument" in the philosophical language of chemistry had a very different meaning from "agent", "cause" or "principle" of chemical changes. Fire or heat and solvents (*menstrua*) were therefore regarded as true active, universal and primitive instruments of chemists.[20] The expression "apparatus" was closely related to the term "instrument", and often both terms did not differ from one another. Furthermore, in the English-speaking world during the seventeenth and eighteenth centuries both terms were used interchangeably.[21] Incidentally, the entry *Appareil* in the *Encyclopédie* had no chemical meaning.

---

[17] These were instruments for measuring quantities (rulers, balances, clocks, barometers, thermometers, calorimeter, electrometer, and chemical balance) or for surveying, navigation and astronomy (sundials, compasses). According to Heilbron (1990, pp. 5-6), this class of instruments might justifiably be called "measurers". A second class, "explorers", includes instruments, such as the air pump and the electrical machine, which intervened in nature to produce artificial phenomena for demonstration, research and measurement. A third class, "finders", comprises those instruments that were employed as measurers and explorers for practical purposes (telescopes, chronometers or theodolites).

[18] Crosland & Smith, 1978, pp. 3-4; Crosland, 1994, pp. 22-25; Hankins & Silverman, 1995, pp. 3-5; Lundgren, 1990, pp. 258, 262.

[19] Klein & Lefèvre, 2007, pp. 21-28.

[20] Venel, 1766, Vol. 8, p. 802. Luigi Cerruti (1998, p. 42) coined the expression "epistemic trinity" for the interaction among instruments, reagents and techniques in the practice of chemistry. The fact that chemistry incorporated an arsenal of reagents along with the inseparable tandem of instruments and techniques marked a major difference from other experimental disciplines.

[21] Warner, 1990, p. 83.

For Venel there were two 'essentially chemical instruments': the aforementioned heat-fire and menstrua-solvents. Chemical instruments acted directly on the corpuscles of the constituent parts of bodies. The idea of menstrua-solvents and heat as true instruments of chemistry clearly reflects the influence of the instrumental theory of chemistry proposed by Boerhaave, which was moreover an alternative approach to the traditional system of the chemical principles (salt, sulphur, mercury as well as phlegm and earth) for interpreting chemical composition and changes.[22] Mechanical instruments were used in preparatory operations and acted on tangible, observable bodies consisting of aggregates of corpuscles. In general, these mechanical instruments were not regarded as true instruments of chemistry, but were frequently referred to as "tools" or "utensils", and even "apparatus". The most common of these were furnaces, vessels and lutes, among others. However, these components and their arrangement in a chemistry laboratory would undergo an evolution from mid-eighteenth century onwards. Two factors would be decisive in these developments; the interest aroused by the study of gases and the approach to experimental physics via chemistry.

The instrumental view of the chemical approach to experimental physics

The studies undertaken by the English experimental philosophers on "airs" had their effect on the instrumental ware of chemistry laboratories in the eighteenth century. One of the most striking examples of this growing interest in the different kinds of air was the article "Gas", written by the chemist Pierre-Joseph Macquer (1718-1784) in his *Dictionnaire de Chymie*. The entry in the 1766 first edition of the dictionary barely filled half a page, whereas in the 1778 second edition it took up nearly one hundred pages.[23] Among the latest instruments to emerge for the study of gases, Macquer highlighted the pneumatic trough. This was an apparatus that would become indispensable in a chemistry laboratory for research into gases. Other accessories such as siphons, retorts and small funnels accompanied these pneumatic troughs.

This second edition of Macquer's dictionary was also crucial in regarding the fact that chemistry laboratories were equipped with various machines and instruments that hitherto could only be found in the cabinets of physics as a natural result of the combined approach between chemistry and

---

[22] Powers, 2012, pp. 70-77.

[23] Macquer, 1766, Vol.1, p. 550; 1778, Vol. 2, pp. 240-407.

experimental physics.[24] These devices included mercury thermometers, hygrometers, barometers, pyrometers, balances with their weights, magnifying lenses, magnetised steel rods, a microscope, a pneumatic pump and an electric machine. The inclusion of these additions in Macquer's article on gases in the second edition of his dictionary was one of the reasons for its much greater length.[25] It is also worth noting the quantitative nature of some of these instruments, such as thermometers, hygrometers, barometers, pyrometers and balances, which reflected the growing quantification of chemistry. These instruments, made with the intention of providing quantitative results, actually contributed to the introduction of numerical measurements into chemistry.[26] With regard to the quantification of chemistry, the pneumatic kit of chemistry laboratories contributed substantially to making the balance an essential instrument of the chemical practice.[27] It was Lavoisier who followed up on Macquer's endeavours by foreseeing that the progress of chemistry was bound to adhere to the methodological standards of mathematics and experimental physics. And so it turned out, since experimental physics exerted a profound influence on Lavoisier's way of organizing, interpreting and presenting his chemical researches.

Just as it had done in the second edition of Macquer's dictionary of 1788, the expansion of studies on gases also left its imprint on the equipment used in chemistry laboratories. In Antoine-François Fourcroy's (1755-1809) article, *Appareil,* published in the second volume of the *Encyclopédie Méthodique,* he associated the changes in appearance of chemical apparatus to the increasing interest in the gaseous state.[28] In actual fact,

---

[24] Macquer described it as follows: 'Finally, as chemistry and physics are now only one science [...]'; 'Enfin, comme la chimie et la physique ne sont plus presentment qu'une même science [...]'. The term *"physique"* should be understood in the context of *la physique expérimentale.* (Macquer, 1778, Vol. 2, pp. 2, 6). Translations into English hereafter are by the author, unless stated otherwise.

[25] Macquer, 1778, Vol. 2, pp. 6-7. In the same article, *Laboratoire,* which appeared in the first edition of the dictionary, no mention was made of either the pneumatic trough or other philosophical instruments.

[26] Lundgren, 1990, pp. 260, 265.

[27] In this respect, it is worth remembering that Joseph Black was able to identify the fixed air (carbon dioxide) by detecting the loss of weight when heating magnesi alba (Bensaude-Vincent, 1992, pp. 219-220: Crosland, 2000, pp. 82)

[28] In 1780, Louis-Bernard Guyton Morveau (1737-1816) was commissioned by the publisher Panckouke to write the chemistry volumes for a new encyclopaedia, the *Encyclopédie Méthodique,* which was to be a thematically organized work. (It must not be forgotten that the articles in the *Encyclopédie* were arranged alphabetically,

the difference between the foregoing period in chemistry, considered by Fourcroy himself as the time when elastic fluids counted as zero in chemists' calculations, and the present state of chemistry did not lie so much in the creation of new devices, but rather in the redesign and conversion of the available equipment in accordance with the new insights into the nature of substances to be analyzed or combined, and the expected results.

Fourcroy understood any chemical device as a coupling of different parts or recipients, which could either be directly or by means of intermediate parts such as glass or metal tubes. Sometimes this connection was intimate – such as a bottle with its cap - although this was often not the case, and some external cement such as mastic was needed to make the device airtight. This special care taken to maintain control over the material environment of experiments has to be placed in the context of a quantifying spirit in the practice of chemistry. Chemists wanted to simplify operations in order to understand them better, to bring them under the control of the balance and to make this instrument a guarantee of their results. This idea led chemists to design devices in which losses could be minimized, thereby enabling them to obtain the quantitative results of their analysis and synthesis with much greater accuracy.[29] Unlike Venel, Fourcroy wrote no articles devoted to instruments in the fourth volume of the *Encyclopédie Méthodique*.

A new place to locate physico-chemical instruments in the chemistry laboratory

The instruments, such as balances, that Macquer wanted to move from the cabinets of experimental physics to the chemistry laboratory were not be stored within the laboratory itself, but in a dry place to prevent their deterioration due to the vapours released from most chemical operations.[30] All these instruments would eventually find their own place within the chemistry laboratories, and by the early nineteenth century they all had

---

irrespective of subject). Guyton completed only the first volume, because in 1790 he was elected to an important administrative post in Dijon. As a result, Fourcroy took over from Guyton for the task of completing the work. The second volume was published in 1792 and the third in 1796, while the fourth and fifth volumes did not appear until 1805 and 1808, respectively. After Fourcroy's death, Vaquelin was engaged to write the sixth volume, which was not published until 1816.

[29] Fourcroy, 1792, Vol. 2, pp. 350-351.

[30] Expensive items (balances) were kept ready to hand rather than in a separate room, despite the potentially damaging effects of fumes. Such an arrangement required space and not all the laboratories were large enough (Morris, 2015, p. 58).

their allotted location. The territorial demarcation of these instruments would be described in Fourcroy's article, *Laboratoire*, which he wrote for the fourth volume of the *Encyclopédie Méthodique* of 1805. He explained in great detail the configuration and equipping of what he termed a universal or polyvalent laboratory for all kinds of work and research, and appropriate for philosophical chemists as well as teachers and consultants.[31]

The main room in this polyvalent laboratory had to be equipped with water fountains in order to perform operations requiring a plentiful supply of water, access to a warehouse and also a yard to carry out open-air detonations and malodorous experiments, and an intermediate specific room adjoining the central room.[32] The great novelty of this initial description is the inclusion of the adjoining room. This room was particularly - but not exclusively – designed for instruments of physics consisting of metallic components that might be damaged by the acid vapours released in operations conducted in the laboratory itself. This adjoining room came to be regarded as an indispensable storeroom, extremely useful in a laboratory of great activity, and it would become an established feature of the chemical approach to experimental physics. The importance of this new room was evident in Fourcroy's detailed description of the main chemical recipients and tools that were kept there. Along with a long list of glass containers of all shapes and sizes (vessels, bottles, vials, cylindrical glassware, funnels, capsules, retorts, stills, round-bottomed flasks, bell jars, curved and capillary tubes), Fourcroy also included the instruments of physics grouped into different categories corresponding to the variety of disciplines belonging to experimental physics.[33] These categories were: electrical instruments (electric machines with a battery of Leyden jars, large electrophorus, electrometers and a Coulomb's balance); galvanic instruments (galvanic piles of approximately sixty plates of copper and zinc; silver, gold and platinum threads needed

---

[31] Industrial laboratories also existed for different manufacturing processes: pottery, ceramic, glass, enamel, foundries, metallurgy, dying, extraction (salt works) or paper.

[32] Specialized rooms apart from the main laboratory were already an important feature of early laboratories. Andreas Libavius envisaged different kinds of special rooms in the design of his "chemical house" in 1606, but this laboratory was never constructed (Morris, 2015, pp. 21, 32-33).

[33] In the main room of the laboratory - outside the intermediate site - copper and iron instruments could be found (stills, basins, mortars, sand baths, retorts, crucibles and other equipment for cutting, grinding and scraping), which for their size and/or regular use were not kept in the intermediate site, although this was where they were normally stored (Fourcroy, 1805, Vol.4, pp. 568-569).

for various applications of galvanism in chemistry, such as electrolysis); optical instruments (concave metal mirrors and mobile magnifying lenses with their pedestals); meteorological instruments (barometers, thermometers and hygrometers); pneumatic instruments (pneumatic pumps with their auxiliary vessels); magnetic instruments (a magnetic stone, artificial powerful magnets and magnetic needles); measuring instruments (simple and hydrostatic balances for high accuracy assays with platinum weights), and finally instruments for measuring and analyzing gases (copper gasometers, round flasks with tubes and taps for collecting and weighing gases, and eudiometers, especially those of Volta, Fontana, Guyton and Séguin). Having acquired its own place in the laboratory, it is clear that the eudiometer had become a standard instrument in chemical practice.

A large number of these instruments had certainly been discreetly transferred to the chemistry laboratories from the cabinets of physics, and therefore the impulse behind this trend was not to make their incorporation explicit, but rather to provide them with a physical space that guaranteed their role in chemical experimentation. However, some of these instruments may not have been found in the former cabinets of physics, since they were employed in galvanism and the measurement and analysis of gases.[34] Firstly, galvanic instruments, which developed from 1800 after the discovery of voltaic electricity, had their origins in animal physiology and were included because they belonged to the new and separate science of galvanism. Furthermore, it was competent and methodical chemists and not physicists who were working most closely on galvanism. Secondly, gasometers were one of Lavoisier's (together with Meusnier) most spectacular contributions to the control and weighing the volumes of gases in experiments. Lastly, all eudiometers had their basis in a chemical reaction involving oxygen consumption.

---

[34] Cabinets of physics were not entirely impermeable to chemical equipment. Jean Nollet was inclined to include chemical demonstrations (experiments on fermentation and distillation and on the uses of burning lenses) in the courses on experimental physics he gave in the late 1760's and early 1770's. Sigaud de la Fond and Brisson also followed this tendency (Beretta, 2014, pp. 203-204).

1.

The nitrous air test
and the first eudiometers

The question of the goodness of air

The interest in determining the goodness of air originates in a hygienist tradition that attributed to atmospheric air the capability of bearing harmful exhalations for health. Putrefying animal remains, rotting vegetation and stagnant water, presumably exuded these poisonous exhalations. Bad environments engendered bad air, which then turned stinking, thereby explaining the spread of disease; firstly, by a climatic causality that predisposed living beings towards illness, then followed by the presence of casual miasma[1] (bad air) that could affect a weakened humoral constitution.[2] This was a belief rooted in the Hippocratic tradition that urged physicians to diagnose diseases by considering the influence of meteorological and environmental factors.[3]

The English physician Thomas Sydenham (1624-1689) brought this Hippocratic tradition up to date in the seventeenth century, when he distinguished between acute diseases due to noticeable qualities of the air, and epidemic diseases that generated hidden and unexplained changes in the atmosphere. The seasonality of some illnesses led him to conceive a project for the systematic natural history of disease that would provide the basis for a program of collective observation of the air. This program was developed during the eighteenth century with the aim of establishing correlations between the occurrence of disease and local climate

---

[1] From the ancient Greek word "Μίασμα".

[2] From Hippocrates (fifth century BC) to Galen (second century AD) humoral medicine reinforced the analogies between the four Aristotelian elements and the four bodily humours (blood, phlegm, choler or yellow bile and black bile), whose balance determined health. Hygiene was vital for maintaining a balanced constitution, and the role of medicine was to restore this balance when disturbed.

[3] Beretta, 1995, pp. 17-18.

characteristics.[4] In general, the main concern of medical practice was to determine the effects on the bodies of persons and animals of the airs that permanently or accidentally constituted the atmospheric air. In addition, physicians aspired to find those combinations of airs[5] that affected animal health both in order to lessen their effects and prevent them from causing disease.

Whatever approach is adopted to the goodness of air, it cannot neglect the theoretical views on the nature of aerial substances, i.e. the modern gaseous state. As Maurice Crosland states, the history of gases is anything but simple, since it consists of a combination of complementary confluences. Ancient chemical operations, distillations above all, provided evidence of the incoercibility of gases; that is, the inherent characteristic that revealed the extreme difficulty of collecting gases, which also raised an epistemological barrier: what cannot be handled may not be understandable. Apart from considering gases as products of chemical changes, and therefore as essentially incoercible, they were also perceived by natural philosophers as basically atmospheric air. Thus, Robert Boyle (1627-1691) characterized air for its transparency and its property of being compressible and expandable; in other words, its elasticity. The different gases, or kinds of air, would be atmospheric air in varying degrees of purity, containing saline or acid particles that altered its elasticity.[6]

This purity of atmospheric air was also associated with those exhalations or miasmas contained in common air, which not only made it more or less impure and inelastic but also harmful for health. In this respect, small animals such as mice and birds became the usual detectors of the goodness (i.e. respirability) of any air. Boyle used his air-pump, described in the *New Experiments Physico-Mechanical, Touching the Spring of the Air, and its Effects* of 1660, to show that air was necessary for respiration. He placed birds and mice in the air-pump receiver and observed how they expired when air was removed. During the Civil War, the city of Oxford became a refuge for Royalist exiles from London. Boyle arrived at Oxford in the 1650s, and among his assistants at this time were Robert Hooke (1635-1703) and John Mayow (1641-1679). The latter fully developed the nitro-aerial theory, conceived in an atmosphere of interest surrounding chemical questions that was encouraged by Boyle in his Oxford circle. The

---

[4] Hannaway & Hannaway, 1977, pp 182-183.

[5] For instance, the volatile alkali contained in the air latrines, smoke and fumes (tobacco, rubber, resins, etc.) and various perfumes.

[6] Crosland, 2000, pp. 80-86.

nitro-aerial theory assumed that air was a universal solvent containing certain nitro-aerial particles that were also found in the nitre used in gunpowder.[7] During burning and animal respiration in closed vessels, a fraction of the air composed of these nitro-aerial particles was used up. In his work *Tractus Quinque Medico-Physici* of 1674, Mayow contrived different devices for experimenting with mice (Figure 1.1) in an attempt to confirm his hypothesis that the air given out from the lungs of animals diminished in its elasticity due to the loss of its nitro-aerial particles.[8]

Figure 1.1 Two of the experiments by Mayow on respiration in a confined volume of air.

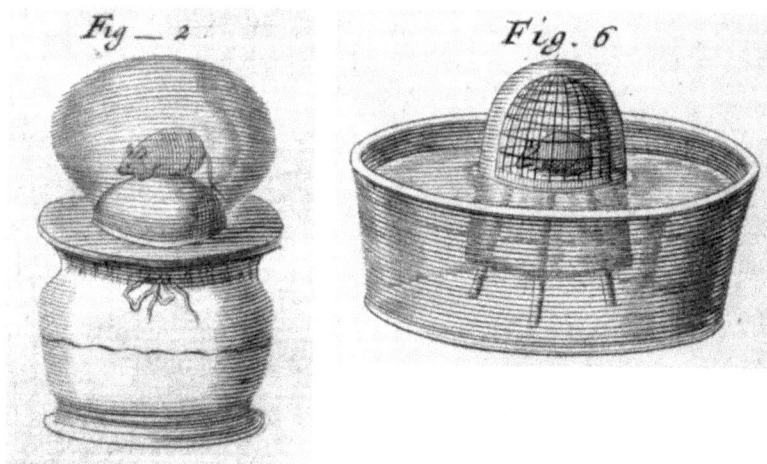

From John Mayow, *Tractus Quinque Medico-Physici* (London, 1674)

Stephen Hales (1677-1761) repeated Mayow's experiment on the portion of air absorbed by the breath of animals enclosed in glass vessels. This was the first of a number of experiments on the effects that the respiration of animals had on the air, the findings of which were published in his *Vegetable Staticks* of 1727. Hales held that all gases were no more than air whose most interesting property was not their composition but their elasticity. In accordance with Newton's ideas, air particles repelled each other, which provided an explanation for its elasticity. However, air could

---

[7] The theory was based on a gunpowder analogy that linked thunder and lightning to the explosion and flashing of gunpowder. Sulphur and nitre were the components of gunpowder, therefore, by analogy, a clap of thunder and its lightning could be explained as a reaction between nitrous and sulphurous particles in the air.

[8] Mayow, 1674, pp. 103-104, 169-170, plate V, *figs. 2* and *6*; 1907, pp. 72-73, 17.

also contain alien particles from acids or salts that attracted the air particles and altered their elasticity. Hales performed his experiments on respiration using different kinds of animals (mice, cats and dogs), including himself. From his experiments on respiration he deduced that some of the elasticity of the air was destroyed when it was inspired because a number of air particles were converted from an elastic-repulsive to a strongly attractive state.[9] Hales ideas prompted Alexandre-Julien Savérien (1720-1805) to make an instrument for measuring the goodness of air.

Figure 1.2 The main parts of Savèrien's *queynomètre*.

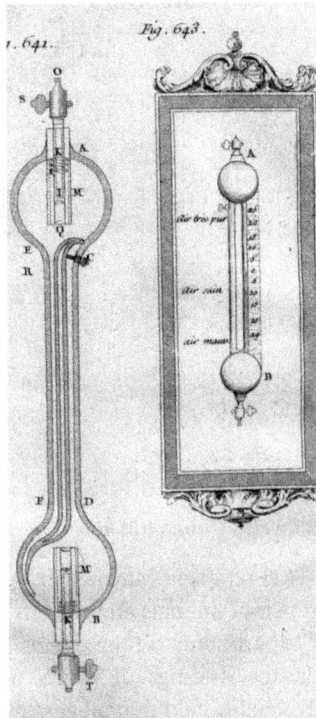

From *Dictionnaire universel de mathématique et de physique* (Paris, 1753), plate 19.

Savérien was a French engineer dedicated to research on navigation who first devised an instrument to measure the purity of the air. He was inspired by Hales' ideas of how air could lose its elasticity when it was impregnated by particles harmful to health. Since the air purity was conditioned by its elasticity, and this was proportional to its

---

[9] Hales, 1727, p. 139.

compressibility, Savérien designed an instrument, the *queynomètre*,[10] for the purpose of determining the compressibility of the air (Figure 1.2). This compressibility of the air, i.e. its purity, was measured on a scale ranging from "bad air" to "healthy air" to "pure air".[11] Savérien's innovation received little attention at the time but was retrieved after Landriani presented his own instrument in 1775, to which he gave the long-lasting name of "eudiometer", as well as in subsequent contemporary narrations on eudiometry.[12]

During the last quarter of the eighteenth century, the early speculations on airs as variations of atmospheric air evolved in terms of the phlogiston theory. Gases were not the only subtle and invisible fluids; phlogiston also appeared on the list with magnetism, electricity, heat and gravity. Chemists believed that phlogiston was an intangible fluid released in combustion and was impossible to handle. With regard to animal respiration, the prevailing belief was that expired air conveyed the phlogiston released from the lungs during breathing and evacuated it from the body. Phlogisticated air (i.e. saturated with phlogiston) was consequently unable to absorb more phlogiston and would not be appropriate for breathing, that is, it would be unhealthy or mephitic air. On the other hand, dephlogisticated air was deprived of more than its normal allocation of phlogiston and therefore was more wholesome.[13] In other words, the healthier an air, the less phlogiston it contained.

As Joseph Priestley (1733-1804) pointed out in his landmark paper of 1772, the use of animals in experiments to determine whether they were able to live in any kind of air occasioned some inconvenience; for example, keeping a sufficient stock of them. Additionally, care was needed to ensure that animals (mice for Priestley) were maintained under suitable conditions. Priestley's solution was to place the mice in glass receivers that were open at both top and bottom and then stand them on perforated plates to allow fresh air to enter. The receivers also contained a quantity of paper or tow, which had to be changed and the recipient washed every two or three days (Figure 1.3). Moreover, it was necessary to keep the mice at a controlled temperature, since either much heat or much cold could

---

[10] This term stems from two Greek words meaning "salubrity" (υγιεινό) and "measure" (μέτρο).

[11]Savérien, 1753, Vol. 2, pp. 468- 469, plate xix, figs. 641, 642 and 643.

[12] Servières, 1777; Sigaud de la Fond, 1779, pp. 200-201.

[13] Atmospheric air consisted basically of a mixture of phlogisticated or mephitic air (later called nitrogen) and dephlogisticated or vital air (later called oxygen).

kill them. Priestley usually kept them on a shelf over the kitchen fireplace. Furthermore, because they were required to pass through the water, they needed a considerable degree of heat to warm and dry them. Priestley must have breathed a sigh of relief when he inferred that nitrous air could provide the basis of a promising chemical test to ascertain the goodness of air.[14]

Figure 1.3 Priestley's glass receiver used to keep mice.

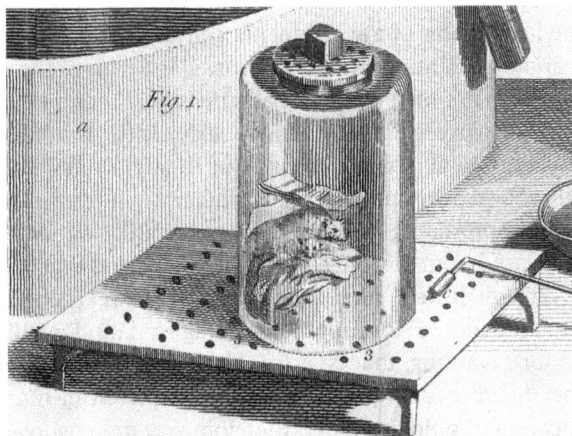

From Joseph Priesley, *Experiments and Observations of Different Kinds of Air* (London, 1775), plate to face the title.

The phlogiston theory virtually dominated the world of chemistry when the first chemical test to determine the goodness of air emerged. However,

---

[14] Priestley, 1772, pp. 214, 249-251; 1775, pp. 9-10, 114-115. Priestley furnished detailed information about the care taken of the mice in order to facilitate the replication of his experiments. He was conscious of the need to treat animals humanely (Priestley, 1775, p. 9):

'For the purpose of these experiments it is most convenient to catch the mice in small wire traps, out of which it is easy to take them, and holding them by the back of the neck, to pass them through the water into the vessel which contains the air. If I expect that the mouse will live a considerable time, I take care to put into the vessel something on which it may conveniently fit, out of the reach of the water. If the air be good, the mouse will soon be perfectly at its ease, having suffered nothing by its passing through the water. If the air be supposed to be noxious, it will be proper (if the operator be desirous of preserving the mice for farther use) to keep hold of their tails, that they maybe withdrawn as soon as they begin to show signs of uneasiness; but if the air be thoroughly noxious, and the mouse happens to get a full inspiration, it will be impossible to do this before it be absolutely irrecoverable.'

it was clear that phlogiston was an immaterial principle whose presence could not be measured directly. It was therefore necessary to find methods to provide gradations of the effects that phlogiston was supposedly able to produce, assuming that these gradations were proportional to the amount of hypothetical phlogiston that caused them. The case was to look for visual effects revealing how much phlogiston an air sample was able to absorb up to the point of saturation and thereby arrive at an indirect measurement of its original phlogiston content.

In the last quarter of the eighteenth century, doubts began to be cast on the simple nature of atmospheric air, and the idea that only a part of the common air was breathable started to gain acceptance. Then, in parallel with the hygienist approach to atmospheric air, a more analytical-quantitative approach to determine the composition of common air was adopted, mainly in regard to the uncertainty about its proportion of respirable or vital air. This approach involved procedures for determining the vital air content of the air in places as diverse as streets, gardens, wetlands, grocery stores, cemeteries, hospitals, prisons and theatres. The supposed correlation between the vital air content and the salubrity of a particular location came into question because of discrepancies in the proportion of vital air found in the atmosphere of different places.

Priestley's experimental device

The goodness of air was considered to be inversely proportional to the content of phlogiston or, alternatively, proportional to the amount of phlogiston which common air was able to capture. Accordingly, testing the goodness of air was equivalent to testing its phlogiston content. In his paper of 1772, Priestley proposed a nitrous air test to replace mice and birds for checking the goodness of air:[15]

'It is exceedingly remarkable that this effervescence and diminution, occasioned by the mixture of nitrous air, is peculiar to common air, or air fit for respiration; and, as far as I can judge, from a great number of observations, it is at least very nearly, if not exactly, in proportion to its fitness for this purpose; so that by this means the goodness of air may be distinguished much more accurately than it can be done by putting mice or any other animals to breath it'

---

[15] Priestley, 1772, pp. 214. Victor Boantza (2013b) has mapped the rise and fall of the nitrous air eudiometer in relation to Enlightenment ideals of quantification and challenges in contemporary experimental philosophy and practice.

For this reaction of nitrous air, Priestley was quoting from Stephen Hales' work *Statical Essays.*[16] However, the reaction had already been observed by John Mayow and described by Stephen Hales himself in his previous work *Vegetable Staticks.*[17] This nitrous air test relied upon the contraction in volume that an air sample underwent when mixed with nitrous air (nitric oxide).[18] Hales attributed this diminution in volume not only to a loss of

---

[16] Hales, 1738, Vol.1, p. 224; Vol. 2, pp. 280-284.

[17] Mayow, 1674, pp. 162-163; Scherer, 1782, p. 5; Hales, 1727, p. 125, Experiment XCVI.

Hales produced nitrous air from a mixture of powdered Walton pyrites (iron-cupper sulphides) with diluted nitric acid:

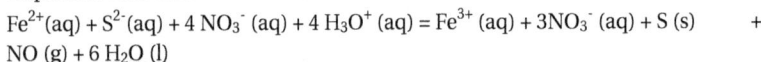

$Fe^{2+}(aq) + S^{2-}(aq) + 4 NO_3^- (aq) + 4 H_3O^+ (aq) = Fe^{3+} (aq) + 3NO_3^- (aq) + S (s)$ $+ NO (g) + 6 H_2O (l)$

In the spring of 1772, Cavendish indicated to Priestley sources other than Walton pyrites. Priestley began this research the following summer by reacting different metals (iron, copper, tin, mercury, nickel and gold) with the appropriate aqueous solutions of nitric acid. In the case of gold, nitric acid was to be mixed with hydrochloric acid (1:3).

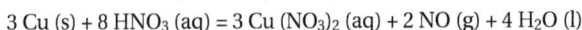

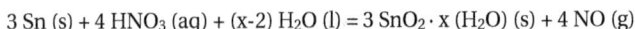

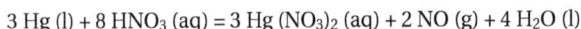

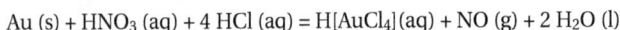

$Fe (s) + 4 HNO_3 (aq) = Fe (NO_3)_3 (aq) + NO (g) + 2 H_2O (l)$

$3 Cu (s) + 8 HNO_3 (aq) = 3 Cu (NO_3)_2 (aq) + 2 NO (g) + 4 H_2O (l)$

$3 Sn (s) + 4 HNO_3 (aq) + (x-2) H_2O (l) = 3 SnO_2 \cdot x (H_2O) (s) + 4 NO (g)$

$3 Hg (l) + 8 HNO_3 (aq) = 3 Hg (NO_3)_2 (aq) + 2 NO (g) + 4 H_2O (l)$

$Ni (s) + 4 HNO_3 (aq) = Ni (NO_3)_2 (aq) + 2 NO_2 (g) + H_2O (l)$

$Au (s) + HNO_3 (aq) + 4 HCl (aq) = H[AuCl_4] (aq) + NO (g) + 2 H_2O (l)$

[18] The relevant reactions taking place in the nitrous air test carried out over water, are equilibrium processes that do not occur independently. They are outlined in the following chemical equations (Usselman *et al*, 2008, p. 107):

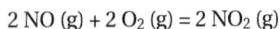

$2 NO (g) + 2 O_2 (g) = 2 NO_2 (g)$	Relatively slow and goes essentially to completion within minutes.
$2 NO_2 (g) = N_2O_4 (g)$	Rapidly achieve equilibrium.
$NO (g) + NO_2 (g) = 2 N_2O_3 (g)$	Rapidly achieve equilibrium.
$N_2O_4 (g) + H_2O (l) = HNO_3 (aq) + HNO_2 (aq)$	Fast and irreversible.
$N_2O_3 (g) + H_2O (l) = 2 HNO_2 (aq)$	Fast and irreversible.

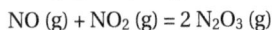

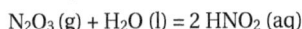

Additionally, oxygen dissolved in the water can play a role.

elasticity of the common air particles, but also to the fact that they were fixed by certain unhealthy sulphurous particles. It was this idea that inspired Priestley to prepare the ground for a test to determine the goodness of air that would undergo diverse instrumental and procedural design modifications throughout the rest of the eighteenth century.[19]

After living for six years in Leeds (from 1767 to 1773), Priestley moved from his post as Dissenting minister in this city to become the librarian and companion of William Petty, Lord Shelburn. Over the next seven years (from 1773 to 1780), he was living between the village of Calne, Wiltshire County, two miles from Shelburn's summer residence in Bowood, and at Shelburn's house in London during the winter season. Priestley's most creative period in the field of chemistry was between 1772 and 1780 under the patronage of Lord Shelburn, during which time he published his major works in chemistry; in particular, his research on gases in three papers published in *Philosophical Transactions* for the years 1774, 1775 and 1776; the work *Experiments and Observations on Different Kinds of Air*, published in three volumes (1774-1775,[20] 1776 and 1777) and its continuation *Experiments and Observations Relating to Various Branches of Natural Philosophy* (1779, 1781 and 1786), as well as the last three-volume abridged edition of the preceding six-volume series *Experiments and Observations on Different Kinds of Air and other Branches of Natural Philosophy* (1790).

Upon contact of NO with atmospheric oxygen, a brownish gas [NO_2] forms immediately and the colour dissipates within a few minutes as NO_2 dimerizes to N_2O_4 and dissolves in water completely and irreversibly, giving a mixture of HNO_3 and HNO_2. Simultaneously, NO and NO_2 react and form N_2O_3, which also dissolves in water to give additional HNO_2. Consequently, a contraction in volume of the gas mixture is observed. In stoichiometric ideal conditions the global process would be:

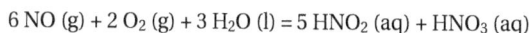

$$6\,NO\,(g) + 2\,O_2\,(g) + 3\,H_2O\,(l) = 5\,HNO_2\,(aq) + HNO_3\,(aq)$$

Therefore, the only residual gas with a common air sample in such conditions would be nitrogen.

[19] Maurice Crosland highlighted the moral background of Priestley's obsession by the goodness of the airs he collected. The idea was that air was injured by the respiration of human beings and animals and it was a matter of human concern to restore it (Crosland, 1983, pp. 235, 237)

[20] The first edition of this first volume was presented to the Royal Society in May, 1774. The second edition of 1775 has been used as a reference work.

The first results of Priestley's experiments on the remarkable phenomenon observed when mixing nitrous[21] and common air were published in 1773 in his paper *Observations on Different Kind of Airs* (*Philosophical Transactions* for the year 1772). Priestley regarded the mutual annihilation of both airs as a promising observation on which to establish a test for determining the goodness of air. In his correspondence with Benjamin Franklin and Torbern Bergman during the last half of 1772, Priestley expressed his high expectations in this test.[22] In fact, 1772 was the year that marked the start of the proliferation of different kinds of devices and instruments designed to measure how healthy air was.[23] Priestley was lavish in his praise ("amazing", "agreeable", "surprising" or "conspicuous") when referring to this particular phenomenon.[24] Notwithstanding, he believed that its most useful and remarkable property was its ability to preserve animal products from putrefaction and of restoring those that were already putrid,[25] although later he was to express his disappointment about these supposed antiseptic properties of nitrous air.[26]

Priestley was of the opinion that nitrous air appeared to consist of the nitrous acid vapour combined with an excess of phlogiston, perhaps also with a small portion of the metallic calx used to obtain the nitrous air. The appearance of a reddish color when mixing nitrous and common air was nothing more than the usual colour of the nitrous acid vapour, which became disengaged from the superabundant phlogiston with which it was combined in the nitrous air. Phlogiston was united with the acid principle of the common air,[27] while a fixed air that it contained was precipitated out, and the water in which the mixture was made absorbed the acid of the nitrous air. These two latter phenomena caused the contraction in volume of the air mixture. Nitric acid and nitrogen oxides would become one of Priestley's obsessions throughout their chemical research.[28]

---

[21] Priestley coined the term "nitrous air" because it came from the same spirit of nitre of the nitrous (nitric) acid.

[22] Schofield, 1966, pp. 104, 110.

[23] Priestley, 1772, p. 214.

[24] Priestley, 1772, p. 211; Priestley, 1775, pp. 111, 115.

[25] Ibid., 1775, p. 123.

[26] Ibid., 1779, p. 69.

[27] Priestley thought that marine acid (hydrochloric acid) was the basis of common air.

[28] Priestley, 1775, pp. 271-274. On Priestley's commitment to nitrous air, see Boantza (2013a, pp. 153-161).

'I do not know any inquiry more promising that the investigation of the properties of *nitre*, the *nitrous acid*, and *nitrous air*. Some of the most wonderful phenomena in nature are connected with them, and the subject seems to be fully within our research.'

Priestley gave the first known description of his nitrous air test in the first volume of *Experiments and Observations* (1774), where in the second section of the introduction he presented the instruments for the forthcoming experiments.[29] It should be noted that there was no description of this particular test in his 1772 paper, neither in the section on nitrous air nor in the final part in the description of the instrumentation included in plate IX.[30] This description, referring to the parts of the experimental device, barely took up a page. In the operating procedure he made a distinction between a sizeable and an exceedingly small air sample.

> In the first case, Priestley put two measures of the air sample into a jar standing in a pneumatic trough containing water. After marking the exact place of the boundary between air and water on the glass, he introduced one measure of nitrous air into the jar. After waiting for a suitable length of time, he noted the quantity of the observed diminution in volume of the air mixture.[31]

> In the second case, if the quantity of the air sample was sufficiently small as to be contained in a part of a glass tube (a) (Figure 1.4), Priestley first measured the length of the column of air in the upper part of the tube, the remaining part (b) being filled with water. He then introduced a wire of appropriate thickness (b), bent to a sharp angle, into the tube. After that, the whole of this little apparatus was passed through the water into a jar of nitrous air, and then the wire was drawn out. In consequence, the nitrous air from the jar occupied the place left by the wire. He then measured the length of the column of nitrous air that had been passed into the tube, in order to determine the exact length of both columns (air sample and nitrous air). Next, while holding the tube underwater, he brought

---

[29] Priestley, 1775, pp. 20-21; plate facing the last page.

[30] Priestley, 1772, pp. 210-225, 250-252.

[31] Priestley observed that if the mixture was made over mercury, the redness remained for a very long time, and the diminution in volume was not so great as when it had been made over water (Priestley, 1775, pp. 112).

the two separate columns of air into contact by means of a small wire. After remaining together for a sufficiently long time, Priestley measured the length of the final air column and compared it with the length of both the columns as noted before.

Figure 1.4 Priestley's glass tube used in his nitrous air test.

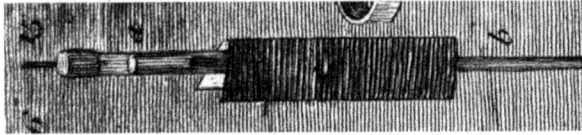

From Joseph Priestley, *Experiments and Observations of Different Kinds of Air* **(London, 1775), plate facing the last page**

These experimental procedures do not appear to be easily replicable. The weakest feature of both procedures was the determination of the endpoint of the test. Expressions such as 'a proper time' or 'sufficient time together' were completely unhelpful in obtaining reliable results. Furthermore, the execution of the second procedure appears to be quite contrived. It would not have been an easy task to introduce the little apparatus through the water into the jar of nitrous air, or, as Priestley himself recognised, to push the wire into the tube to allow the necessary amount of to enter without loss or excess.[32] For the designers of future instruments for measuring the goodness of air, Priestley's description of his nitrous air test constituted more a source of inspiration than a material model.

Landriani's nitrous air eudiometer and the imagined "Priestley eudiometer"

Marsilio Landriani (1751-1815), a professor of experimental physics at the Scuole Palatine in Milan, realized the potential of the contraction in volume observed when common air was mixed with nitrous air as an analytical tool for determining the purity of air. Landriani was aware of Priestley's test from his paper in the *Philosophical Transactions* of 1772 (or from the corresponding translations into French and Italian published in 1773 and 1774, respectively),[33] as well as from the first volume of Priestley's *Experiments and Observations,* published in 1775.[34]

---

[32] Levere, 2000, p. 111.

[33] The French translation was published in the *Observations sur la Physique et Mémoires, sur l'Histoire Naturelle et sur les Arts et Métiers* (1773, Tome I, Partie IV, pp. 292-325; Partie V, pp. 394-419). The Italian version appeared in 1774 in the booklet *Osservazioni del dott. Priestley sopra differenti specie d'aria, tradotte dall' Inglese, da*

The question of the goodness of air had already been raised in enlightened circles of Milan since 1756, regarding the noxious aerial exhalations coming from rice paddies in the outskirts of the city. Therefore, there was a genuine concern in Lombardy to address the social problems that the advances in agriculture were causing due to the air quality in the environment of that region. Jointly with Pietro Moscati, a professor of surgery at the Ospedale Maggiore in Milan, Landriani had already since 1772 been engaged in the search for a method to analyze gases for therapeutic purposes.[35] The chemistry of the air had aroused Landriani's curiosity from mid-1774, and it was in that year when, together with Moscati, he undertook a project for examining the purity of air. Thanks to Priestley's work, Landriani was aware of the key role that nitrous air could play as an indicator of the goodness of air.[36] In this context, in 1775 he published a small booklet, entitled *Ricerche fisiche intorno alla salubrità dell'aria* (*Physical researches on the salubrity of the air*), devoted to an instrumental view of the phenomenon of the contraction in volume when mixing nitrous with common air, the aim of which was to measure the salubrity of the air to prevent infections and epidemics. The publication of this work was announced in a letter published by Landriani in the *Observations sur la physique* in October 1775, accompanied by an abridged description of a new instrument with the corresponding plate.[37] This new development by Landriani has been recognized in some quarters as the first scientific and socially relevant contribution of the functions and aims that the chemical analysis of atmospheric was able to provide.[38]

Despite Landriani's recognition of Priestley's researches, he found that Priestley's nitrous air test was neither sufficiently accurate nor suitable enough for measuring sudden minor variations in the atmospheric air.

Gio. Francesco Fromond, coll' aggiunta di varie annotazioni consultate coll' Autore, Milano, Appresso Giuseppe Galeazi. The first volume of Priestley's *Experiments and Observations* (1774-1775) was not published in French until 1777 (*Expériences et observations sur différentes espèces d'air*, 1777, Paris, Nyon). The corresponding Italian version had to wait until 1784 (*Sperienze ed osservazioni sopra diverse specie di aria, Opera del Signor Priestley Dottor in Legge, Membro della Società Reale di Londra*, 1784, Napoli, Stamperia di Amato Cons)

[34] Landriani, 1775b, pp. 2-3.

[35] Beretta, 2000, pp. 49-50.

[36] Ibid., 1775b, pp. v-vi.

[37] Landriani, 1775a.

[38] Beretta, 1995, p. 5.

Landriani used the term "apparatus" (*apparato*) to refer to the experimental device used by Priestley to conduct his nitrous air test. On the other hand, Priestley did not use the terms "instrument" and "apparatus" but referred to his experimental device as a "test". In his *Ricerche*, Landriani took the opportunity in the *Appendice* (Annex), firstly, to insist on the lack of knowledge behind Priestley's apparatus, and secondly, to affirm that this apparatus could be redesigned in a more suitable form (Figure 1.5). As a result, he devoted part of the annex to a better description of Priestley's apparatus than that provided on the opening pages of the *Ricerche*.[39] Although Landriani insisted 'that this was the apparatus by Priestley that I rightly believed superior to that by Lavosier', there is no evidence that Priestley had built such an apparatus, as far as is known.[40] Actually, Landriani converted an idealized version of Priestley's experimental nitrous air test device into an instrument.[41] Priestley devoted the second part of the final 1790 edition of his *Experiments and Observations on Different Kinds of Air* to the properties of nitrous air. The first section was retitled *Of nitrous Air as the Test of the Purity of Air*, and the description of the test itself, as well as the accompanying experimental device, were faithfully reproduced without any allusion to a eudiometrical instrument or an apparatus.[42]

After describing the redesigned version of Priestley's putative apparatus, Landriani felt able to present his own instrument as an evolved and

---

[39] Landriani, 1775b, pp. 4-5, 76-78, tavola 2; Belloni, 1960, p. 139.

[40] Landriani, 1775b, p. 78: 'Questo è l'apparato di Priestley che a ragione egli crede superiore a quello di Lavosier'.

[41] Schaffer, 1990, p. 282. Marco Beretta added a caption to each of the three plates in his transcription of the *Ricerche*, although there were no captions for these plates in Landriani's original work of 1775. In particular, the caption to the second plate reads as follows: *Eudiometro di Priestley perfezionato da Landriani* (Priestley's eudiometer perfected by Landriani). Thus, this caption reinforced the misleading idea that such an apparatus by Priestley really did exist (Beretta, 1995, p. 107). In fact, the description of an apparatus in a printed text or manuscript does not necessarily mean that the object was ever constructed or used (Anderson, 1985, p. 219).

[42] Priestley, 1790, Vol. 1, pp. 354-365. The fact that Priestley never actually designed and constructed a benchmark eudiometrical instrument for his air nitrous test may owe something to his ideal of simplicity and economy in the design of experiments. Priestley's vision of the public culture of science was closely bound up with his moral imperative that for knowledge to be established and rendered useful it had to be widely replicated. Lack of resources should not hinder those who wished to confirm discoveries or put them to use. (Golinski, 1992, pp. 114, 117).

improved version of the said apparatus. Thus, in his opinion, Priestley had failed to fulfil the expectations he had aroused for determining the salubrity of the air. For instance, Landriani disagreed with Priestley's conjecture that the presence of fixed air (carbon dioxide) in common air was a cause of the salubrity rather than the insalubrity of air. Consequently, common air could hold a large quantity of fixed air, which because it was unhealthy gave an unexpectedly remarkable positive nitrous air test. On the other hand, a healthier air subjected to the same test could give a less positive result.

Figure 1.5 Eudiometer attributed to Priestley by Landriani.

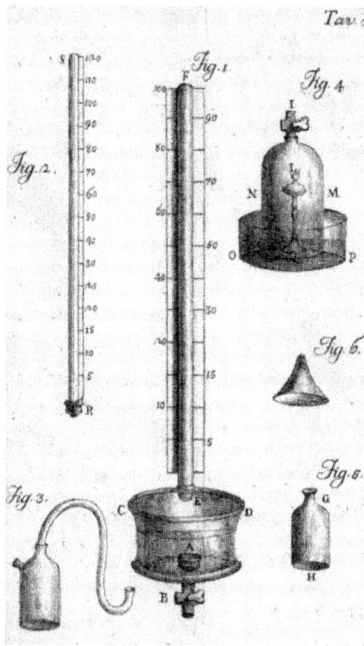

From Marsilio Landriani, *Ricerche fisiche intorno allà salubrità dell'aria* (Milano, 1775), plate 2.

This is the context in which Landriani's proposal for a new, more suitable, sophisticated, and apparently simpler instrument should be placed, which for the first time he named the *eudiometro*.[43] The purpose of the instrument

---

[43] This term is derived from two Greek words. The first part of the term (εὔδιος) means "clear or mild weather", but also with the implication of good air, because (διος) - stemming from Zeus - can mean "weather" or "air". The second part of the term (μέτρο) means "measure" (Landriani, 1775b, pp. 3, 6).

was to measure the salubrity of the air and not to indicate the causes of a vitiated air, which, given the limited knowledge of physics and chemistry at the time, was an unsolvable problem; the aim was rather to determine with some precision the main modifications undergone by air with regard to the principal functions of the human body. Landriani's eudiometrical procedure can be summarized as follows (Figure 1.6):

> The instrument consisted of a glass flask (AB, *Fig.1*) of known capacity that had already been filled with water. The upper neck (A) of the flask was internally threaded. A glass or ivory tap (stopcock)[44] (BC, *Fig.* 2) was placed in the lower neck (B) of this flask. The eudiometrical tube (CD) with the same capacity as that of the flask (AB), was connected to the lower neck (C) of the tap (BC). Beneath this tube was fixed a small brass cylinder (HL, *Fig.* 4) containing a helical spring at whose end was attached a brass strip (L) covered with a small leather bag filled with soft wax (I), which was used to seal the lower end (D) of the tube. By opening the tap (BC) water was poured from the glass flask (AB), thus filling the tube (CD).
>
> A double-necked flat-bottomed glass bottle (*Fig.5*) was used to obtain nitrous air. One end of a curved tube (TV) was inserted into the perforated cork stopper in the upper neck (T) of the bottle and sealed with soft wax. To the other end of the tube was attached a wooden pipe tube (V), internally threaded to easily screw the ivory tap (stopcock) (NO, *Fig.* 3) in which a bladder (P) – previously emptied of atmospheric air - was attached to receive the nitrous air. A certain amount of iron filings and water was poured through the side neck (S). After gently stirring the mixture, an amount of *aqua fortis* (nitric acid) was also

---

[44] A letter from Landriani to Priestley dated November 17th, 1776, proves that he sent an example of his eudiometer to Priestley (Priestley, 1776, pp. 380-381). The instrument was included in the inventory of Priestley's instruments in a claim for damages after the destruction of his home in 1791 by a violent mob protesting against his support to the French Revolution. The writer of the claim described the instrument as 'Landriani's eudiometer with Glass Cocks sent from Italy' (McKie, 1956, p. 122).

Actually, in the *Ricerche*, Landriani described the taps of his eudiometer as being similar to those used by Jean-André De Luc (Figure 1.7) in his barometer, and bearing a close resemblance to ordinary pneumatic taps, except that they were made of ivory instead of metal, and that the key was made of highly compacted cork and perforated by a small steel blade to prevent it from twisting when turned. In addition, there was a section of a goose quill inside the hole bored through the key so that compression on the cork did not close its internal cavity (De Luc, 1772, Vol.2, pp. 7-12; Landriani, 1775b, p. 7).

poured into the bottle through the same side neck, so when the effervescence began the neck was stoppered. When the whole device was purged of common air, the tap (NO) of the bladder was screwed into the pipe (V) to inflate it with nitrous air, which once completed, the tap (NO) was closed.

The bladder was then removed from the bottle and screwed into the upper neck (A) of the glass flask (AB). By removing the brass strip (L) and opening the taps (NO) and (BC), the nitrous air was introduced into the flask (AB) while water flowed into the glass bowl (E) below under the effect of gravity. This bowl could be raised or lowered at will by means of the pressing screw (G).[45] When the taps were closed and the glass bowl was lowered, the atmospheric air, whose salubrity was to be ascertained, entered the tube and displaced the remaining water. The bowl was then positioned so that the level of water it contained coincided with the zero of a scale divided into twenty-four graduations. To measure the quality of the air, the tap (BC) was opened to bring the two airs into contact in the flask (AB). This triggered the reaction between the atmospheric and the nitrous air, thereby causing a contraction in volume. Driven by atmospheric pressure and sucked by the depression generated by this reaction, the water rose up the tube, and the air quality could be evaluated by measuring its rise on the scale.[46]

---

[45] Alternatively, by removing the small spine of ivory (M) in the tap (BC), atmospheric air also entered the tube and the water drained out of it. The problem was that ivory was very sensitive to changes in moisture and dryness, and therefore the spine was not appropriate for pneumatic use. For that reason, Landriani preferred to proceed directly to lowering the bowl (E). However, if a narrower eudiometrical tube was used, lowering the bowl would not be sufficient to drain the water completely, so Landriani suggested the use of a gold-tipped metal spine, since gold impervious to attack by nitrous air.

[46] Landriani, 1775b, pp. 3, 6-9, 70-71; tavola 1. This example of Landrian's eudiometer is kept at the Museo Galileo in Florence (Figure 1.8, left), signed by the craftsman Saruggia in 1776, and is similar to the one described in the *Ricerche*. It is quite likely that Fontana had acquired this example when he was director of the Gabinetto di Fisica in Florence (Beretta, 1995, p. 29). Landriani possessed another version of his eudiometer for performing the test on mercury instead of water. The difference was that water could absorb the fixed air of common air, whereas mercury could not. The disadvantage with mercury was that it could react with the nitrous air. Figure 1.8 (middle and right) shows a modern replica of Landriani's eudiometer.

Figure 1.6 Landriani's eudiometer.

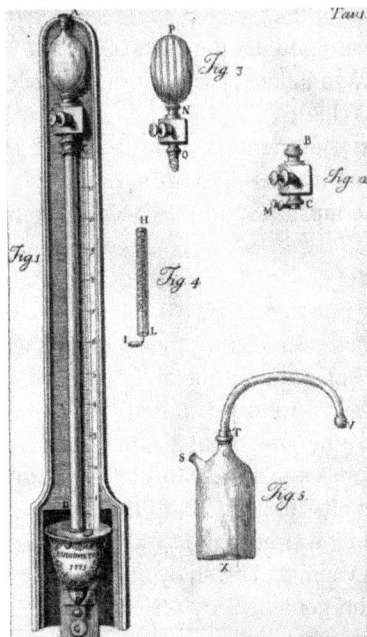

From Marsilio Landriani, *Ricerche fisiche intorno allà salubrità dell'aria* (Milano, 1775), plate 1.

The chief advantage of this eudiometer was probably that of reducing the manipulations required, both beneath the water and with both airs.[47] Regarding the reacting ratio of the airs, Landriani used a ratio of 1:1 of nitrous and common air, given the identical capabilities of the flask (AB) and the tube (CD). However, this was not the proportion at the point of saturation. When establishing this point of saturation Landriani ran into the problem of standardizing the nitrous air, which to a large extent depended on both the process used to obtain it and on the reagents.[48] This is a convenient juncture at which to state that the meaning of saturation refers to the point beyond which the nitrous air and the dephlogisticated air (pure, or as a part of the atmospheric air) could neither receive nor retain a greater quantity of each other in combination. The endpoint of

---

[47] Capuano *et al.*, 1998, p. 26.

[48] Eventually, Landriani set the reacting ratio at 2:5 of nitrous and common air as an average value deriving from different experiences, while warning that this would change the ratio between the capabilities of the flask and the tube of his eudiometer. Landriani, 1775b, pp. 31-32.

the test was reached at this point of saturation, and it should be detected after achieving the greatest contraction in volume of the air mixture. A different matter was the combination or reaction ratio of nitrous and dephlogisticated or common air used in a particular test, since it may or may not have coincided with the ratio at the point of saturation.

Figure 1.7 Profile view of De Luc's barometric tap (stopcock), which was the basis for the design of Landriani's eudiometrical taps.

From *Recherches sur les modifications de l'atmosphere* (Genève, 1772), Vol. 2, plate 1.

With regard to the effective use of this eudiometer, Landriani employed the version on water as well as mercury. He also used a more accurate version of his 1775 eudiometer during an extensive tour he made through different regions of Italy between the late summer and early autumn of 1776 when in particular he analyzed air samples from the mountains near Pisa and in his ascent of Vesuvius.[49] Landriani expected that his eudiometer would be able to measure the air salubrity, and therefore prevent infections and epidemics. In an excess of enthusiasm, he came to believe that his eudiometer would pave the way to the creation of a new

---

[49] Priestley, 1776, pp. 380-381; Proverbio, 2007, p. 81, note 188.

discipline - aerial medicine – by means of which the therapeutic action of different types of air on the human body could be controlled.[50]

Figure 1.8 (Left) Landriani's eudiometer, at the Museo Galileo. (Middle and right) Modern replica of Landriani's eudiometer with the bladder and the bottle to obtain nitrous air (above) and the mixing flask (below), at the Centro Studi Lazzaro Spallanzani, Scandiano.

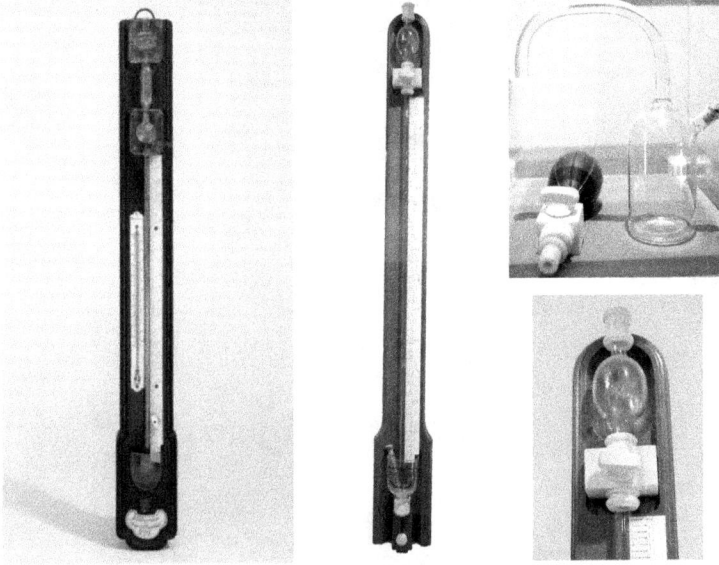

Photo Franca Principe, Museo Galileo, Firenze. Courtesy of Centro Studi Lazzaro Spallanzani, photographs by the author.

Landriani began to conceive an initial version of the instrument at the beginning of 1775, and in March of that year he presented the definitive instrument to Count Firmian, counsellor of state of Habsburg Lombardy. In August he wrote Alessandro Volta (1745–1827), with whom he maintained a cordial friendship, announcing his project of an aerial medicine. Volta's reply, a few days after, is probably the first assessment of Landriani's eudiometer. Volta recognized that the instrument was a suitable, elegant and successful invention that could be used safely and immediately. However, he was quite sceptical about Landriani's project of an aerial medicine. According to Volta, the most controversial issue was the acceptance of eudiometric measures as indications of the air salubrity. That is, it was necessary to make a distinction between the breathability

---

[50] Beretta, 2000, p. 51.

and wholesomeness of air, whereas the eudiometer indicated the air respirability (i.e. its dephlogistication) but not its salubrity.[51] In short, as Priestley had remarked about his nitrous air test, the function of the eudiometer was limited to indicating the degree of phlogistication of the air. After Landriani, studies on the salubrity of the air approximated more closely to chemistry and medicine, while moving further away from meteorology.[52]

Fontana's first generation of nitrous air instruments

In that same year, 1775, Landriani was aware of the publication of the work *Descrizione ed usi di alcuni stromenti per misurare la salubritat dell'aria* by Felice Fontana, in which the author described as many as eight versions of an instrument also designed for measuring the salubrity of air. Fontana sent a copy of his work to Landriani, which the latter received when had almost completed his *Ricerche*. Nevertheless, Landriani was still in time to use the *Appendice* to claim priority for his own eudiometer.[53]

The Abbé Fontana (1730-1805) was appointed as the court physician of the Granduca Pietro Lepoldo of the Tuscany in 1766. However, in the previous year, which saw the publication of his *Descrizione,* he had also been appointed director of the Reale Gabinetto di Fisica e Storia Naturale of Florence as well as publishing the essay *Ricerche fisiche sopra l'aria fissa, (Physical Research on the Fixed Air),* which was his first contribution to the chemistry of gases. At the end of this work, after pointing out that the authorities responsible for public welfare should focus their concerns on breathable air, Fontana announced the invention of an instrument that would determine the quality of air and indicate its different degrees of purity and wholesomeness as if it were a new thermometer.[54] He eventually fulfilled this commitment in the *Descrizione.* Although Priestley's nitrous air test was the prescribed qualitative reference for this instrument, the thermometer, an instrument for the quantification and graduation of the effect of fire on bodies, became his own quantitative benchmark for grading

---

[51] VO, Vol. 6, pp. 5 -10.

[52] Beretta, 1995, p. 35; 2000, pp. 54-55.

[53] Landriani, 1775b, pp. 75-76.

[54] Fontana published the French version of the *Ricerche fisiche sopra aria fissa,* in which he announced his new instrument, in the same volume of the *Observations sur la physique, sur l'histoire naturelle et sur les Arts* of 1775, in which Landriani had set out the appearance of his eudiometer (Fontana, 1775a; *Observations,* pp. 288-289).

the goodness of air and its influence on the health of people and animals.[55] Nevertheless, in this work Fontana failed to provide or reproduce any theoretical reflection on the contraction in volume in the nitrous air test.

Figure 1.9 The eight "macchinette" of Felice Fontana.

From Felice Fontana, *Descrizione ed usi di alcuni stromenti per misurare la salubritat dell'aria* (Firenze, 1775), plates 1-8.

When Fontana wrote the *Descrizione*, Landriani had yet to publish his *Ricerche*, and therefore Fontana was unaware of the term "eudiometer" proposed by Landriani. At first he assigned no particular name to his different apparatus, referring to them by the terms *stromento*[56] or *macchina*. In the *Descrizione*, he presented a first generation of eight instruments, four with a gravimetric basis and the other four with a volumetric basis (Figure 1.9). Two of the gravimetric instruments were

---

[55] Schaffer (1990, p. 300) considers that Fontana envisaged eudiometry as part of a broad program of social and economic reforms promoted by the Granduca of the Tuscany. Golinski (1992, p. 118) also points out that Fontana and Landriani promoted eudiometry as a part of the government improvement programs. The objective was for surveillance of the quality of the air in a certain region to become a means of controlling crop productivity and epidemic diseases. Other have argued that eudiometers became part of the socio-political context due to nineteenth century capitalism, when the salubrity of the air associated with the productivity of the labour force (Beretta 1995, pp. 42-49; 2000, p. 52, note 13). In any case, the practical and theoretical implications of the nitrous air eudiometer were far too complex to be reduced to a sociologically determined process (Magiels, 2010, p. 227).

[56] The present Italian spelling of the word is *strumento*.

made of glass, while the other two were made of wood with glass windows. The four volumetric instruments were equipped with a glass tube - three vertical and one horizontal – with an attached scale for reading the measures. In general, the iconography deployed by Fontana was impressive and was intended to convince the reader of the potential of his instruments.[57]

> The four gravimetric instruments followed the same design pattern: two chambers for storing each of the two airs (common and nitrous), directly connected to each other, or with a third one for the mixing of both airs. Despite Fontana's considerable efforts, it appears that some difficulty was involved in ensuring that the instruments could be tightly sealed. The nitrous air test was performed on mercury instead of water,[58] with a single addition of equal volumes of both airs (i.e. a 1:1 combination ratio). Fontana's quantification of the goodness of air would eventually require a balance sensitive enough to detect small but significant differences in the weight of the mercury added in order to compensate for the contraction in volume after the mixing of the air. His failure to mention the specifications of such a balance, and the absence of experimental data in the *Descrizione*, raise serious doubts about whether all these instruments had actually been built, and above all if and how they could have been used.[59] However, evidence exists that at least one of Fontana's devices was in fact constructed. The Museo Galileo in Florence houses an incomplete example of the fourth instrument (Figure 1.10, left).[60] A description of how it functions is given in Appendix I. [61]

---

[57] Boantza, 2013b, p. 387.

[58] Fontana gave no explanation for using mercury instead of water, even knowing that the contraction in volume over mercury was slower. In fact, it is blatantly obvious the absence of theoretical background in the *Descrizione*.

[59] Knoefel, 1984, pp. 168-169; Schaffer, 1990, p. 302; Proverbio, 2007, pp. 77-78

[60] The exemplar in the Museo Galileo was found by Marco Beretta at the storerooms of the Galleria degli Uffizi in Florence. Its size is 240 x 75 x 115 mm. Figure 1.10 (right) shows a modern replica of the same fourth instrument (Levere, 1999, p. 60).

[61] Fontana, 1775b, pp. xxvii – xxx.

With regard to the four volumetric instruments, the main differences
were, firstly, that the two airs were mixed in a glass tube initially full of
mercury and, secondly, that the goodness of air was determined by
reading the length of the resulting column of mercury in an attached
graduated scale. The test was also performed with a single addition of
equal volumes of both airs (i.e. 1:1 combining ratio). The absence of
information about the size of the instruments or the experimental data
makes their replication unreliable. Only the account of the fifth
instrument provides operational details that would hardly be remarked if
not actually executed. For this reason, the following description of the
instrument together with the operating procedure is given below (Figure
1.11).

The instrument, *Fig. I*, consisted of a long glass tube (MoM) and two
glass balls (C, D). The tube as well the balls had known capacities.
The two balls were connected to with each other using a key (B) that
closed tightly, and each ball had two openings that could be
stoppered (a, c; b, d). Two openings (h, i) with the corresponding
stoppers (f, g) are depicted in *Fig. III*. The key (B) fitted a large
opening (c), *Fig. II*, in the middle of a wooden section (A) with two
holes (d,b) that allowed connection between both balls when passing
through the key. A third hole (not marked in the plate), perpendicular
to the other two, went into the middle of the key, connecting with the
lateral holes (d, b). Each ball, *Fig. III*, had a short glass neck (lm) that
fitted into the aperture (a) of *Fig. II*, firmly cemented with mastic. The
end (M) of the glass tube was attached with mastic to the hole (n) in
the piece (A). In this way, the tube was connected through the key
with the two holes (db), and therefore with the two balls. These were
attached to the piece (A) with mastic and also by two strong steel

bands (ef). The whole device was seated on a wooden board (EEEE). Along the tube there were two ivory blades divided into 100 degrees. The opposite end (o) of the tube passed through a thin ivory ring (n) attached to the wooden board. A stopper (p) was inserted into the opening (o) of the tube up to the ring (n) where the divisions began. The whole device was placed on a pedestal with a bucket partially filled with mercury (Figure 1.12).

Once the two stoppers (a, c) of the ball (C) were removed, a continuous flow of nitrous air was forced into and out of the ball so that all the original air in the ball was consumed, a process that lasted for less than a minute. The same operation was carried out on the other ball (D) with respirable air. The key (B) had to be manipulated in such a way that its opening did not connect with the tube (MoM). After turning the instrument upside down, the tube was completely filled with mercury through the opening (o) by using a long, thin glass tube with a funnel-shaped end, and then closed with the stopper (p) to ensure that no air remained above the stopper and that the mercury inside the tube was not too compressed. Both of these requirements could be fulfilled after two or three attempts. The key was then turned to connect the tube with the two balls, and the mercury descended into the balls, leaving the connection between the tube and the nitrous and respirable air open. Simultaneously, the end (o) was submerged in a bucket partially filled with mercury, the stopper (p) was removed, and the instrument was carefully lifted in the bucket so that the ring (n) remained beneath the surface of mercury, which was easily achieved by adjusting the pedestal. Both airs mutually destroyed on mixing, thereby increasing the imbalance between the external atmospheric air (i.e. pressure) and the internal air of the instrument. As a consequence, the external air pushed the mercury in the bucket up to a certain height in the tube, thus providing a way of measuring the amount of air annihilated inside the instrument by means of the graduated blades, and therefore the goodness of respirable air.[62]

---

[62] Ibid., 1775b, pp. xxx – xxxiv. Procedural aspects such as that one of the balls could be filled with nitrous air by removing the original common air 'in less than a minute', or that it was possible 'after two or three attempts' to evacuate the remaining air in the tube without excessive compression of the mercury it contained, suggest that this fifth "macchina" may indeed have been built. For Fontana, this device would be able to serve as a model for others by changing a few non-essential things, such as placing the balls over one another.

Figure 1.11 The fifth "macchina" of Felice Fontana.

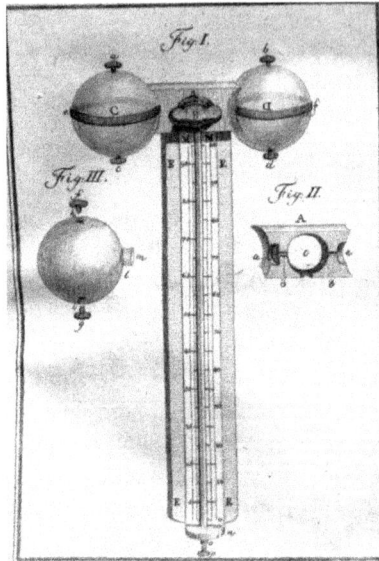

From Felice Fontana, *Descrizione ed usi di alcuni stromenti per misurare la salubritat dell'aria* (Firenze, 1775), plate 5.

Unlike Landriani, Fontana included some remarks about the need to establish a scale of salubrity in his instruments and the difficulties involved therein. The main problem of establishing such a scale was the setting for its two extreme points, i.e., the minimum and maximum values of the air salubrity. The minimum value had to be set with a lethal air sample; that is, with a sufficient sample of unhealthy air so as to ensure the performance of a negative nitrous air test at the very instant of the death of an animal breathing this air. The maximum value of the scale was to be set with a sample of air as salubrious as possible in order to ensure a better positive nitrous air test. The point was to locate and standardize this kind of excellent air. Fontana's proposal was to assign the value of 100 to the maximum air salubrity and to divide the scale into 100 parts or degrees. He was well aware, however, of the extreme difficulty involved in establishing the two points of reference for the scale in order to obtain absolute values of the salubrity of the air. He was therefore satisfied with the indications given by his instruments about the relative salubrity of the air from different locations.[63]

---

[63] Ibid., 1775b, pp. viii-xvii.

Figure 1.12 Modern replica of Fontana's fifth "macchina" placed on a pedestal with a bucket, at the Centro Studi Lazzaro Spallanzani, Scandiano.

Courtesy of Centro Studi Lazzaro Spallanzani, photography by the author.

Instrumentalizing the nitrous air test. The contributions of Magellan and Gérardin

In February 1776, Priestley published the second volume of his *Experiments and Observations on Different Kinds of Air*, in which he made no mention of the brief description of the nitrous air test provided in the first volume. The only reference to this experimental device is to be found in the introduction to the work, where Priestley recommended some time-saving strategies, should it be necessary to conduct a series of tests. Thus, in order to circumvent the need to collect all the air measures separately, he first suggested using a number of vials of proportional capacities (Figure 1.13; f, f, f), each vial having a measure of twice the preceding one, and secondly a cylindrical container (g) on which all these measures were engraved in order to indicate where to mix the two airs. In the case of amounts of air samples that were too small, Priestley proposed a series of smaller tubes (h, h, h) with the same proportions as the larger vials, and a long tube (i) on which all these small measures were engraved for the mixture of the airs.[64]

---

[64] Priestley, 1776, pp. xliii – xliv, plate I.

Figure 1.13 Priestley's glass vials used for eudiometrical measurements.

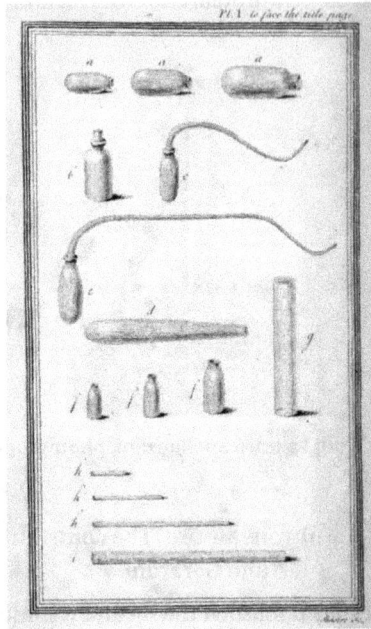

**From Joseph Priestley, *Experiments and Observations of Different Kinds of Air*
(London, 1776), plate I**

When dealing with dephlogisticated air and the constitution of the atmosphere, Priestley was faced with the problem of establishing the saturation ratio between nitrous and dephlogisticated air. He had already addressed the question of the proportions involved in the contraction in volume between nitrous and common air in the sixth section of the first volume of *Experiments and Observations*.[65] In the second volume of his work of 1776, Priestley concluded that on dosing nitrous air in two measures of dephlogisticated air, the contraction in volume ceased to be observed after the addition of practically five measures of nitrous air. Furthermore, he determined that the dephlogisticated air used in the experiment was between four and five times as good as common air.[66] He also took the opportunity to clarify misunderstandings about some aspects of his previous publications. He was especially concerned about the interpretations made by some of his non-native, English-speaking colleagues. In particular, his clarifications referred to the mistakes

---

[65] Priestley, 1775, pp. 110-112.

[66] Priestley, 1776, pp. 47-48.

observed in Lavoisier's and Landriani's works, the *Opuscules Physiques et Chimiques* and the *Ricerche*, respectively. For instance, Priestley regretted that Landriani had taken for granted that he had considered the fixed air in the atmosphere as the principle of salubrity. He recognized that this misunderstanding might have stemmed from his observation about the lethal nature of unmixed fixed air when breathed in, but not being harmful when mixed with common air. However, this was far from admitting that fixed air was the principle of salubrity.[67]

The third volume of *Experiments and Observations*, published in the spring of 1777, contained no relevant contribution to the execution of the nitrous air test. However, Priestley was proactive in his explorations into the purity of the air from various sources. For instance, in a letter to Matthew Boulton prior to November 1777, Priestley asked him to collect air samples from different parts of Birmingham and send them to him so he could ascertain their salubrity. Priestley also gave him instructions about how and where to collect the air samples. The procedure was very simple: fill a clean vial with about eight ounces of clear water; empty it in the place chosen to collect air; tightly seal the vial without opening it again; note the weather conditions, and send a box containing the vials to Priestley's address in London. The samples were to be collected in places where a low air quality was suspected: factories emitting smoke and fumes or with many manual workers living in narrow, crowded streets.[68] The practice of field eudiometry had enthusiastic followers in England, especially among the medical fraternity. In a letter to William Cullen, dated June 1777, the physician William Falconer stated that he had constructed a eudiometer according to Priestley's plans. Another physician, William White, conducted up to twenty-seven trials between August and September 1777, to check the purity of the air in various circumstances.[69] The paper he wrote on these trials was read at the Royal Society the following year, and has the peculiarity of containing a concise description of the tools and procedures used:[70]

> The air sample filling a one-ounce vial (i.e. capable of containing an ounce of water) was conveyed by means of a glass funnel submerged in water into a barometric tube with a large bore and graduated in

---

[67] Ibid., p. 311.

[68] Schofield, 1966, pp. 161-162.

[69] Golinski, 1992, pp. 119-120.

[70] White, 1778, p. 197.

inches and decimals. The air sample occupied on average 134 decimal parts of an inch. Then, immediately upon a further addition of a half-ounce vial of nitrous air by the same method, the space occupied by them both as well as the time were noted down, as well as the time. After standing for half an hour (except in some cases), the space occupied by the residual mixture was also noted, which on being deducted from the first space gave the result of the diminution in volume.

White described his device as being a very simple apparatus which, while less ostentatious, was perhaps more accurate than more complete instruments (he was probably referring to Landriani's and Fontana's instruments). The simplicity of its design reflected that of the ideal instrument advocated by Priestley and was probably based on the experimental design that Priestley was currently using in his nitrous air test. The paper concluded with a fierce defence of the procedure for ascertaining the goodness of air when olfactory detections were not possible:[71]

'Hence I do not hesitate to declare, that in jails, hospitals, and other crowded places, we ought not by any means to estimate their wholesomeness by the absence of disagreeable smells alone. The principle of disease may lurk therein unperceived by our limited senses. The method used in these experiments is the only true one by which we may judge with some degree of safety.'

Despite Priestley's limited contribution to the nitrous air test in the third volume of his *Experiments and Observations*, the work provided three eudiometrical references that would later attract much more attention. The first appeared at the end of the preface, where Priestley remarked on the darkening of a solution of green vitriol (iron sulphate) by the nitrous air.[72] The other two items concerned the publication of two letters; the first one sent to Priestley on November 30th, 1776, by Magellan, announcing the construction of three new eudiometers based on the nitrous air test. In the second letter, dated December 10th, 1776, Alessandro Volta informed Priestley about the discovery of a new kind of inflammable air associated with marshes.[73]

[71] White, 1778, p. 205.

[72] Priestley, 1777, p. xxxiii.

[73] Ibid., pp. 349-350, 381-383.

Magellan's contribution

João Jacinto de Magalhães (1772-1790) was educated by the congregation of Augustinian monks at the Monastery of Santa Cruz in Coimbra, of which he later became abbot. At the age of 31, he formed a close friendship with the French army officer Gabriel de Bory when he visited Portugal for the observation of a solar eclipse. Probably as a result of this friendship with Bory, Magellan requested a temporary secularization, which lasted from 1754 to 1758. During this time he embarked on a philosophical tour, visiting some European countries and eventually settling in London at the end of 1763. Here he came to be known as John Hyacynth Magellan and was soon acknowledged as an expert, not only in the use of astronomical and philosophical instruments but also in mechanical devices such as steam engines and cranes. The amicable and academic relationships he established with Priestley and Benjamin Franklin, among many others, paved the way to his appointment as a Fellow of the Royal Society in 1774. Magellan maintained a regular and extensive correspondence with colleagues all over Europe: instrument makers, mathematicians, astronomers, physicians, chemists, philosophers and politicians. He also maintained a close relationship with Priestley, and both men frequented the scientific circle of Franklin and the same London coffee houses, such as the Chapter Coffee House and Slaughter's Coffee House. Franklin and Priestley signed the certificate nominating Magellan to the Fellowship of the Royal Society in 1774. It was against this background that Magellan began to address issues concerning pneumatic chemistry in his correspondence in the early 1770s, and went on to become a leading advocate on the continent of Priestley's experimental findings.[74]

The three eudiometers referred to by Magellan in his letter to Priestley were described in the second part of his work *Description of a Apparatus for Making Mineral Waters, like those of Pyrmont, Spa, Seltzer, &c, Together with the Description of Some New Eudiometers*, published in 1777. In principle, the three eudiometers were able to perform the test over water or mercury, although mercury was inadvisable due to its reactivity with nitrous air. Likewise, although water could also interfere with the dissolution of the nitrous air, it was also preferred because it was lighter and more affordable than mercury.[75] Magellan's descriptions of his eudiometers and the operating procedures were detailed enough to

---

[74] Mason, 1991, pp. 155-157, 159, 161; Malaquias, 2008, pp. 255-256, 258-260, 270-272.

[75] Magellan, 1777, pp. 20-21.

enable the test to be reproduced (Figure 1.14; first eudiometer:[76] *Figs. 11, 12, 14, 16* and *17*; second eudiometer: *Fig. 15*; third eudiometer: *Fig. 8*). Magellan regarded his third eudiometer as the most similar to Priestley's and the best of all the three. Its description may therefore constitute the first more comprehensive approach to Priestley's eudiometrical device.

Figure 1.14 Magellan's eudiometers.

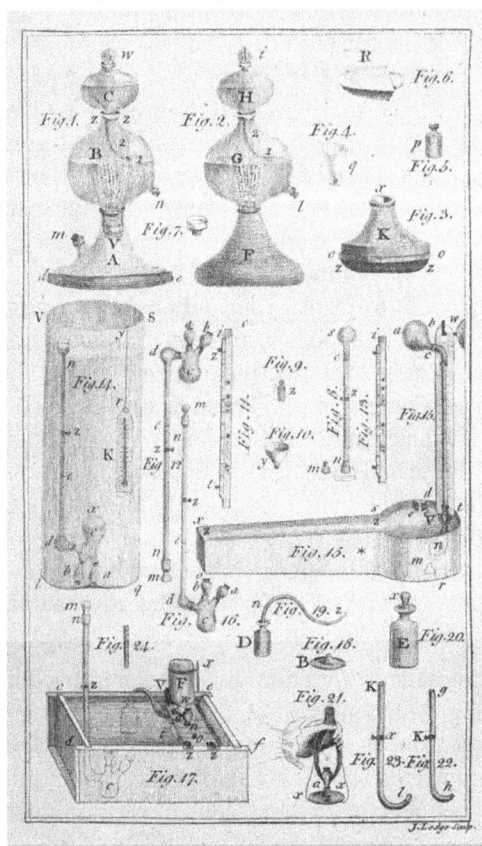

From John Hyacynth Magellan, *Description of a Apparatus for Making Mineral Waters, like those of Pyrmont,* **Spa, Seltzer, &c, (London, 1777)**

---

[76] The incomplete copy of Magellan's first eudiometer (Figure 1.15), kept at Teylers Museum, was probably made by W. Parker, London.

Figure 1.15 Magellan's first eudiometer.

Teylers Museum, Haarlem, The Netherlands.

This third eudiometer consisted of a glass tube (en, *Fig. 8*) approximately 2-3 feet long with a large ball (s) on the top and a glass stopper (m) fitted air-tight to the funnel-shaped mouth (n). There was also a glass funnel (y, *Fig. 10*) and a small vial (z, *Fig. 9*). The content of this vial occupied the third part of the ball (s), and when introduced into the glass tube (ns) took up no more than half of its length. Lastly, the instrument had a scaled ruler (*Fig. 13*) and a glass funnel, which was ground to the mouth (n) of the instrument when this was not wide open. The first step of the test consisted of filling the tube with water and placing it in a vertical position with the mouth (n) under the surface of the water in a trough (*Fig. 17*). The vial was then filled with nitrous air, introduced into the tube by means of the glass funnel (y), which was ground to the mouth of the eudiometer. The same vial was again filled with the air to be tested, which was also introduced into the tube. On reaching the stationary moment at which the air mixture was at the point of its greatest diminution, the eudiometer was placed horizontally beneath the water for a few minutes in such a manner that no part of the enclosed air could escape. Afterwards, the mouth (n) was sealed with the glass stopper and the instrument was inverted with the mouth (n) facing upwards. Lastly, the space occupied by the residuum of the diminished air was measured by applying the scaled ruler.[77]

Magellan's description of his eudiometers provided fully detailed and precise operating procedures that were likely to make the replication of the tests feasible. Despite all the precautions, Magellan accepted that the results of his experiments, even those with common air, had not matched

---

[77] Magellan, 1777, pp. 39-41.

his expectations. He suggested several reasons for this shortcoming, such as not having a nitrous air of standard quality to enable reproducible results. The accounts of these three Magellan eudiometers, especially the first, encouraged Gérardin to submit a project for his own eudiometer

Géradin's eudiometer

René-Louis Gérardin (1753-1808) was captain of the guard of the Duke of Lorraine until 1766. In 1762, he inherited the 800-hectare estate of Ermenonville (a commune in the Oise department of northern France), where he settled in 1766 to engage in landscape gardening and agriculture. He was a strong supporter of the values of Enlightenment and closely wedded to Rousseau's ideal of the return to nature. He was until the end of his life a passionate devotee of country life. Once ensconced in Ermenonville, one of his first undertakings was to drain some marshy lands because they were a source of miasma harmful to animal life.[78] These activities may well have led Géradin to his interest in eudiometry, which would have enabled him to ensure the salubrity of the air in order to make such places habitable. In 1778 he published a paper in which he described Magellan's first eudiometer, and subsequently the design of a new eudiometer intended to be as easy to use as the thermometer and the barometer. See its description in Appendix II.[79]

Gérardin suggested repeating the eudiometrical test every day for a year in the location where the salubrity of the air was to be tested. By taking a mean of the results of each season, in order to compensate for accidental differences in temperature and the quality of nitrous air, it would be possible to ascertain the salubrity of the air *in situ*. In this procedure, both airs were mixed in a fixed combination ratio of 1:1, without addressing the problem of the endpoint (i.e. point of saturation) between both airs. The device would not have been easy to transport, and its use cumbersome, and Gérardin was aware of the need to improve and simplify it, and even to make it portable. There is documentary evidence that this eudiometer was in fact built. Nevertheless, Gérardin was still confident that the time it took for a flame to be extinguished, or the life duration of an animal exposed to an air sample, might still serve as a basis for future eudiometers.[80]

---

[78] Marti-Decaen, 1912, pp. 14-15, 19.

[79] Gérardin, 1778, pp. 252-253, plate 2.

[80] Gérardin, 1778, p. 253.

Summary

The nitrous air test was envisaged as a chemical procedure to replace animals for checking the goodness of air. This test was founded on in the striking diminution in volume that occurred when mixing nitrous and atmospheric air in a closed vessel. It is worth mentioning the contributions made by Mayow, Hales and Cavendish to the development of the chemical trial undertaken by Priestley in the summer of 1772 and that culminated so successfully in May 1774. In the first edition of his work *Experiments and Observations on Different Kinds of Air* (1774-1775), Priestley presented the experimental device he had conceived for conducting the nitrous air test, and which was soon to become a source of inspiration for Landriani and Fontana, the first devisers of instruments for this test.

Landriani insisted that the device he presented in his *Ricerche* of 1775 in the form of an imagined eudiometer was an idealized version of Priestley's apparatus and that this apparatus really did exist. This was a rhetorical strategy designed to persuade his colleagues by referring to Priestley's authority while announcing that he was going to present his own eudiometer as an improved version of the instrument apparently devised by Priestley himself. Fontana, like Landriani, also made the most of another rhetorical resource. The visual impact of the elaborate iconography of the eight machines in his *Descrizione* was aimed at convincing others of their potential. This visual presentation was underlined in the text by the distinction made between the description and the use of each machine, thereby resembling the operating manual of any modern appliance. Fontana's notable contribution was his proposal of grading the goodness of air in the same way that a thermometer measured heat in bodies. On the whole, serious doubts exist about the construction of most of these machines.[81] After the pioneering eudiometers of Landriani and Fontana, the instrumentalization process of the nitrous air test remained in the hands of Magellan in England and Gérardin in France. Meanwhile, practice in the emerging field of eudiometry was gaining adepts in England. Most of them were physicians who, far from instrumentalizing the test, conducted their eudiometric tests according to Priestley's experimental plans.

The construction and use of the first eudiometers brought some of the perennial problems dogging the nitrous air eudiometers to the fore: (1) the imprecise detection of the endpoint of the test, which was intimately

---

[81] The *Ricerche* and the *Descrizione* have been regarded as examples of the interplay between the practical and the impractical (Boantza, 2013b, p. 385)

related to; (2) the determination of the reaction ratio of nitrous and dephlogisticated air at the point of saturation; (3) the purity of the reagents, nitrous and dephlogisticated air; (4) the execution of the test over water or mercury, and (5) the technological constraints concerning the performance of materials used in their construction. Regarding the latter issue, Sigaud de la Fond indicated that inventive ingenuity was not always enough when it came to constructing machines.[82] The experimenter was often at a disadvantage due to the limitations of the nature of the materials he was obliged to use when conducting his experiments. This was certainly the case of the taps in Landriani's eudiometer. Taps had to be made of ivory, or preferably of glass, because they were resistant to attack by nitrous air, or at most it did so slowly, while on the other hand it corroded metals. Furthermore, since ivory was very expensive, as well as being sensitive to moisture and dryness, and because glass taps were not commonly used,[83] it was possible to replace them by taps made of dense, oven-dried wood.[84]

---

[82] Sigaud de la Fond, 1799, p. 214.

[83] Glass taps were probably found to be unsatisfactory on account of leakage. However, ground glass stoppers were introduced into chemistry laboratories in the 1770's (Smeaton, 2000, p. 221).

[84] Landriani, 1775b, p. 12.

2.

The inflammable air test
and Volta's eudiometers

As mentioned in the preceding chapter, Volta sent a letter dated December 10th, 1776, to Priestley informing him about his discovery of a new kind of inflammable air associated with marshes. Priestley included an extract of the letter as an appendix in the third volume of his *Experiments and Observations*.[1] This year, 1776, in which Volta discovered, described and analyzed the inflammable air released from marshes, can be taken as his official entry into the field of chemistry.

The choice of this new line of investigation, the chemistry of airs, is an example of Volta's wide-ranging interest in the field of natural philosophy. The premises for Volta's early development as a natural philosopher in the 1760's and 1770's came partially from an informal network of acquaintances and friends and partially from his self-taught training in the classical tradition, as well as by the reading of seventeenth and eighteenth-century classics, notably Newton, Musschenbroek, Dufay, Nollet, Franklin and Beccaria.[2] In the beginning, Volta developed his scientific work privately, his friendship with the cannon Giulio Cesare Gattoni being of priceless assistance. From 1765 onwards, and for many years, Volta made use of the well-equipped physics laboratory that Gattoni had set up at his home in Como, Volta's birthplace.[3]

Volta's approach to the chemistry of airs

Volta's interest in airs must be placed in the context of his scientific correspondence. It is worth mentioning that Volta lived on the periphery of the powerful scientific centres of London and Paris. However, since 1763 he had been in touch with some leading figures in natural philosophy,

---

[1] Priestley, 1777, pp. 381-383.

[2] Pancaldi, 2003, pp. 27-28.

[3] Gigli, 2002, pp. 54-55.

especially expert electricians, in an attempt to make himself known on the scientific stage, and during this time he also started some correspondence with Giambatista Beccaria in Turin and the Abbé Jean-Antoine Nollet in Paris, which remained somewhat intermittent. But from 1767 things improved when he expanded his circle of correspondents, which included Joseph Priestley in England and Lazzaro Spallanzani in Pavia. Before March 1772, Volta wrote to Joseph Priestley, the author of the highly successful *History and Present State of Electricity* (1767), with the intention of establishing an intellectual relationship. On this occasion, things went better, and Volta persisted with the correspondence in the hope that his electrical discoveries might be included in a future edition of Priestley's *History*.

In October 1774, Count Firmian appointed Volta to the position of superintendent of public education in Como. This job did not absorb all of Volta's energy and left him time to progress from an amateur to an expert in natural philosophy. He made an important addition to his credentials in this field in 1775, when he successfully built his innovative electrical machine, the electrophorus. This discovery impressed Count Firmian and in November 1775 Volta was appointed permanent teacher of experimental physics at the *ginnasio* (grammar school) of Como.[4]

Volta announced the discovery of his electrophorus in an open letter to Priestley in June 1775, published in the *Scelta di opuscoli interessanti sulle scienze e sulle arti*, but it was not until almost a year later that Priestley responded to Volta, when he realized that Priestley was more interested in his researches on airs than on issues relating to electricity. Volta felt that his connection with the European scientific community was weakening and it was probably no coincidence that at this time he also switched his attention to the investigation of airs.[5] Another line of contact with England was through Magellan, with whom he established a fluid correspondence that shows how he furnished Volta with apparatus to support his research and teaching. The extraordinarily widespread nature of Magellan's correspondence made him the chief scientific agent of the late eighteenth century; he supplied his many correspondents with the books and instruments they requested, Volta's forthcoming eudiometer among them.[6]

---

[4] Home, 2000, pp. 117-119; Pancaldi, 2003, pp. 27-28, 31-32.

[5] Home, 2000, p. 121.

[6] Volta's correspondence with Magellan was very rewarding for Volta (Ibid., pp. 123-131).

Inventing the inflammable air eudiometer

Volta's observation and collection of marsh air,[7] obtained by stirring up the mud at the bottom of stagnant pools in marshes at Angera around Lake Maggiore,[8] are documented in the seven letters to his friend Father Carlo Giuseppe Campi, written between November 1776, and January 1777.[9] Volta observed that this air was different from the inflammable air obtained by dissolving metals in acids, since it burned slowly with a glowing flame,[10] for which coined the name of *aria infiammabile native delle paludi* (inflammable air native to marshes). While comparing both kinds of inflammable airs, Volta observed how after blowing up a mixture of inflammable and atmospheric air, ignited by an electric spark, its volume diminished. Early in 1777, Volta published those seven letters to Father Campi in the book *Lettere del Signor Don Alessandro Volta sull'aria infiammabile native delle paludi* in which he described these observations (Figure 2.1).[11]

---

[7] Marsh gas is a common name for gas in wetlands, whose principal component is methane, produced naturally within some geographical marshes. The surface of marshes is initially porous vegetation that rots to form a crust that prevents oxygen from reaching the organic material trapped below. This is the condition that allows anaerobic digestion and fermentation of any plant or animal material, which incidentally also produces methane.

[8] Lake Maggiore is located on the south side of the Alps. The lake and its shoreline are shared between the Italian regions of Piedmont and Lombardy and the Swiss canton of Ticino.

[9] Meanwhile, Volta was writing a treatise on this air (*Proposizioni e sperienze di aerologia*), to be published in 1776, Giuseppe Jossi (one of his students) being the official author.

[10] In general, this inflammable air, as it was known at that time, could be identified as an individual gas, or mixtures of gases such as hydrogen and methane. Other kinds of inflammable air, such as carbon monoxide and volatile hydrocarbons coming from the distillation of organic materials, were distinguished later.

[11] Volta, 1777a. A summary of these seven letters was published in the *Scelta di opuscoli* (1777, pp. Vol. 28, pp. 43-78; VO, Vol 6, pp. 19-102)

Figure 2.1 Frontispiece of Volta's first letter to Father Campi showing the collection of marsh air samples.

From *Lettere del Signor Don Alessandro Volta sull'aria infiammabile native delle paludi,* (Milano, 1777)

These letters were widely distributed in Italy and abroad. For example, between 1777 and 1778 four German translations were issued (*Briefe über die entzündbare luft der sümpfe: nebst drey andern briefen*), and the French translation appeared in 1778 (*Lettres de Mr. Alexandre Volta sur l'air inflamable des marais, auxquelles on a ajouté trois lettres du même auteur, tirées du Journal de Milan).*[12] Volta considered that the three letters to the Marquis Franscesco Castelli, written between April and May 1777,[13] were a sequel to the preceding ones to Father Campi, and decided to include them in the German and French translations of the original book in Italian. He had already begun to think about a new apparatus that would be useful for both observations and experiments with inflammable air. Thus, in his second and third letter to Francesco Castelli he revealed the invention of an electric pistol (Figure 2.2) that worked by triggering an electric discharge in a sample of inflammable air obtained by dissolving metals (such as iron) in acids.[14] According to Volta, the reading of

[12] Volta, 1778a.

[13] These three letters were first published in the *Scelta di opuscoli* (1777, Vol. 30, pp. 86-96, 97-109; Vol. 31, pp. 3-24; VO, Vol. 6, pp. 125-150)

[14] This pistol, which was also used for recreational purposes, became a teaching resource and was known by other names such as *moschetto ad aria infiammabile, pistola elettrico-flogopneumatica, pistola elettrico-aereo-infiammabile, pistola elettrico-infiammabile.*

Priestley's *History of Electricity* inspired him to use an electric spark to ignite the air mixture.[15]

Volta designed this pistol with different aims in mind: first, as an alternative to gunpowder as a weapon of war; second, as a device for transmitting distant signals and, third, as an instrument for measuring the degree of inflammability of airs that could also help to determine the nature of inflammable air.[16]

Figure 2.2 Volta's musket and electric pistol.

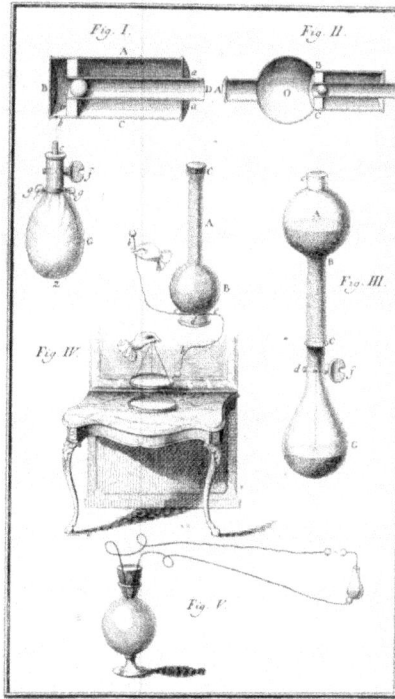

Fig. 1 and *Fig. 2*, show drawings of the musket charged with a mixture of common and inflammable air from a bladder. *Fig. 3* shows the electric pistol charged with inflammable air from a brass flask. The charged pistol could be fired by an electric spark triggered from an electrophorus (*Fig. 4*) or a Leyden jar (*Fig. 5*). From *Lettres de Mr. Alexandre Volta sur l'air inflamable des marais, auxquelles on a ajouté trois lettres du même auteur, tirées du Journal de Milan* (Strasbourg, 1777).

---

[15] Volta, 1777b, p. 98, note; 1778a, p. 157, note c; VO, Vol. 6, p. 133, note a.

[16] Beretta, 2000, pp. 55-56.

In his experiments with the electric pistol, Volta observed that when the device was filled with inflammable and common air, the mixture diminished in volume after the detonation.[17] In order to simplify the experiment and make it more remarkable, Volta redesigned the electric pistol and began developing a prototype instrument of his later eudiometer without displaying any image of it.[18] Finally, on September 2nd, 1777, he presented a perfected model of this instrument in an open letter to Priestley, published in the *Scelta di opuscoli*. This new apparatus measured the reduction in volume that occurred when a mixture of the air in question and a certain amount of inflammable air was fired by means of an electric spark.[19] Volta's detailed description of the apparatus was followed by an explanation about his complete ignorance of Waltire's experiments.[20]   However, it was not until April 1778, after he had successfully repeated his experiments before the Académie des Sciences, that an account of Volta's work was published in the *Observations sur la physique*.[21] A French translation of Jean Senebier of Volta's first letter to Priestley was published in the *Observations sur la physique* some months afterwards.[22] This paper included a reproduction of the plate published in the *Scelta di opuscoli* depicting the first example of a Volta eudiometer.[23]

---

[17] Volta, 1778a, p. 182; VO, Vol. 6, pp. 146-147.

[18] Volta 1778a, pp. 183-185; VO, Vol. 6, pp. 147-148. This prototype was described in the third letter to Francesco Castelli of May 1777. It consisted of a long glass tube that widened at the base and was closed at the upper end by a stopper fitted with two wires. The tube was placed in water and filled with a mixture of common and inflammable air. After the mixture had been exploded, the space occupied by the residue was measured.

[19] Volta, 1777c; VO, Vol. 6, pp. 175-184.

[20] Another appendix in the third volume of Priestley's *Experiments and Observations* included a letter from John Waltire, an English lecturer in natural philosophy. Waltire described an experiment in which he had also observed the diminution in volume when burning inflammable air in a closed recipient. Volta stated that he had been unaware of this earlier experiment by Waltire and claimed that his own experiments were more varied and accurate.

[21] Volta, 1778b; Le Roy, 1778. The Baron Philippe-Frédéric de Dietrich performed these experiments after having repeated them with Volta himself.

[22] The Genevan botanist and naturalist Jean Senebier (1742-1808) had a literary and theological education and was ordained a minister in 1765. After a stay in Paris to finish his studies, he returned to Switzerland in 1769 and became pastor of a church in Chancy, where he remained until 1773, when he was appointed city librarian of Geneva. He soon formed part of the intellectual circle consisting of Charles Bonnet,

The development of the inflammable air eudiometer

Volta's first portable eudiometer

Volta described the perfected prototype instrument developed from his original electric pistol.[24] The new instrument was extremely simple, ingenious and easy to use (Figure 2.3).

> It consisted of a graduated glass tube at least 1 inch in diameter and 14 to 15 inches long (*Fig.1*, A). Two small brass wires, each ending with a ball (d, d), were inserted into a cork stopper sealed with mastic that closed off the upper end of the tube. The lower end of the tube was funnel-shaped (E), and a small gap (c) was left between the ends of the two wires inside the tube. The device was complemented with a graduated auxiliary tube (ef), which was narrower and longer than the tube (A). Its open end (e) was lined with leather on the outside and fitted into the mouth (E) of the tube (A).[25] Firstly, the tube (A) was filled with water, then inverted and dipped into a trough of water (C). Several measures of inflammable[26] and common air were introduced by means of the tube (ef) so that the volume of the air contained in the tube reached the highest degree of the scale (B). With the help of a Leyden jar, a spark was discharged through the gap between the two wires and ignited the mixture inside the tube. The air expanded and the water rose up until it reached a level on the scale indicating the diminution in volume of the gas mixture.[27]

Abraham Trembley and Horace-Benedict de Saussure. Volta established a regular contact with Senebier in the course of September, 1777, during his first trip to Switzerland (Pancaldi, 2003, pp. 115-116, 153-154).

[23] Volta, 1777c, p. 12; Ibid., 1778c, planche III.

[24] For an early extensive study of Volta's development of the inflammable air eudiometer, see Osman (1958).

[25] This tube (ef) is not drawn to scale.

[26] It was the inflammable air of metals (i.e. hydrogen). All the results reported by Volta were obtained with inflammable air prepared by mixing iron with diluted vitriolic acid.

[27] Volta, 1778c, p. 366, planche III, Fig. 1; VO, Vol. 6, pp. 177-178.

The main substance formed in the explosion of the gas mixture is water:

$$H_2 \text{ (g)} + O_2 \text{ (g)} = H_2O \text{ (l)}$$

Figure 2.3 Volta's prototype of an inflammable air eudiometer.

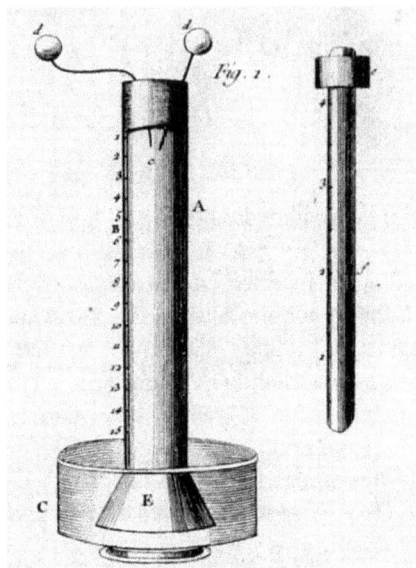

From *Observations sur la physique, sur l'histoire naturelle et sur les arts* (1778),
plate 3.

Volta summarized the results of a large number of experiments relating the ratio of inflammable to common air with the different types of explosion (very small, small and very large). He stated that these results were repeatable, provided that the strength of spark (triggered by an electrophorus or an electrical machine), the quality of inflammable air and the goodness of common air remained constant. For Volta, this device could be also used for the nitrous air test. In principle, he was mostly interested in the nature of inflammable air rather than in testing the respirability of common air, and he adapted his new instrument to the needs of this particular research. However, only at the end of his letter to Priestley does Volta associate the new instrument with the eudiometer when he seeks Priestley's approval, asking rhetorically:[28]

However additional gaseous substances - such as ozone and different species of nitrogen oxides (NO_x) - are produced in the atmospheric air as side products during the combustion of hydrogen. Furthermore, the presence of water vapour in the air provides the formation of nitric acid from the mono-nitrogen oxides and, as a consequence, the water produced turns acidic.

[28] Ibid., 1778c, p. 80; VO, Vol. 6, p.182. 'Che ne dite, Signore, di questa novissima foggia di *Eudiometro*? Non lo è difatti?'

'What do you say, Sir, about this newest form of *Eudiometer*? Is it not, in fact so?'

Volta provided the answer himself by placing the instrument in a utilitarian context by saying that giving it the name "eudiometer" was not only because it was a question of testing the air respirability but also of portability.[29] Indeed, it was to this purpose that he devoted the final part of the letter to describing a portable version of the instrument. Portability was a necessary requirement for an instrument designed for determining the respirability of common atmospheric air, which could vary in different locations and according to diverse circumstances such as meteorological conditions. Volta's description of his portable eudiometer (Figure 2.4) was important because it constituted the first step forward in the development of further versions of his inflammable air eudiometer.

The main section of this portable eudiometer consisted of a small oval jug (*Fig. 2*, A) which could contain four ounces of water. Each of its two ends had a brass tap (stopcock) (D, C). The tap (C) ended in an extension (B) that could contain about an ounce of water. Two brass wires were implanted in the metallic profile of both taps, and converged towards the centre of the jug closer than one line apart. The wire coming from the tap (D) had to be longer than the other wire to ensure that it was out of the water when performing the test (Figure 4, right).[30] The other section of the device was a jug (E) with a copper tap that contained inflammable air. When in use, the two sections of the device were fitted together. Firstly, the jug (A) and its extension (B) were filled with water, and then the water contained in (B) was drained out after closing the tap (C). The neck (e) of the extension (B) was then inserted into the mouth of the jug (E). On opening the key (C) the extension (B) filled with water falling from (A) and the air contained in (B) gurgled upwards to (A). After closing the tap (C) and opening the tap (E), the water contained between the two taps of the extension (B) fell in the jug (E) and was replaced by an equal amount of inflammable air coming from the jug (E). On closing

---

[29] Later, in 1790, Volta reinvented the new term *aparato universale per l'infiammazione delle arie in luogo chiuso* (universal apparatus for ignition of airs in closed spaces) for his eudiometer, as he was convinced that the apparatus was unsuitable for measuring the salubrity of the air (Seligardi, 2000, pp. 33-34).

[30] These two wires are not depicted in the French version of the letter published in the *Observations sur la physique*. On the other hand, they were depicted in the *Scelta di opuscoli* and in *Le opere di Alessandro Volta* (Figure 2.4, right).

the tap of the jug (E) and opening the tap (C), the inflammable air rose in the jug (A) displacing an equal amount of water, as before. Two equal measures of common and inflammable air were thus introduced into the main body of the device. The latter series of operations were then repeated, so that two measures of inflammable air and one of common air were introduced into the jug. Since this mixture was not yet ignitable, additional measures of common air had to be introduced. To that end, the upper part (AB) was then detached and inverted so that the tap (D) faced downwards and the small amount of water remaining in the jug (A) covered the inner part of the tap (D). This tap (D) was not perforated completely, but had two spherical or conical holes that did not communicate directly. With each half rotation of this tap, one of these holes faced the water contained in (A). This water descended, filling the hole and displacing the air, which rose in the form of one or two bubbles. On repetition of this operation, a small amount of common air was added to the mixture, and the respirability of the air could be measured in terms of the number of the half rotations needed to make the mixture inflammable.[31] Although Volta did not describe how the mixture was ignited, this was presumably done with the help of a portable electrophorus, which discharged a spark between the two brass wires in the centre of the main jug.

Figure 2.4 Volta's portable inflammable air eudiometer.

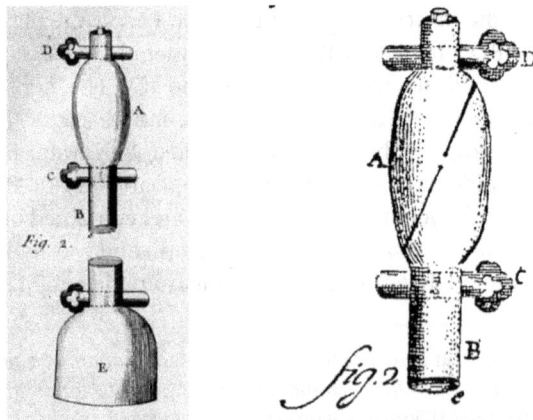

(Left) From *Observations sur la physique, sur l'histoire naturelle et sur les arts* (1778), plate 3. (Right) From *Scelta di Opuscoli* (1777), p. 121.

[31] Volta, 1778c, pp. 371-373, planche III, Fig. 2; VO, Vol. 6, pp. 182-183.

Volta was aware of some of the drawbacks in his new eudiometer; namely, its lower accuracy and sensitivity in comparison with other existing eudiometers (i.e. nitrous air eudiometers); the need for an electrophorus and the difficulty of operating it correctly, and the tedium involved in bubbling common air repetitively until explosion of the gas mixture could be achieved. Nevertheless, Volta adopted a proactive approach to promoting his instrument: it was easy to make; unlike the nitrous air eudiometers it did not need glass taps, and the test was performed over water instead of mercury. In addition, it was easier and more affordable to obtain inflammable air of more comparable purity than nitrous air, which altered too easily. Finally, inflammable air did not have such an unpleasant odour as nitrous air.

Interpreting what occurred inside the eudiometer

A follow-up paper on the combustion of inflammable air, also written in the form of a letter to Priestley in January 1778, was again translated by Senebier and published in the *Observations sur la physique* in 1778.[32] Volta reported the results from a large number of experiments with mixtures of inflammable air and airs at the opposite extreme of phlogistication, i.e. dephlogisticated and phlogisticated airs, as well as common air. The experiments he found most interesting were those using mixtures of inflammable and dephlogisticated air, the latter obtained from red lead.[33] At that time, Volta saw himself as the inventor of a new instrument and was eager to explore the extreme air mixture proportions with which his eudiometer was able to work.[34] He found, for instance, that from one to one hundred parts of dephlogisticated air could be mixed with eight parts of inflammable to obtain an explosion.

In this letter, Volta returned to an issue that he had already addressed in another letter to Landriani of July 1777, in which he expressed his conviction about the close analogy between the actions of inflammable and nitrous air on dephlogisticated air regarding the contraction in volume observed in both cases.[35] At the beginning of 1778, Volta

---

[32] Ibid., 1779; VO, Vol. 6, pp. 185-215. Volta considered that Senebier had made a bad translation of this second original letter (Osman, 1958, p. 230).

[33] Red lead, also called "minium", is lead (II, IV) oxide [2 $PbO \cdot PbO_2$]. When heated to 500 °C, it decomposes to lead (II) oxide and oxygen:

$2 Pb_3O_4$ (s) = $6 PbO$ (s) + O_2 (g)

[34] Volta, 1779, p. 279; VO, Vol. 6, p. 188.

[35] Ibid., pp. 159-160.

supported the idea that these analogous actions also implied analogous compositions in both airs. Accordingly, if one of the components of the two airs was phlogiston, and the other ingredient of nitrous air was nitrous acid, then a pure acid should also form part of the composition of inflammable air. He argued that no acid had been detected in the decompositions of inflammable air performed so far, due to the minute proportion of this acid in the inflammable air.[36]

Actually, Volta's first thoughts on the nature of inflammable appeared in the letters he wrote to Father Campi in November 1776. On the basis of the combustibility of both inflammable air and sulphur, Volta had proposed that inflammable air was a species of sulphur, a *solfo aereo* (aerial sulphur) resulting from the combination of an aerial acid with phlogiston.[37] Later, in January 1777, he abandoned this interpretation and defined inflammable air as a compound of phlogiston with an undescribed aerial salt. When inflammable air was ignited, phlogiston was disengaged from that aerial salt and transferred continuously into common air. Only a beautiful flame could be expected to result from the transference of phlogiston from one air to another.[38] Experimenting with mixtures of inflammable and common air in a closed tube to find the best mixture for his electric gun provided Volta with an opportunity to conjecture further on the nature of inflammable air. In the third letter to Francesco Castelli of May 1777, Volta pondered the possibility of collecting the other principle of inflammable air that might have been precipitated after the ignition of the mixture. He hoped that any vapour deposited on the inside of the tube, or whatever solid or liquid had been formed, would be observable, but in the event he was not able to detect anything. He believed that it was the small quantity of missing air that had made the detection difficult.[39] Nevertheless, in a letter to Senebier of December 1777, he revived this hope again by planning to work with larger volumes of both airs in order to obtain greater diminutions in volume, and also by using mercury instead of water.[40] He took up this idea again in a second letter to Priestley

---

[36] Volta, 1779, p. 287; VO, Vol. 6, p. 195. Using a thought experiment, Volta conjectured that in equal volumes of both airs, the amount of nitrous acid in nitrous air was nine times greater than the amount of the acid of inflammable air (Volta, 1779, pp. 291-292; VO, Vol. 6, pp. 202-203; Holmes, 2000a, pp. 97-98).

[37] VO, Vol. 6, p. 41.

[38] Ibid., p. 82.

[39] Ibid., pp. 149-150.

[40] Ibid., pp. 252-253.

written early in January 1778, convinced that inflammable and dephlogisticated air did not destroy each other but changed their original aerial form into another different one. It therefore follows that almost a year later Volta was greatly interested in what may have proved to be the counterpart of phlogiston in inflammable air.[41]

It would have been no easy task to detect the formation of water, because the burning of the air mixture was carried out in a closed tube above water, and the condensed vapour would have simply been confused with the existing water in the tube. Alternatively, the new water drops formed on the inside of the tube would have merged with many such others sprayed as a result of the shaking of the tube after the explosion. Replacing water by mercury might have overcome these drawbacks and helped Volta to detect the formation of water drops, but he himself stated that he had never had enough mercury to effect this replacement.[42] Had he used larger volumes of both airs, as Laplace and Lavoisier would do later in 1783, he may well have been successful. However, with this large-scale experiment Volta could have obtained a condensed acidic material,[43] which is what Lavoisier, Laplace and Cavendish[44] would detect in 1785. It is quite likely that he would have also tested this acidity, as he did in 1777 in his experiments with the electric pistol, thus confirming his belief that an acid was the companion of phlogiston in the inflammable air.

In a manuscript assumed to an extract of a letter written in French circa 1785, Volta described a crucial experience which he regarded as the key to all others of the same kind. He ignited a mixture of two measures (200 degrees) of a very pure dephlogisticated air with half a measure (50 degrees) of a very pure inflammable air in his eudiometer. He then repeated successive additions of half a measure of inflammable air to the remaining residues obtained after the corresponding ignitions. He was able to achieve up to an eighth addition, which left a final residue of 8-10

---

[41] Volta, 1779, pp. 286-287; VO, Vol. 6, pp. 196-197.

[42] This Volta's claim of 1783 may be understand as a mere self-justification for his inability to discover the formation of water (Ibid., pp. 410-411; Seligardi, 2000, p. 39).

[43] See note 27.

[44] In June 1785, Cavendish read the paper *Experiments on Air*, from which he concluded that phlogisticated air (nitrogen) was enabled to combine with the dephlogisticated air (oxygen) by means of the electric spark, the phlogisticated air being thereby reduced to nitrous acid (see note 27). These experiments also served Cavendish to establish that common air consisted of one part of dephlogisticated air mixed with four of phlogisticated air (Cavendish, 1785, pp. 376, 79, 382)

degrees of phlogisticated air. Volta wondered whether from these experiments he was entitled to conclude that inflammable and dephlogisticated air mixed in a proportion of 100 to 48-50, respectively, completely destroyed themselves by ignition, and whether the small residue of phlogisticated air was already mixed in one air or the other, since it was impossible to obtain absolutely pure airs. The manuscript ended abruptly with a further hesitation:[45]

'I wondered in my first experiments what became of these two airs that disappeared entirely? Having observed a nebulous vapour filling the container while airs languished [...]'

Volta had been able to ascertain the volumetric proportion for a complete combination of both airs, but unable to identify this 'nebulous vapour' as being anything other than phlogisticated air. His main concern was with the phlogistication of airs, and this was an epistemological barrier to conceive the synthesis of water.[46]

Completing the device of the inflammable air eudiometer

Leaving aside his theories and conjectures on the nature of airs, by January 1778 Volta had returned to the question of the diminution in volume after ignition of the air mixture. He noticed that this diminution depended on the goodness of the common air. It was quite clear that the largest decrease occurred when mixing inflammable air with dephlogisticated air instead of common air, and that a lesser decrease was observed when mixing the same inflammable air with vitiated air coming from extinguishing candles or respiration. Volta definitely stated that inflammable air could behave in the same way, or even better, than the nitrous air when measuring the degree of respirability of different airs using the diminution in volume that they underwent.[47] In the inflammable air test, airs were mixed in a fixed proportion (usually in equal parts), the observed contraction in volume being the final outcome of the test. On the other hand, in the nitrous air test, airs were mixed in variable proportions until a maximum contraction in volume was reached, which was taken as the test result. On the basis of the

---

[45] VO, Vol. 7, pp. 221-222.

'Je me demandais dans mes premières expériences qu'est-ce que donc que deviennent des deux airs qui disparaissent entièrement? Ayant ensuite observe une vapeur nébuleuse qui remplissait le récipient à mesure que les airs se consommaient [...]'

[46] Seligardi, 2000, pp. 38-46; Holmes, 2000a, pp. 110-111.

[47] Volta, 1779, p. 296; VO, Vol. 6, p. 207.

former portable version of his eudiometer, Volta took another step forward in the devising of his apparatus.

He adapted the device described in his first letter to Priestley (*Fig.2*) by using the same jug (A), the tap (D) now being superfluous and replaced by a cork stopper sealed with mastic with two metallic wires (as in *Fig. 1* of the same letter). Alternatively, the tap (D) could also be replaced by a brass stopper transfixed by a glass tube through which a brass wire was passed and turned towards the inside against the brass stopper itself. This assembly was to be sealed with mastic to make it airtight. The procedure described for the portable version was followed until one measure of common and one measure of inflammable air had been introduced into the jug. One measure was the amount of air contained in the prolongation (Be) of the jug (A). The air mixture was then ignited with the help of a portable electrophorus. After that, the extension (B) was filled with water; the mouth (e) was closed with a finger tip and immersed in a glass vessel containing water. Afterwards, the finger was removed and the tap (C) opened. Consequently, water rose rapidly in the jug (A) to occupy the place left by the missing volume of air. The tap was then closed and the jug was removed from glass vessel.

For the next step in the test, a long tube (ABB) with a diameter of four lines was used (Figure 2.5). The tube was calibrated so as to contain a little more than two measures of air. The mark (B) on the lower part of the tube corresponded to one measure. The next upper mark (C) corresponded to two measures. A graduated scale was attached to the tube. This scale was divided into 90 degrees from (C) to the mid-point (B), and further equal degrees from this point (B) to the top point (A). Once filled with water, this tube was attached to the extension (B) of the jug (A) so that it fitted with its mouth (e).[48] This two-part assembly was turned upwards so that on opening the tap (C) the remaining air in (A) rose to the top of the tube, while some of the water in the tube fell into the jug. The level of the water in the tube therefore indicated directly the contraction in volume undergone by the air mixture (55 units or degrees in the drawing), which had to be proportional to the goodness of the air sample examined. This proportion was indicated on the graduated scale by the degree of respirability.

---

[48] Figure 2.5 shows that Volta used a crooked brass wire in the jug (A) as part of the device conceived to trigger electric sparks.

Figure 2.5 Sketch drawing of a Volta-type eudiometer (1778).

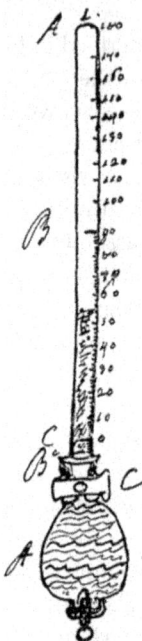

From *Le Opere di Alessandro Volta.* (Milano, 1918), Vol. 6, plate 7.

At that time, in Volta's mind the concepts of "goodness" and "respirability" of the air were apparently interchangeable. He regarded this new version of his apparatus as being not only manageable and portable but also versatile, since it proved itself able to function as a eudiometer in two different ways. Firstly, it could count the number of the common air bubbles needed to make the air mixture inflammable, as explained in the first letter, and secondly, it could measure the reduction in the volume of the air mixture with the aid of the graduated scale. He doubtless preferred the second version, because it converted the eudiometer into a comparable instrument. The disadvantage with the inflammable air eudiometer was that it did not work with poor respirable airs, whereas the nitrous air test was effective with both rich and poor respirable airs.[49]

---

[49] Volta, pp. 296-297; VO, Vol. 6, pp. 207-208.

Samples of common air with this eudiometer ordinarily gave between 50 and 55 degrees, while the best common air never yielded a result as high as 60. The air in which a candle had been extinguished gave a value between 30 and 35. Volta noticed that an excessive amount of inflammable air neither burnt nor decomposed, but remained in the final residue. This meant that one measure of inflammable air surpassed the amount necessary to saturate an equal measure of common air with phlogiston. For this reason, the residual air after ignition was greater than the measure of common air. Therefore, if a new measure of fresh common air was introduced in the jug (A), after a second ignition of the mixture the residual air would be less than the two measures of common air. To illustrate this situation, Volta drew attention to the results of a number of experiments in which successive measures of inflammable air were added to a mixture of equal measures (four instead of one) of common and inflammable air, the mixture being ignited after each addition and the residual air measured (Table 1). These results were only one part of a series of eight experiments with mixtures of common and inflammable air.[50]

Table 1
Results of the sixth series of experiments based on equal parts of common and inflammable air.[51]

Measures of common air	Measures of inflammable air	Burning	Residue
4	4	Medium strong	5
	Add 1	Null	6
	Idem.	Null	7
	Idem.	Null	6 ½
	Idem.	Mediocre	7 ½
	-	Null	-
	2	Weak	
	Add 2	Strong	
	Idem.	Null	
	Add several bubbles	Mediocre	
	Add 2	Null	
	Idem.	Null	
	Idem.	Weak	About 8 saturated with phlogiston

---

[50] Volta, 1779, pp. 283-284, 299; VO, Vol. 6, pp. 193-194, 211.

[51] The results from the fifth to the eighth series of experiments were incorrectly printed in *Observations sur la physique*. The print in *Le Opere di Alessandro Volta* is the correct version (VO, Vol. 6, pp. 193-194).

Addressing the synthesis of water

In September 1777, during Volta's first trip to Switzerland, he had shown Senebier how to use his new eudiometer and made arrangements to have a eudiometer designed for him so he could repeat the experiments.[52] This was constructed in Geneva and turned out to be better than Volta's own. Back in Como to resume his teaching obligations, in April and May 1778, Volta received two letters from Senebier containing the results obtained with this improved version of the Volta eudiometer. These results revealed the difficulty in obtaining inflammable air without any traces of common air. Actually, Volta applied the nitrous air test to check the purity of the inflammable air, since this test should show no contraction in volume if no common air was present. The purity of the inflammable air was a critical factor for obtaining the greatest possible reduction in volume on ignition of inflammable air in common air.[53]

In two letters he sent to Senebier in the second semester of 1778, Volta described some modified versions of his eudiometer. In the first letter, written after 20 June 1778, he outlined a simple eudiometer (Figure 2.6, left) for measuring the amount of inflammable air in common air, since he was sure that there was always some present.[54] In the second letter, dated September 10th, 1778, Volta expressed his satisfaction for the delivery of Senebier's version of his eudiometer, which had been sent by Senebier himself. The rest of the letter was almost entirely devoted to suggesting a new design for his eudiometer made completely of glass, which he intended to use with mercury instead of water in order to determine again what was precipitated during the burning, although this eudiometer prototype never actually entered into operation. Volta's eagerness to improve the accuracy of his eudiometer led him to design an accessory for the instrument that consisted of a tube graduated from 0 to 100 (Figure 2.6, right). The capacity of the tube was almost equal to the diminution in volume observed after the ignition of a mixture of three measures of inflammable air and two measures of good common air.[55]

---

[52] Holmes, 2000a, p. 95; VO, Vol. 6, pp. 167-169.

[53] Holmes, 2000a, pp. 99-100; VO, Vol. 6, pp. 257-260, 261-266.

[54] Ibid., p. 294.

[55] Ibid., p. 299; Osman, 1958, pp. 236-240.

Figure 2.6 (Left) Sketch of a Volta-type eudiometer sent to Senebier in 1778. (Right) Sketch of an accessory tube for improving the accuracy of the eudiometer (1778).

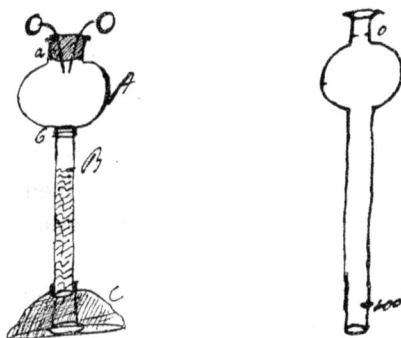

From *Le Opere di Alessandro Volta* (Milano, 1928), Vol. 6, pages 294 and 299.

Between 1779 and 1790, Volta temporarily interrupted his eudiometrical publications, since during the 1780s his attentions were concentrated on other affairs. He managed to change his position as a secondary school teacher in Como for a chair of natural philosophy at the University of Pavia. Regarding his work on electricity, he was busy writing a paper on conductors, which included the theory of the electrophorus as a case study.[56] During that time, he was faced with a serious setback in his chemistry career. In April 1782, he visited Paris on the first stage of a lengthy scientific tour. With the *condensatore*, his new machine conceived in 1780 and capable of detecting very weak electrical charges, he contributed to Lavoisier and Laplace's experiments to demonstrate that when liquids evaporated or were in effervescence they gave signs of electricity. While in Paris, Volta presented his eudiometer to the chemist Jean D'Arcet and other French scientists. It is unlikely that Volta would show his eudiometer to D'Arcet and not to Lavoisier.[57] At the end of 1783, Landriani informed Volta that Lavoisier and Laplace had that June been able to verify that what was condensed after the ignition of a mixture of inflammable air and oxygen with an electric spark was pure water.[58] On receiving this news Volta felt

---

[56] Pancaldi, 2003, pp. 115-121.

[57] Beretta, 2001a, pp. 45-47. In the unpublished inventory of Lavoisier's laboratory, one *eudiomètre de Volta*, one *pistolet en cuivre* and one *pistolet de Volta en fer blanc* appeared (Ibid., 2000, p. 63).

[58] Gaspard Monge independently achieved the same result in June-July, 1783, in the laboratories of the École du Génie of Mézièrs. He used a version of Volta's

demoralized. He had believed himself to be close to this discovery and claimed some part in it, albeit without denying credit to Lavoisier and Laplace for their achievement. However, he was far from accepting the theoretical consequences of this finding. In October 1783, he was still of the opinion that water was a simple element and not a compound of the inflammable and dephlogisticated air. It was the water that was contained in these airs and not these airs that were contained in the water.[59]

Volta continued his eudiometrical reflections when the chemistry and botany professor at the University of Pavia, Giovanni Antonio Scopoli, requested him to contribute to the Italian translation of Macquer's *Dictionnaire de chymie*. In 1783, he was able to benefit from the notes he prepared for this work; he contrasted his own eudiometer with the family of nitrous air eudiometers known to date, while in the notes for the article on the nitrous air he warned about the uncertainty associated with this air for the great difficulty involved in always obtaining it with equal strength. The other drawback was that the nitrous air test was more time-consuming than his inflammable air test, in which the decrease in volume occurred suddenly. Furthermore, the process of making the nitrous air was much more inconvenient, and the smell of this air was also more unpleasant than that of the inflammable air. In Volta's opinion, these were good and sufficient reasons for recommending the inflammable air eudiometer without hesitation, although it required a more complex device.[60] However, Volta's major contribution to the Italian version of Macquer's dictionary was the drafting of the new entry *Eudiometro*. This article served without doubt to distance the eudiometer further away from medicine. The instrument was declared to be incapable of judging the salubrity of the air, but only its respirability by determining the relative amount of dephlogisticated air.[61] No significant contribution to the design of the instrument was included in this article.

Volta's return to eudiometrical publications

In the early 1790s, after a hiatus of almost eleven years, Volta finally resumed his eudiometrical publications. Between 1790 and 1791, he published his last contribution to the inflammable air eudiometer in three

eudiometer to analyse the residual gas resulting from the ignition of a mixture of inflammable air and oxygen (Monge, 1786, p. 85).

[59] VO, Vol. 6, p. 411; Holmes, 2000a, pp. 108-111.

[60] Volta, 1783, pp. Vol. 2, pp. 227-228; VO, Vol. 6, pp. 420-421.

[61] Volta, 1783, Vol. 4, pp. 118, 121; VO, Vol. 7, pp. 64, 70.

articles, for which he chose to abandon English and French journals in favour of the Italian journal *Annali di chimica ovvero raccolta di memorie sulle scienze, arti e manifatture ad essa relative.*[62] Volta declared that his intention was to describe the additions and improvements that had been gradually incorporated into his eudiometer. The fact is that he spent little time in tracing the history of the instrument, but hastily announced its novelties by providing a detailed description of both the eudiometrical device and procedures for making the test replicable (Figure 2.7).

Figure 2.7 Final versions of Volta's eudiometer.

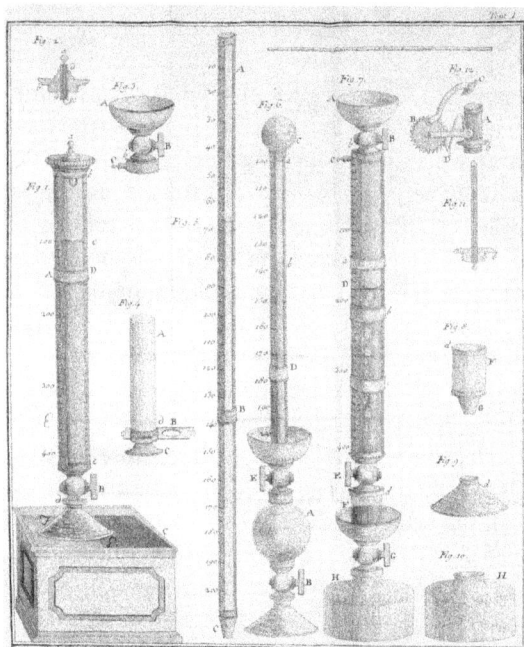

From *Annali di chimica ovvero raccolta di memorie sulle scienze, arti e manifatture ad essa relative* (1790), plate p. 297.

---

[62] This journal was also known as the *Brugnatelli Annali*. These three articles were reprinted in VO (Vol. 7, pp. 175-213). The third paper concluded with the announcement of an unpublished continuation (*Sarà continuato*). The choice of this journal to publish these articles was probably a matter of interior policy. In 1778 Count Firmian, Volta's patron, had already pressed Volta to publish his paper on conductors in the Lombard journal *Scelta di opuscoli* instead of in French or English journals (Pancaldi, 2003, p. 116).

The main body of the instrument described in Volta's article of 1790 consisted of a glass tube 15 or 16 inches long with a scale graduated in 400 degrees (*Fig. 1*), (see also Figure 2.8). The inner diameter of the tube was not to be less than 10 lines in order to facilitate the mixture of the two airs, while the thickness of the glass thickness was to be of at least 2 lines at least to resist the force of expansion due to the explosion of the gas mixture. The upper end of the tube was closed off by a brass lid (a), screwed by means of an interposed greased leather ring into another brass ring (b) sealed with onto the glass with mastic. This lid (*Fig. 2*) was actually a device designed to discharge an electric spark into the tube and was similar to that conceived by Volta as an alternative to the two-wire system described in his second letter to Priestley in 1778. It consisted of a brass wire curved in (c), and ended in a little ball (a). This wire was covered by a thin glass tube passing through the brass lid from top to bottom (dd), so the metallic wire (ac) remained insulated. The lower end of the wire (c) was bent upwards so that its tip was roughly one line away from the brass piece towards which it was pointed. An electrical spark triggered from an electrophorus on the ball (a) should excite another one at the tip (c) and ignite a mixture of inflammable and respirable air.[63]

At the lower end of the cylindrical container there was a brass component (cBdF) consisting of three sections that could be adapted into a single component or divided for ease of work into three sections which joined at (c) and (d). The ring (c) embraced the contours of the tube exactly, and like the other upper ring (b) it was sealed with Spanish wax or another good quality mastic. The tap (stopcock) was equipped with an airtight key (B), a quarter rotation of which revealed an aperture sufficiently wide for the simultaneous passage of air and water. Volta paid close attention to the making of this tap because it needed a craftsman of skill to drill a relatively large hole of at least 4 lines in diameter. This was because air had to rise through water, and the water had to drain without clogging. The foot (F) was funnel-shaped in order to introduce measures of air easily underneath.[64] Volta remarked on the extreme importance of working

---

[63] It often happened that the second spark failed because of a drop of water remained, filling the gap between the tip (c) and the brass piece. In such a case, a stronger spark from a Leyden jar charged with an ordinary electric machine or the electrophorus itself was needed (Volta, 1790, pp. 225-227; VO, Vol. 7, p. 196).

[64] Volta, 1790, pp. 178-181; VO, Vol. 7, pp. 177-178.

with an airtight instrument, and advised the interposition of greased leather rings in (a, b, c) as well as checking the instrument for leaks before starting the eudiometrical test. This part of the eudiometer resembles the perfected prototype instrument that Volta developed from his electric gun and which he described in his first letter to Priestley in September, 1777.

Figure 2.8 A copy of the Volta-type eudiometer depicted in *Fig. 1* (ca. 1790).[65]

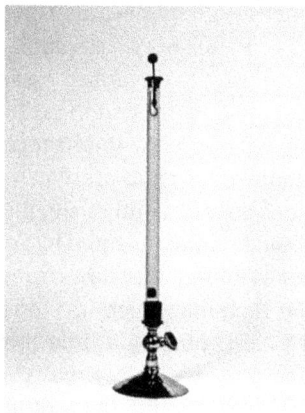

Photo Franca Principe, Museo Galileo, Firenze.

The first step in the eudiometrical procedure consisted of filling the tube with water from the pneumatic trough (C). To this end, the tube was laid down and immersed horizontally in this vessel, which had to be larger than that shown in *Fig.1*, with the tap (B) opened. The tube was then inclined so that the water from the vessel rose into the tube, displacing all the air inside. After that the eudiometer was stood upright on the shelf (G) so that the water in the vessel surpassed it by one or two inches. Volta proposed another and perhaps more convenient way of filling the recipient with water. This was to unscrew and remove the lid (ab), open the tap (B) and plunge the whole foot (F) into the water above the tap, or at least the ring (c). In this way the water entered freely up to the point of immersion. Then, after closing the tap, water was poured into the tube through the open upper end. Once the tube was filled, the lid (ab) was screwed on firmly to prevent any air bubbles from remaining inside. A mixture of

---

[65] The glass tube of this eudiometer is not graduated, perhaps because it is a modern replacement.

one measure of inflammable air and one or two measures of common air was then introduced into the tube, displacing the water. After exploding the mixture with an electric spark, the water rose back up the tube to a given height, which determined the respirability of the air sample.[66]

Volta's concern with accuracy reflected his commitment to an exact determination of the decrease in volume of the air mixture, and he gave warning about the existence of complications in this respect. Firstly, the need to introduce equal amounts of an air sample on several occasions. It was not enough always to use the same flask as a measure to ensure that this was done, since the difficulty resided in preventing the emergence of air bubbles in the measure. Secondly, the heat released to the air in the measure by contact with the hands also presented a problem.[67] In order to overcome both difficulties as well as other significant errors, Volta used a measure (*Fig. 4*) very similar to that invented by Fontana for his nitrous air eudiometer.[68] The capacity of this measure corresponded to 100 degrees of the scale engraved on the tube, which was in turn subdivided into 100 parts. In this way, the tube contained a little more than four measures. With regard to the graduation of this scale, Volta had to address the problem of ensuring the correct calibration of the eudiometrical tube, an issue that had already arisen in the case of the nitrous air eudiometers.[69] To remedy any parallax error when reading the position of the concave meniscus of water in the tube to determine the test endpoint, Volta implemented the same device that Magellan had adopted for his nitrous air eudiometer of 1783.[70] He fitted a ring (AD) to the tube, which could slide to the point where the top edge of the ring was tangent to the bottom of the concave surface of the water.[71] He devoted his first 1791 article to continuing the description of what he had changed or added to the apparatus to make it more manageable and adaptable to a greater number of tests.

The accessory device depicted in *Fig.3* could replace the brass lid (a) of *Fig.1*, as shown in *Fig. 7*. This was a brass component consisting of

[66] Volta, 1790, pp. 181-189; VO, Vol. 7, pp. 179-181.

[67] Volta, 1790, pp. 189-191; VO, Vol. 7; p. 182.

[68] Volta, 1790, pp. 192-199; VO, Vol. 7, pp. 183-185.

[69] See chapter 3; Tiberio Cavallo and the assessing of the nitrous air eudiometers.

[70] See chapter 3, note 45.

[71] Volta, 1790, pp. 199-214; VO, Vol. 7, pp. 185-191.

three sections: the bowl-shaped upper part (A), with a tap (B) that was screwed below into the ring (b), and the small section (C), quite similar to that in the device shown in *Fig. 2*, for the purpose of triggering an electrical spark within the tube. The whole component was multifunctional; firstly, for filling the tube easily without having to dip it entirely into the pneumatic trough or removing the brass lid; secondly, for measuring the reduction in volume after the ignition of the air mixture by letting water into the bowl (A) to replace the missing volume. Nevertheless, Volta considered that the most significant advantage of this device was the ease with which the residual air remaining in the tube after ignition could be transferred to another tube for further examination.[72]

Particularly advantageous was the transference to a narrow graduated glass tube 30 inches long (*Fig. 5*) in order to measure the amount of air over a much larger scale with the help of the sliding ring (B). The only open end of this tube finished with a small brass conical tube (C) that fitted into a similar conical hole at the bottom of the bowl (A, *Figs. 3, 7*). Once this tube full of water had been inserted into the bowl and the tap (B) had been opened, a column of air rose up through the tube, while the water drained along its walls. Since it was not always easy to find such a long, strong glass tube, it could be replaced by a shorter tube with a spherical or oval recipient (C) in the upper end (*Fig. 6*; Figure 2.9, left). Eudiometrical tests using one measure (100 degrees) of inflammable air and one or two measures (100 or 200 degrees) of common air, always left a residue higher than 100. Accordingly, for testing common air samples, the long tube could be shortened by replacing the volume of the tube corresponding to the first 100 degrees with a bulb (C) of the same capacity.[73]

The eudiometer depicted in this *Fig. 6* is a perfected version of the device sketched by Volta in Figure 2.5 (see Figure 2.9, right). The jug (A) in Figure 2.5 has been replaced by the glass globe (A) in *Fig. 6*, and a funnel-shaped foot of the same figure replaces the extension (Be) in Figure 2.5. The open end of the graduated tube fits at the bottom of a bowl in *Fig. 6*, whereas in Figure 2.5 it is directly attached in the extension (B) of the jug (A).

---

[72] Volta, 1791, Vol. 2, pp. 161-174; VO, Vol. 7, pp. 199-204.

[73] Volta, 1791, Vol. 2, pp. 174-186; VO, Vol. 7, pp. 204-209.

Figure 2.9. (Left) A copy of the Volta-type eudiometer depicted in *Fig. 6*, 115 cm height, with the narrow glass tube of *Fig. 5*. (Right) A view of the lower part of the same instrument.

Tempio Voltiano, Musei Civici di Como, Comune di Como.

The second 1791 paper was mostly devoted to the calibration of the graduated tube mentioned above. It is important to point out two notable particulars in Volta's last contribution to his inflammable air eudiometer. First, the absence of any theoretical commitment explaining what occurred inside the eudiometrical tube during and after the air mixture ignition. Second, the abandonment of the portability of the instrument. The last version of the eudiometer was intended to be a laboratory bench instrument rather than a portable one.

Volta's two letters to Priestley in 1777 and 1778 were not the only textual references he made to his eudiometer in the 1780s. Subsequent updates of Volta's eudiometer were already known from the description of the latest version of the inflammable air eudiometer that appeared in Guyton de Morveau's article "Air", written for the *Encyclopédie Méthodique* published in 1789. The fact is that working versions of the instrument were already circulating before the publication of Volta's last articles in 1790 and 1791. By May 1778, Senebier had been provided in Geneva with an example of the instrument that he had ordered to be made, a copy of which he later sent to Volta. Some time between 1782 and 1784, Lavoisier used a version of Volta's eudiometer to replicate the experiments on the synthesis of water.[74] In 1783, Martinus van Marum and Paets van Troostwijk submitted a eudiometrical paper to the Royal Danish Academy of Sciences and

---

[74] Beretta, 2001a, p. 47.

Letters, based on their own version devised to correct faults in Volta's eudiometer.[75]

In June 1785, Berthollet read a paper at the Académie des Sciences in Paris that awakened expectations for the analytical potential of Volta's eudiometer. Berthollet was able to determine the composition of ammonia with the aid of this instrument. He first decomposed a sample of dry gaseous ammonia by means of an electrical discharge. The resulting gas mixture was introduced into the eudiometer by adding successive measures of vital air (oxygen) and igniting the final mixture after each new addition. Berthollet found that the volumetric composition of ammonia was of nearly 3 parts of inflammable gas (hydrogen) to 1 part of atmospheric mophete (nitrogen).[76] Berthollet also used the same eudiometer to determine the quantities of different kinds of inflammable air from marshes, and those obtained from the distillation of charcoal, silk, sugar and oil, that were needed to mix with one hundred parts of vital air to achieve an ignitable mixture.[77]

Guyton de Morveau had in his possession the latest version of the eudiometer made by the Parisian instrument maker Pierre Bernard Mégnié.[78] Guyton had ordered the instrument for use in his lecture on hydrogen as part of the course of chemistry he was to give at the Dijon Academy (February–May 1789). He described this eudiometer as a part of the experiment XXXIV in the article "Air" that appeared in the *Encyclopédie Méthodique*. According to this description, which took the form of a simple drawing (Figure 2.10, left), the eudiometer can be seen as an apparatus constructed with different parts deriving from some sections later depicted in the plate published in the first volume of the *Annali di chimica* of 1790.[79]

---

[75] Turner, 1973, p. 245. John Cuthbertson probably was the instrument maker.

[76] Berthollet, 1788a, pp. 323-326. Berthollet also determined that the weight proportions for nitrogen and hydrogen were 0.85 (0.82) and 0.14 (0.17), respectively. The figures in parenthesis are the actual values.

[77] Berthollet, 1788b, pp. 340-344.

[78] Along with Nicolas Fortin and Étienne Lenoir, Pierre Bernard Mégnié was a leading instrument maker in Paris at the end of eighteenth century. Mégnié also made a barometer and a pair of gasometers for Lavoisier in 1787 (Beretta, 2014, p. 207).

[79] Guyton, 1789, pp. 718-719. The eudiometer described in the *Encylopédie Méthodique* was dimensioned differently than Volta's instrument described in the *Annali di chimica* of 1790. Thus, the cylinder (A) was 8-9 inches long with an inner diameter of 20-24 lines, and the thickness of the glass was 4-5 lines (see also Figure 2.10, right).

The lowest end of the instrument, which would correspond to the component (BdF) in *Fig. 1*, consisted of a funnel-shaped glass foot (C) with a tap (B) and a screw knob (i). This component was made of a platinum wire coated with greased leather and was used as a quick closing valve (stopcock). The middle section consisted of a cylinder (A), a graduated metallic scale (n), a piece (G) designed to trigger an electrical spark, a tap (D) and a bowl (E). This central section and the upper half of *Fig.7*, from the bowl at the top to the end of the tube, were almost identical. The upper part of the eudiometer consisted of a narrow tube with a graduated metallic scale attached. The tube was open at the bottom and ended with a top bulb of one measure of capacity. This part of the instrument was also practically identical to the upper half of *Fig.6*, from the bulb at the top to the end of the narrow tube. This tube was intended to give more accurate readings of the diminution in volume after igniting the air mixture. For that purpose, the tube was removed, filled with water and put back in the bowl (E), also filled with water, by covering the tube hole with a finger. Afterwards, the tap (D) was opened, and the gaseous residue that remained in the cylinder after the ignition rose in the tube. A part of the residue was collected in the top bulb, while the remainder occupied an upper segment of the tube. The length of this segment could be read in terms of hundredths of a measure.

Figure 2.10 (Left) Sketch of Volta's eudiometer described in the *Encyclopédie Méthodique*. From *Recueil des planches du Dictionnaire de chimie et de métallurgie, faisant partie de l'Encyclopédie Méthodique par ordre de matières* (Paris, 1813), Huitième classe, planche xxvi. (Right) A copy of a Volta-type eudiometer[80] at the IES Francesc Ribalta.

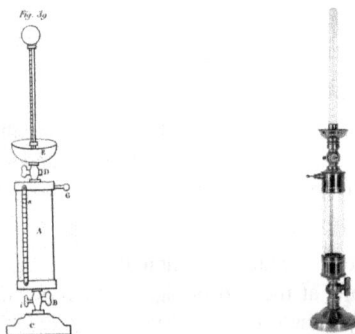

Courtesy of the IES Francesc Ribalta, Castelló de la Plana, Spain.

---

[80] The upper narrow tube of this eudiometer has no top bulb.

Later versions of the instrument had fittings made entirely of iron instead of brass. Iron, unlike copper, did not form an amalgam, so mercury could be used instead of water.[81] While the inflammable air eudiometer was being developed during the 1780s, the nitrous air eudiometer was enjoying a remarkable rise; this heyday is described in the following chapter.

Summary

Volta's eudiometer evolved from his electric pistol and traced an investigative pathway in which trial and error played a role in the process. The initial stages of this pathway were closely associated with Priestley's prestige as a leading natural philosopher at that time. Volta was seeking Priestley's intellectual backing and authority when he wrote to him for the first time describing his electrical investigations. Priestley had written his *History of Electricity*, as well as his *History and Present State of Discoveries Relating to Vision, Light and Colours* (1770), on the basis of previous writings on electricity and optics, repeating some of the experiments described therein. His authority was therefore grounded in his literary researches. However, from 1770-71 onwards, when he turned to the study of airs, due probably to financial difficulties, his researches were based on original experiments.[82] Volta's interest in collecting marsh air and in the nature of the inflammable air of metals was probably not unconnected with Priestley's turn towards his research work on airs. The study of airs became a common field of concern that enabled Volta to maintain contact with Priestley. Furthermore, Volta called on Priestley's authority again in this field, as well as in electricity, when he presented the prototype of his inflammable air eudiometer by means of the letter written to him in September 1777.

Volta highlighted the simplicity of his first prototype of eudiometer, which was in line with Priestley's device for the nitrous air test, and did not even discard his instrument for the performance of Priestley's test.

---

[81] Smeaton, 2000, p. 217. Guyton used this mercury eudiometer in his lecture to prove the composition of water. Once the eudiometer was filled with dry mercury and the two gases passed through, a sheet covered with calcined potash (potassium carbonate) was introduced into the tube. This sheet remained there for 24 hours. The sheet was then removed and the mixture ignited. After a cooling-off period, water drops were collected on precisely weighed sheets of paper and the entire weight of water was calculated. Letter from Louis-Bernard Guyton de Morveau to Martinus van Marum, February 26th, 1789 (Lefebvre & Bruijn, 1976, p. 139).

[82] Crosland, 1983, pp. 230-235.

Nevertheless, he recognized some of the drawbacks of his inflammable air test when compared with the nitrous air test: above all, it was inferior in accuracy and sensitivity as well as involving some difficulty when obtaining pure inflammable air. Even so, Volta convincingly furnished a list of positive features of his instrument. Apart from its portability, most of these features concerned the general characteristics of the inflammable air in contrast to that of the nitrous air. In summary, and despite its differences from the nitrous air test, the main advantage of Volta's inflammable air test was that a certain endpoint for the test did indeed exist. On the other hand, its major disadvantage was its ineffectiveness with air mixtures containing low proportions of dephlogisticated air (oxygen). Nonetheless, Volta was confident that his test was clearly able to compete with the nitrous air test in measuring the respirability of a common air sample.

The development of the inflammable air eudiometer was not the work of only one person. Volta was its creator and first developer whose work was later taken up by other scientists such as Monge, Senebier, Van Marum and Guyton de Morveau. Furthermore, the evolution of the instrument owed much to the inestimable collaboration of instrument makers and artisans (glassmakers, ironsmiths and boilermakers), who all played a part in the peculiar development of an apparatus that was characterized by a modular conception enabling innovative and alternative versions to be made. The basic modules of the inflammable air eudiometer were the 1777 prototype (Figure 2.3) and the portable (Figure 2.4) designs. The following year, Volta modified the jug of the portable eudiometer by making it a globe-shaped recipient, while also adopting the crooked wire ignition device and the narrow graduated tube (Figure 2.5). Later in the same year, the globe-shaped recipient with the ignition device was incorporated into the prototype design (Figure 2.6). During the 1780s, the ignition device was attached as a part of a bowl-shaped section (Figure 2.7, *Fig. 3*), and narrow graduated tubes with a bulb on the top end were used to improve the measurement accuracy (Figure 2.7, *Figs. 3, 6*). The loss of the cylindrical tube in the last version of the prototype eudiometer, with the ignition device incorporated into the globe-shaped recipient, signified the merger of the two original basic modules of the inflammable air eudiometer (Figure 2.9). Guyton's eudiometer (Figure 2.10) is a remarkable case of the modular nature of the inflammable air eudiometer. Regarding the use of Volta's eudiometer in chemical research, it is worth noting that during the 1780s the instrument was employed in experiments conducted simultaneously by Monge and Lavoisier on the nature of water, and also that Berthollet demonstrated its potential as a gas mixture analyser in his experiments on the composition of ammonia.

Although Volta's first intention for his eudiometer was to determine the respirability of air in different locations, the research work in which he used the instrument were focused on exploring the nature of the inflammable air and the range of proportions of the mixtures of inflammable with common or dephlogisticated air, both areas in which his eudiometer was effective. In the first area, Volta was obsessed with the idea that an acid could join phlogiston in the composition of inflammable air, and despite his plans for a glass version of his eudiometer, in which mercury was to be used instead of water to detect that acidic component, he may not have been successful. Volta's theoretical commitment to phlogiston in his interpretation of the nature of the different kinds of air would probably have prevented him from seeing the formation of water.

3.

Nitrous air eudiometers at work

Priestley published the first volume of the new series *Experiments and Observations Relating to Various Branches of Natural Philosophy* (the continuation of his *Experiments and Observations on Different Kinds of Air*) in 1779, in which he described the procedures and the equipment involved in his nitrous air test in a more detailed and understandable manner than he had done previously:[1]

> He used a vial (capable of containing an ounce of water), called 'the air measure', to collect an air sample for examination. This air sample was introduced into a low jar and then a measure of nitrous air was added. If after two minutes the reduction in volume of the mixture was found to be insufficient, another measure of nitrous air was added. The residual air mixture was transferred into a glass tube (3 feet long and $1/3$ inch wide) graduated according to the air measure and divided into tenths and hundredths parts. The tube was then immersed in a trough of water until the water inside the tube was at the same level as the water on the outside. The result of the test was expressed in measures occupied by the final air mixture .

First of all, this description makes it clear that the test was carried out using an instrumental device rather than a compact instrument that was recognizable as Priestley's eudiometer with all its assembled parts. Secondly, no a priori criterion was established to determine the end of the test, unlike in Landriani's and Fontana's eudiometrical tests. The final point of the test was rather tentative since an air measure could require up to two equal measures of nitrous air to be saturated.[2]

In the same year, 1779, the relationship between Priestley and Lord Shelburne began to deteriorate. Between July and December 1780, the Priestley's moved to Birmingham, where Priestley had accepted a position

---

[1] Priestley, 1779, pp. xxx-xxxii.

[2] Ibid., pp. 245-247.

as a preacher. It was in this city that in 1781 he published the second volume of the new series of *Experiments and Observations*. In the preface to the book, Priestley suggested that the most important part of the work concerned the experiments with the 'green vegetable matter',[3] which was considered responsible - with the necessary collaboration of light – for the purification of rarefied air. At this point, Priestley's research and the development of the nitrous air test overlapped with the researches that Fontana and Ingenhousz had already initiated.

Fontana-Ingenhousz's second generation of nitrous air eudiometers

After publishing the *Descrizione* in the autumn of 1775, Fontana set out on a journey to Paris, where he arrived in January 1776. His stay in Paris would have its turning point in August of 1777 when he met up again with Ingenhousz, whom he had first met in Florence in 1769. Ingenhousz himself arrived in Paris in August 1777, for a meeting with Benjamin Franklin, with whom he had maintained a good relationship since their meeting in London in 1767.

Jan Ingenhousz (1730-1799) began work as a doctor in Breda (his hometown) in 1757, although he had completed his studies in London (John Pringle being his mentor) and Edinburgh, where he attended the lectures of William Cullen. On his return to London, he became interested in the treatment of smallpox and learned the technique of inoculation. Since 1768, Ingenhousz had been the court physician of the Empress Maria Theresa of Austria in Vienna, where he had himself inoculated the royal family against smallpox.[4]

Once in Paris, Fontana and Ingenhousz became involved in the polemic between Priestley and Scheele about the respirability of inflammable air. Fontana's experiments on this subject began in Paris and continued in London from September 1778; they also served to check the suitability of the nitrous air test to determine the salubrity of the air. There was no mention of his *stromenti* of 1775 in the paper he wrote on this research work, but it did include a description of his own nitrous air test, which prefigured his next eudiometer.[5] While in London, in April 1779 Fontana

---

[3] The formation of the so-called 'green matter' in fresh water was reported to Priestley. This green material is due to the filamentous algae that forms slimy green coatings on the walls of pools, in puddles, slow-flowing rivers and flooded fields.

[4] Magiels, 2010, pp. 9-48.

[5] Fontana, 1779a, pp. 343-344. In a manuscript by Fontana (*Fontana opuscoli*), written after the paper on the respirabilty of inflammable air, there is a section titled

read a paper to the Royal Society concerning certain experiments he had carried out during his previous stay in Paris. Although these experiments were devoted to the testing of the salubrity of air dissolved in water from different sources, the paper was not confined to his experimental results. Fontana expressed his frustration about the suitability of the nitrous air test for determining the salubrity of air, not only for the difficulties involved the standardization of the test procedures but also, and above all, for its intrinsic constraints.[6] Common air could easily be spoiled by vapours accidentally adhering to it, such as particles of arsenic floating in the atmosphere, which were undetectable by the nitrous air test.[7] Just before Fontana left London for Florence in October 1779, his friend and colleague Ingenhousz had already materialized Fontana's ideas for a new nitrous air eudiometer.

Fontana had already given Ingenhousz access to his notes and drawings for a new eudiometer in Paris, in the summer of 1777. The final instrument and its operating protocol were completed in September 1778, when they had met each other in London. Ingenhousz's experimental activity between July and September 1779 in Southall Green, in the west of London, was frenetic and intense. His laboratory notes on the influence of light on plants in relation to its restorative action to spoiled air due to animal breathing reveal up to 353 experiments, although he later spoke of more than 500.[8] Ingenhousz was always grateful to Fontana for his intellectual generosity in giving him leave to anticipate the publication of Fontana's own device and his eudiometrical procedure for testing the different species of air used and produced in his experiments.[9] Shortly after Fontana left for Florence on October 12th, 1779 Ingenhousz published his *Experiments upon Vegetables*, a benchmark work in his scientific career. A third part of the book can effectively be considered as a thorough operating manual of the new nitrous air eudiometer, which

---

Science de l'air, where Fontana expressed his intention to give his coming instrument the name of 'evaerometre' (Fontana, 1779-1780, p. 30). This term derives from two Greek words, the first part of the term (ευάερος) means "airy" in the sense of "well ventilated", while the second (μέτρο) means "measure".

[6] Boantza (2013b, pp. 391-393) stresses the difficult transmission of the tacit knowledge associated with the operating procedures of Fontana's eudiometer.

[7] Fontana, 1779b, pp. 446-453.

[8] Ingenhousz, 1779, p. xlii; Beale & Beale, 2011, p. 275.

[9] Ingenhousz, 1779, pp. xl-xli.

Ingenhousz always referred to as the instrument devised by his close friend the Abbé Fontana.[10]

To gain the confidence of potential users of his eudiometer, Ingenhousz counselled the need to be scrupulous and to follow the same procedure systematically. Fontana had spent the last few years minimizing the procedural sources of error in order to obtain an accurate method. Ingenhousz claimed to have gathered different sources of error of unequal importance from Fontana's manuscripts, and which might even compensate each other. However, at least twenty of such sources were so important that they could make respirable air harmful if they were neglected. The description of the instrument, and the exhaustive list of all the precautions to be taken into account in the operational procedures, resulted in a long exposition of nearly twenty-five pages that included a quantification of the incidence of errors on the final outcome. The two principal components of the new Fontana-Ingenhousz eudiometer were the eudiometrical tube and the measure (Figure 3.1):[11]

> The first component consisted of a cylindrical glass tube 18-20 inches long with a diameter of approximately half an inch (*Fig. I*, CCCC). This tube had divisions marked upon it, each of exactly three inches. Each division was subdivided into 100 equal parts, which for convenience were engraved upon a brass slider moving along the glass tube. The measure (*Figs. II* and *III*) consisted of a glass tube three inches long with a similar diameter and with the eudiometrical tube fixed in a brass socket that had a flat slider at the opening of the tube.

While recognizing that the procedure devised by Fontana was the most accurate, Ingenhousz was more concerned about controlling his experiments with vegetable materials rather than measuring the salubrity of the common air.[12] In addition to this basic reason, the number of experiments he intended to conduct and the large number of air samples to be examined led Ingenhousz to design an abridged procedure. He regarded this shortened method as a combination of Fontana and

---

[10]Ibid., pp. 150-152. The first Dutch, German and French translations of *Experiments upon Vegetables* were published in 1780.

[11] Ibid., pp. 152-155, 293-295.

[12] Ibid., pp. 155-160, 278.

Priestley's procedures, which according to him was surprisingly accurate and reduced the execution time of the test.[13]

Figure 3.1 Fontana-Ingenhousz's eudiometer.

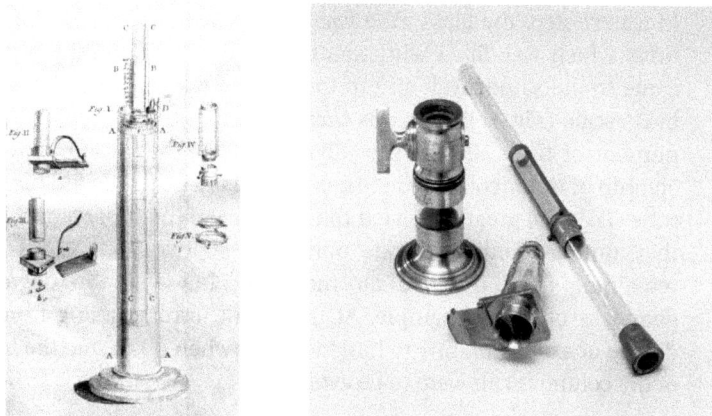

(Left) From Jan Ingenhousz, *Experiments upon Vegetables* (London, 1779), plate p. 294. (Right) Copy with its measure from Charles' cabinet at the Musée des arts et métiers, Paris, © Musée des arts et métiers-CNAM, Paris, photography by M. Favareille.

This abridged procedure entailed using the pneumatic trough to introduce enough air sample into the measure until it was full; holding the brass measure slider under water for exactly fifteen seconds; lifting it until the slider was on a level with the water in the trough, and then shutting the slider to cut off the column of air within the measure. The measure was then inverted under water to let out all the air remaining under the slider, and then the measure of air was allowed to rise up immediately in the eudiometrical tube. All these operations were repeated in the same manner with nitrous air newly made from copper. During the execution of these procedures, any contact with the operator's hands had to be reduced to a minimum, since the amount of air could vary for the same measure due to the air expanding from the heat produced by handling. At the precise moment when the two airs came into contact, the eudiometrical tube was shaken forcibly in the pneumatic trough for

_____

[13] Ibid., pp. 278-287, 296-303.

half a minute. Afterwards, it was placed directly in the brass tube[14] (*Fig. I*, AAAA) and allowed to stand in the middle of the trough for one minute, water being poured on it to bring the temperature of the glass tube up to that of the water.

In a next step, the glass tube was slid up or down within the brass tube, which was filled with water, until the two columns of water came to the same level and to the level of the zero position on the brass scale (*Fig. I*, BB). It was then necessary to observe with what number of the scale the first division of the glass tube above the column of water coincided.[15] The magnifier glass applied to the brass tube (*Fig. I*, D) greatly assisted the accuracy of the observation, since that number indicated not only how many subdivisions were remaining from the two air measures, but also the degree of goodness of the air sample. At this point, errors arising from the degree of expansion due to hand warmth when observing the length of the column of air were to be avoided.[16]

With this abridged method the whole operation was performed in three or four minutes, and its accuracy was such that in ten trials conducted with the same common and nitrous airs the difference in the result rarely amounted to 1/200 of the volume of both airs. Nevertheless, this abridged method did not work with dephlogisticated air samples, since they required additional measures of nitrous air to bring them to full saturation. In the standard unabridged method, one measure of nitrous air was first added to two measures of the air sample. Complementary measures of nitrous air were introduced consecutively until no net diminution in volume of the gaseous mixture was observed.

Fontana-Ingenhousz's eudiometer as an essential instrument for monitoring the first studies on the influence of sunlight on plants

When Ingenhousz changed the direction of his research on the influence of sunlight in the production of dephlogisticated air by plants, he

---

[14] The two copies of this eudiometer kept in the Museo Galileo in Florence have a glass tube instead of brass. Both were built in late XVIIIth century by the English instrument-makers George Adams and Benjamin Martin.

[15] The eudiometrical tube was suspended on a type of stabilizer consisting of two brass rings (Figure 3.1, *Fig. V*), such as were used in common compasses, in order to provide the tube with great mobility and thereby ensuring that it was always in a perpendicular position.

[16] Ibid., pp. 162, 166.

attributed the reasons for this change of course to the impact of Priestley's discovery on the air-purifying capacity of plants,[17] and on John Pringle's presidential address when Priestley was awarded the Copley Medal by the Royal Society in 1773.[18]

It is not possible to recreate the fine structure of Ingenhousz's line of investigation that led him in June 1779 to complete a research program that he meticulously executed during the summer and fall of 1779. Undoubtedly, a key factor in this investigative context was Fontana's intellectual protection and friendly cooperation, without which it would be difficult to understand Ingenhousz's research. Ingenhousz was an adept and thoughtful operator of Fontana's eudiometrical designs, with whom he had shared a great deal of time in Paris and London working together to detect and eliminate any unnecessary components in order to improve the eudiometrical device. Above all, however, Ingenhousz had unshakable confidence in the nitrous air test for determining the salubrity of the air with a precision that far exceeded Priestley's plans.[19] One may surmise that he redevised an instrument that would not only measure the goodness of air but above all would monitor experiments that unequivocally linked sunlight with the production of dephlogisticated air by plants. In the summer of 1779, Ingenhousz himself recognized that he needed a eudiometer as an indispensable instrument for carrying out experiments on which he had spent years meditating. At that time, he possessed a privileged first-hand knowledge of Fontana's eudiometrical device, which he started to use in an attempt to make it more concordant with that of Priestley and which he would have wanted to reproduce exactly if he had possessed the instrument himself.[20]

Ingenhousz planned his experiments with very simple devices: inverted and airtight glass vessels with parts of plants inside them, sometimes submerged in water, to observe and analyse the air released. He experimented with different parts (leaves, stems, roots, flowers, fruits) of a variety of plants (medicinal, culinary, poisonous, exotic) in different circumstances (aerial environment, temperature, weather conditions, light). Stated briefly, these were the main conclusions: green plants released dephlogisticated and absorbed fixed air when they were exposed to sunlight. Dephlogisticated air was mainly released on the underside of

[17] Priestley, 1772, pp. 168-170, 193-199.

[18] Ingenhousz, 1779, p. xv.

[19] Beale & Beale, 2011, pp. 269, 272.

[20] Ingenhousz, 1787, pp. 225-226.

the leaves and its production depended on sunlight intensity, but not on the temperature. Finally, green plants absorbed dephlogisticated air in the dark.[21] More than ninety pages of the eighteen sections of the second part of *Experiments upon Vegetables* were devoted to providing details and data, mostly eudiometrical, of his experiments.[22]

For Ingenhousz, the eudiometer was probably one of the best instruments invented, and one with a great reliability thanks to the procedural protocol he had designed together with Fontana. In a series of ten experiments with the same sample of common air, the differences observed in the residual volume after the addition of three consecutive parts of nitrous air to two parts of common air did not exceed more than 1/500 parts of the total volume.[23] Ingenhousz ran an analytical test with an experimental device that, from the point of view of the explicit as well the tacit procedures, the instrumental equipment (i.e. eudiometer) and the chemicals (i.e. nitrous air) were reasonably standardized.[24] However, despite Ingenhousz's efforts to standardize the eudiometrical procedure, the probability of incurring any of those detected errors was not low enough to ensure the reliability of the instrument in the hands of novice experimenters.[25] Ingenhousz was quite aware that Priestley's nitrous air procedure was the benchmark for those most interested in these matters.

Like Fontana before him, Ingenhousz had to address the question of the suitability of the nitrous air test for determining the air salubrity. This arose when he studied the behaviour of plants in an environment of inflammable air that was harmful to animal life. He used three complementary trials to conduct his experiments; the nitrous air test, a survival test with a chick, and an exploding test with a flame candle. The experimental results on the

---

[21] Ibid., 1779, pp. xxxiii-xxxviii; Beale & Beale, 2011, pp. 277-278.

[22] Ingenhousz, 1779, pp. 185-275.

[23] Ibid., p. 151.

[24] A statistical analysis of the results of twenty-two of the experiments performed by Ingenhousz show the existence of high values for the coefficient of determination between the variation in volume of the gaseous mixture and the volume of nitrous air added to a constant volume of respirable air. Only those experiments providing between six and eight results have been selected in order to provide sufficient data for a statistical analysis. For these experiments, a quadratic regression with a coefficient of determination (R squared) not lower than 0.99 has been estimated.

[25] Boantza (2013b, pp. 389-391) has emphasized that the list of errors proposed by Ingenhousz was an attempt to cover the gap between the tacit knowledge involved in the experimental operations and its effective textual or visual transmission.

capacity of plants to restore the respirability of inflammable air indicated that, according to the nitrous air test, the inflammable air had turned into breathable air with no loss of its explosive force or of its lethal effect on a living animal. Ingenhousz was firmly convinced that the nitrous air test failed entirely to show the degree of salubrity of the restored inflammable air, since by this method it appeared almost to be a dephlogisticated air and yet it was still poisonous. Nevertheless, for Ingenhousz this apparent flaw in the test did not weaken its real value in any way, that is, the nitrous air diminished respirable air in proportion to its salubrity. In other words, the test was entirely appropriate for atmospheric air, which was ultimately the chief object of his experiments.[26] Ironically, this restored inflammable air, which constituted an anomaly for the nitrous air test, had become the key component of Volta's eudiometer.

Tiberio Cavallo and the assessing of nitrous air eudiometers

When Fontana and Ingenhousz continued their experiments on the respirability of the inflammable air in London, in September 1778, they were accompanied by Tiberio Cavallo, who had been obliged to assist Fontana when he fainted as a result of the repeated inhalation of inflammable air. The Neapolitan Tiberius Cavallo (1749-1809) had in 1771 immigrated to England, where for some time he combined business affairs with scientific interests. His regular participation in scientific circles in London distracted him from the commercial activity that had provided him with a certain level of financial self-sufficiency, and this enabled him to devote himself fully to science. The early years of his scientific career between 1775 and 1780 were devoted to the study of electrical phenomena, a subject on which he published papers in the *Philosophical Transactions,* and in 1777 extensive works such as *A Complete Treatise on Electricity.* However, in 1779, and coinciding with his appointment as a Fellow of the Royal Society, he redirected his scientific career. Cavallo began to interest himself in some aspects of terrestrial physics such as pneumatics, hydraulics, geodesy and meteorology. The first work he published in this new stage of his researches was *A Treatise on the Nature and Properties of Air and other Fluids Permanently Elastic. To Which is Prefixed an Introduction to Chymistry* (1781). Throughout the long fifth chapter of this treatise, Cavallo described the different nitrous air eudiometers known at the time, both to highlight the flaws that rendered them inoperative and to propose the necessary modifications.

---

[26] Ingenhousz, 1779, pp. 97-106.

Actually, Cavallo showed little interest in Landrian's eudiometer. He considered that the two main drawbacks were, firstly, the use of glass stoppers which were difficult to make and easy to disable and secondly, the use of mercury instead of water because the former was attacked by nitrous air, which contributed to masking the experimental results. He devoted little time to describing Priestley's contributions, although he acknowledged his merit for developing the nitrous air test. Based on Priestley's latest presentation in 1779, Cavallo in his assessment still referred to a certain 'Priestley's eudiometer' rather than of speaking of an experimental device. On the other hand, he paid much more attention to Magellan's eudiometers, while being highly sceptical of their accuracy.[27]

Cavallo failed to mention any of Fontana's eight 1775 instruments but expressed great appreciation of his eudiometer, which was made known to the public in 1779 in Ingenhousz's work *Experiments upon Vegetables*. It is worth noting that he made no reference to the collaboration between Fontana and Ingenhousz, and in fact did not mention the latter's name at all. Notwithstanding, Cavallo had collaborated with Fontana in his experiments on the respirability of inflammable air in 1778. In this regard, his tacit knowledge of the eudiometrical design and procedures enabled him to formulate his own description of the instrument. From an operational point of view, Cavallo appreciated the fact that Fontana-Ingenhousz's method was based more on the neatness and uniformity of the procedural execution rather than on the particular design of the instrument. As had occurred in the case of barometers, he predicted that the simplest eudiometer would be the best.[28]

In the course of his career, Cavallo had to face technical problems concerning the construction of instruments. He proved himself to be an active and skilful artisan when constructing, for example, an electrometer, a multiplier of electric charges and a micrometre for telescopes. In particular, in his treatise on the properties of air, he presented his own nitrous air eudiometer, accompanied by a method for determining the purity of atmospheric air. The design of this eudiometer (Figure 3.2) was based on the preceding ones he had examined, especially on that by Fontana-Ingenhousz, retaining everything he considered useful and eliminating their drawbacks. He regarded his own eudiometer as being the most expeditious and accurate of all those known to date. Cavallo's interest in instrument-making led him to pay closer attention than other

---

[27] Cavallo, 1781, pp. 315-316, 326-327.

[28] Ibid., p. 315, 333.

authors to the description of technical issues, such as the calibration of the eudiometrical tube. The difficulty of making the correct calibration resided in the fact that successive equal additions of air measures did not correspond to equal increases in volume in the tube. Furthermore, he showed his readers how to deduce the real quantity of air from the apparent space it occupied in a tube, which was partly filled with air and partly with water or mercury. In addition, and in order to avoid the task of calculating, he attached a table to enable that real quantity to be found.[29]

Figure 3.2 Cavallo's eudiometer.

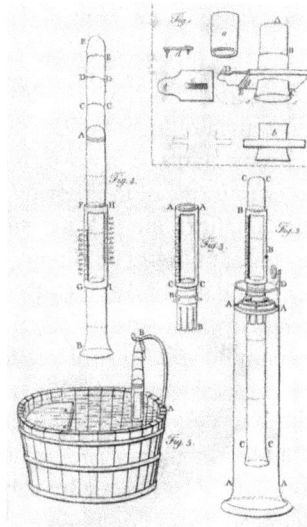

From Tiberius Cavallo, *A Treatise on the Nature and Properties of Air and other Fluids Permanently Elastic.* (London, 1781), plate III.

Priestley's disdain for the Fontana-Ingenhousz eudiometer. The controversy over the effect of sunlight on air purification by plants

In the early summer of 1781 the second volume of the new series of Priestley's *Experiments and Observations* was already on sale. The nineteenth section of the book (*On the mixture of nitrous and common air*) became a source of disagreement and reproaches about Fontana-Ingenhousz's eudiometrical methodology. Without a doubt, the strongest criticism was focused on the irrelevance of the strength of the nitrous air,

---

[29] Ibid., pp. 340, 344-347, 355-359; 378.

stated by Fontana and Ingenhousz, to determine the purity of common air. Ingenhousz had criticized the practice of mixing the nitrous and common airs in a certain proportion at the same time, as had Priestley, since if the nitrous air was of variable quality the results would be uncertain. Ingenhousz argued on behalf of Fontana that the strength of nitrous air was of little importance provided that it oversaturated the common air.[30] Employing a rather sarcastic tone, Priestley expressed his disappointment with Fontana and Ingenhousz's position on the matter. He considered that their arguments were fallacious and extended his dissent to the whole eudiometrical procedure.[31] This disdain of Priestley's towards Fontana and Ingenhousz's contributions has been considered both misplaced and arrogant.[32]

A better way of understanding Priestley's condescending attitude would be to place it within the context of Ingenhousz's discovery on the effect of sunlight in air purification by plants. It is worth recalling that Ingenhousz reassessed his eudiometer in terms of monitoring research unequivocally linking the influence of sunlight to the production of dephlogisticated air by plants. Priestley's dismissal of their eudiometer could therefore also be interpreted as a way of casting doubts on Ingenhousz's merits and that of the instrument in itself in that research. In autumn 1781, when Ingenhousz became acquainted with the publication of the last volume of Priestley's *Experiments and Observations*, he was already aware that Priestley was advocating a new explanation for the appearance of dephlogisticated air bubbles in plants submerged in water. Moreover, Cavallo had already contradicted Priestley by claiming that the bubbles were the result of a chemical process undergone by water with the aid of leaves that favoured the nucleation of bubbles.[33] Ingenhousz was also aware that both Priestley and Cavallo, as Fellows of the Royal Society, had a considerable capacity of persuasion.[34]

Priestley spent the first seven sections of his work setting out his experiments on the role of sunlight in the purifying action of the air by plants, making extensive use of his nitrous air test. His conclusion was that, in the first place, there was no actual production of clean air by aquatic

---

[30] Ingenhousz, 1779, p. 173.

[31] Priestley, 1781, pp. 183-191.

[32] Beale & Beale, 2011, p. 313.

[33] Cavallo, 1781, pp. 823-827.

[34] Beale & Beale, 2011, p. 311.

plants, but only a process of purification or dephlogistication of the air previously dissolved in water. Secondly, since aquatic plants purified the air contained in the water, it was easy to make the analogy that plants could also purify the air around them. At this point, it should be noted that Priestley admitted the originality of Ingenhousz's hypothesis that pure air was produced by the joint action of sunlight with any part of the plant, therefore accepting a certain parallelism between both researches.[35] What happened was that the publication of the new volume of *Experiments and Observations* triggered a dispute over priorities between Priestley and Ingenhousz.[36]

John Pringle, Ingenhousz's great mentor, died in January 1782. Pringle had been promoting eudiometrical practices until the last days of his life, although his academic influence had declined from 1778, the year in which he left the presidency of the Royal Society in the hands of Joseph Banks. It is likely that a sense of loyalty towards Pringle prompted Ingenhousz to continue defending Fontana's eudiometrical procedure despite the growing scepticism surrounding it.[37] In opposition to Priestley's and Cavallo's claims, Ingenhousz proposed to demonstrate that the air bubbles appearing in the plants submerged in water were the result of a chemical process that took place in the leaves of the plants. Once again, Ingenhousz praised Fontana's eudiometer as an instrument for his research and ignoring the discredit surrounding it.

In the winter of 1782, Ingenhousz carried out what he called 'decisive experiments' in a greenhouse of the botanical gardens in Vienna, where he was accompanied by his friends who acted as witnesses (Figure 3.3). The experimental results were conclusive, and Ingenhousz was able to argue against Priestley's views that the bubbles of dephlogisticated air were released from living plants rather than produced by a process of purification of the air dissolved in water.[38] The paper reporting on these experiments was completed in early April and sent to the Royal Society, where it was partly read in June 1782. The fact that the title of the paper was

---

[35] Priestley, 1781, pp. 24, 29-30.

[36] Schofield (2004, pp. 155-156) sustained the hypothesis that Priestley's work was independent from that of Ingenhousz.

[37] Golinski, 1992 p. 124.

[38] Ingenhousz, 1782, pp. 438-439.

scarcely indicative of its contents, and Ingenhousz's absence from London to present it in person, both favoured Priestley's position of primacy.[39]

Figure 3.3 Ingenhousz's device for observing the bubbles of air released from plants under water and in sunshine.

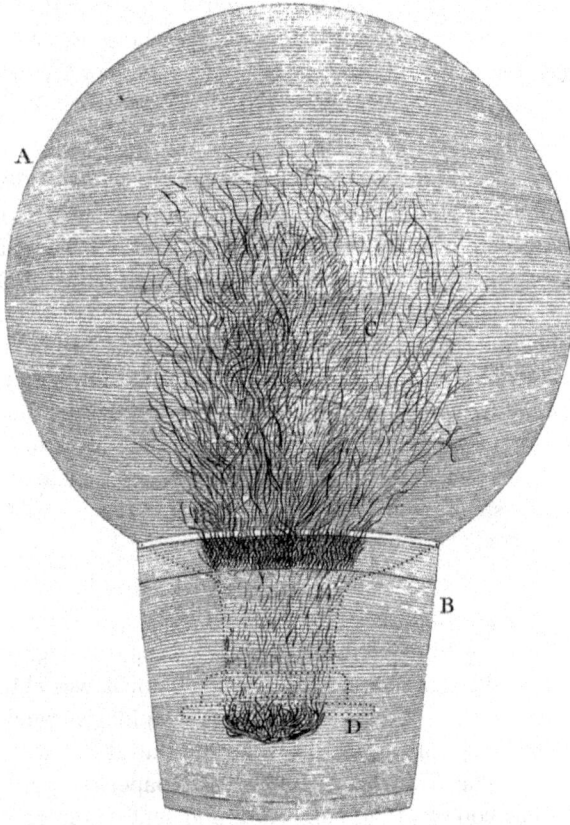

A. *a globular glass vessel containing about 160 cubic inches of water*.
B. *glass vessel fill'd with Mercury in which the orifice of the spherical vessel is plunged*.
C. *a vegetable called Conferva rivularis*.
D. *a piece of wood to which the Conferva rivularis is attached to keep it in its place*.

From *Philosophical Transactions* (1782), Vol. 72, plate XV, p. 435.

---

[39] Beale & Beale, 2011, pp. 317-318.

Cavallo vs. Magellan or Priestleyans vs. Fontanists. Two different approaches to experimental data processing

On November 30th, 1778, Joseph Banks replaced Priestley as president of the Royal Society. No later than 1783, Banks was involved in a personal dispute with Magellan over the latter's role as an independent channel of scientific and technological information with the continent. Meanwhile, Cavallo was in touch with Banks, who in 1781 had appointed him to the Council of the Royal Society to carry out a series of calorimetric researches.[40] Moreover, Magellan maintained a close friendship with Priestley. These institutional relationships provide a better understanding of the feud between Magellan and Cavallo, and ultimately between the eudiometrical methods of Priestley and Fontana, which began in 1783 with the publication of the third edition of Magellan's text: *Description of a Apparatus for Making the Best Mineral Waters, like those of Pyrmont, Spa seltzer, Aix-la-Chapelle & c,Together with the Description of Two New Eudiometers, etc. The Third Edition [...] with an Examination of the Strictures of Mr. T. Cavallo, F.R.S. upon these Eudiometers.*[41] The title of the third edition contains two significant changes; the first was that only two new eudiometers were presented instead of three, the second change was the addition of an analysis of the restrictions that Cavallo had pointed out about Magellan's three original eudiometers in 1779. Magellan suppressed the second of these eudiometers because he regarded it as the most complicated (more apparatus) and therefore the most expensive of the three. The principle underlying this decision was the belief that, all other things being equal, the simplicity of the philosophical experiments and the cheapness of the instruments should be two objectives to be pursued in the research on natural phenomena.[42]

Magellan introduced some procedural modifications in his first eudiometer, such as for instance repeating each experiment three or four times and taking the mean value of the corresponding results as a reference for further comparisons.[43] However, there is no doubt that the procedural observation mentioned by Magellan in a footnote in the first edition of his book, which was embedded in the text of the third edition, was much more

---

[40] Golinski, 1992, p. 124.

[41] A second edition of this work with only few minor variations was published in 1779.

[42] Magellan, 1783, p. 54.

[43] Ibid., pp. 45, 47-48.

far-reaching.[44] The observation concerned the volume of the mixture of nitrous and common air and how within a few minutes it diminished up to a certain level and then expanded again, but well beneath the original volume. This was a crucial observation, since it directly influenced the perception of the endpoint of the eudiometrical test.[45]

In the first edition of 1777, Magellan had left the decision about the moment at which to determine the contraction of volume to the discretion of the experimenter: either as soon as the decrease in volume appeared stationary, or whether to allow time enough for the mixture to settle to a certain volume, which could require a period of as long as twenty-four hours. In the third edition of 1783, Magellan suggested the first option as being the most expeditious and the one preferred by philosophical observers. But he also recommended making every trial as similar to each other as possible in every experiment in order to obtain comparable results.[46] This point formed the basis of his disagreement with Cavallo when he sarcastically remarked that, although this 'Rule of uniformity' was well known even to beginners in experimental philosophy, a 'mighty philosopher' (i.e. Cavallo) had at last been lucky enough to unravel this 'abstruse and mysterious rule of uniformity' in his experiments, thereby introducing a 'wonderful new method' of performing eudiometrical experiments with a margin of uncertainty not exceeding 1/50 of a measure of air with Fontana's eudiometer.[47]

Magellan compensated for the absence of the second eudiometer described in the 1779 edition with an analysis of the restrictions that Cavallo had pointed out about Magellan's original eudiometers in 1779. He devoted an entire section of fifteen pages (*Examination of Mr. Cavallo's strictures of these eudiometers*) to rebutting Cavallo's criticisms about his eudiometers, in which he adopted an extremely hard tone. Magellan believed that Cavallo's rejection of his instruments did not arise from any personal resentment, but rather suspected that Cavallo may have been

---

[44] Ibid., 1777, p. 28; 1783, p. 44.

[45] To ascertain easily this endpoint there was a sliding brass ring (z), which slid in the tube until the instant that the column of water within it appeared stationary (Figure 1.14, F*igs. 12* and *16*). In the 1783 edition, Magellan noticed, however, that the most important use of this ring was avoiding the possibility of parallax error when measuring the height of the surface of water within the tube against the scale (Magellan, 1783, pp. 45, 57, note s).

[46] Ibid., pp. 46-47.

[47] Ibid., p. 47, note t; Cavallo, 1781, pp. 328, 333.

serving certain interests for profit, as a reward for endorsing the merits of Fontana's eudiometer.[48] Magellan also reproached Cavallo for having copied large parts of his text without having taken the trouble to read it, believing him to be guilty of woolgathering (*absence d'esprit*).[49]

The bulk of Magellan's response to Cavallo was focused on his eudiometer. Magellan had been felt obliged to comply with Cavallo's request to witness the performance of the nitrous air test with his first eudiometer. Magellan had conducted four trials with samples of equal parts of atmospheric and nitrous air (132 parts in total) and had found contractions in volume ranging from a maximum of 58 parts to a minimum of 48 parts (the other two measures were 48 and 51). Cavallo subsequently expressed his reservations about the inability of the instrument to differentiate between respirable air and the air resulting from the combustion of a candle. Magellan interpreted this as a form of betrayal on Cavallo's part, suspecting that he had requested to be present at the performance of the eudiometrical tests for ulterior motives rather than to observe the increase in volume of the mixture of nitrous and common air after a certain degree of contraction. Cavallo believed Magellan's observation was simply a mistake and came to the conclusion that it was not possible to achieve a greater degree of accuracy of approximately 1/13 of the mixture of nitrous and common air with Magellan's first eudiometer.[50] Nevertheless, Magellan continued to attribute these findings to what he regarded as Cavallo's absent-mindedness rather than a reproof of his principles.

In actual fact, the issue of instrument accuracy was indeed a problem of experimental data processing, particularly when it came to questioning unreliable data. Common sense led Magellan to accept the existence of erroneous experimental data, thereby rejecting unreliable data and taking the mean value as being representative of accepted data. This contribution to experimental data processing by Magellan enabled him to refute Cavallo's comparison between both eudiometers. Cavallo believed that Fontana-Ingenhousz's eudiometer was much more accurate than Magellan's, while after processing his experimental data Magellan insisted

---

[48] Magellan, 1783, p. 55.

[49] Ibid., pp. 60-61.

[50] Cavallo, 1781, pp. 326-327. In this trial, a measure of air was equivalent to 66 parts of the eudiometrical tube, so each sample of equal parts (measures) of atmospheric and nitrous air was equivalent to 132 parts. Therefore, the variation between the two extreme values was 10 = 58-48, which nearly amounted to 132/13.

on the contrary.[51] Thus, two fundamentally different views of what could be considered valuable and reliable in experimental practice underpinned the approaches to the processing of experimental data. For Magellan, the occurrence of anomalous results was inherent to experimental practice and subordinate to a rule dictated by common sense. Cavallo, however, regarded the emergence of anomalous results as a clear lack of procedural and instrumental efficiency.[52]

Magellan's second eudiometer in the 1783 edition corresponded to the third eudiometer in the 1777 edition, but with the addition of some material and procedural novelties. The procedural developments for this eudiometer reveal that, despite the harshness with which Magellan had countered Cavallo's criticisms, he nevertheless took some of them into account. For instance, he changed the order of the addition of both airs (first the air sample to be analyzed, and then the nitrous air); accepted the necessity of shaking the air mixture, and introduced a procedure to make the surface of water inside the tube level with its external surface in the pneumatic trough. It was for this reason that a new opening (o) was made in the top of the upper ball (s) of that eudiometer (Figure 3.4).

---

[51] The variation between the two extreme values in Magellan's eudiometer (10) amounted to almost 1/7 of a measure of air (10 was roughly 66/7). According to Cavallo, Fontana-Ingenhousz's eudiometer was much more accurate than Magellan's. This was because 1/13 of the mixture of both airs was almost equivalent to 1/7 of a measure of air, which was greater than the margin of uncertainty (1/50) set with Fontana-Ingenhousz's eudiometer (Cavallo, 1781, pp. 328, 333). Magellan believed that it was necessary to apply common sense when ruling out one of the dubious extreme values (48 or 58). Since 48 appeared twice, he excluded the value 58 and chose the mean value (49) of the three remaining data (48, 48 and 51) as being representative of the eudiometrical test, which in his opinion was the most reasonable calculation. After Magellan's new approach to his eudiometrical data, the variations between the extremes and the mean values were 2 = 51 − 49 and 1 = 49 - 48. Magellan chose the minor variation (1) to state that the accuracy of his eudiometer amounted no more than to 1/66 parts of one measure, which was less than the margin of uncertainty (1/50) of Fontana-Ingenhousz's eudiometer. Magellan's eudiometer was therefore apparently more accurate than Fontana's (Magellan, 1783, pp. 63, 65-66).

[52] Boantza, 2013, pp. 397-398.

Figure 3.4 Magellan's second eudiometer of 1783.

From John Hyacynth Magellan, *Description of a Apparatus for Making Mineral Waters, like those of Pyrmont, Spa, Seltzer, &c,* (London, 1783)

But perhaps the most significant development had an epistemological dimension. Magellan considered this renewed second eudiometer not only as the one that conformed most closely to Priestley's but also as one of the best because of its simplicity and its cheapness, two virtues that commended themselves to all kinds of experiments.[53] Magellan and Cavallo's conceptions of instrumental simplicity, cheapness and uniformity reflect two radically different approaches to the experimental philosophy, corresponding to the Priestleyan and Fontanist traditions, respectively. Cavallo's analysis of the eudiometers up to that time, including his own, confirmed his conviction that in spite of any instrumental simplification, the experimental procedure involved many operational factors (tacit knowledge and skills) that made it necessary to follow a strict protocol in order to obtain meaningful results. Simple instruments were always preferred to more complicated ones, even when the latter seemed to have the advantage of greater accuracy over the former, although it was not particularly remarkable. Complex instruments were not only expensive and subject to malfunction, but they also frequently gave rise to error, since the operator was simultaneously busy with other tasks and was thus prone to oversight or mistakes. The fact of

---

[53] Magellan, 1783, pp. 69-70.

the matter was that a simple instrument was preferable to a complex one because it was easier to operate, and therefore more likely to provide uniform results.[54] Stated briefly, a procedural uniformity in the interface between the eudiometer and the experimenter's skills guaranteed the uniformity of results. Cavallo was therefore mainly interested in procedural and manipulative simplicity, whereas Magellan emphasized material simplicity and financial affordability. For Priestleyans like Magellan, instrumental simplicity and cheapness were essential for knowledge dissemination and the promotion of experimental practice before large audiences.[55] This dispute between Magellan and Cavallo over their eudiometers has been viewed as a philosophical conflict materialized in the battlefield of instrumental competition.[56]

Jean Senebier. Criticisms from continental Europe

The reliability of the Fontana-Ingenhousz eudiometer was dependent on the monitoring of a strict procedural protocol, which demanded a fully trained experimenter. The difficulties in acquiring the appropriate training in certain operational skills had constituted the core of most of the criticisms levelled against that eudiometer from the beginning, and further criticisms from continental Europe joined those coming from Britain. This was the case of the Genevan botanist and naturalist Jean Senebier, whose researches in the field of plant physiology were first made known in 1782 with the publication of the three-volume work *Mémoires physico-chimiques sur l'influence de la lumière pour modifier les êtres des trois regnes de la nature et surtout ceux du regne végétal.* This work sparked a controversy between Ingenhousz and Senebier about the importance attached to the discovery of a dephlogisticated air released by the plant leaves under the influence of sunlight.[57] Apart from this personal dispute, Senebier made extensive use of the Fontana-Ingenhousz eudiometer to examine the quality of that air, particularly when the leaves were submerged in water. Senebier apparently followed Ingenhousz's

---

[54] Cavallo, 1781, pp. 716-717.

[55] Boantza, 2013, p. 399.

[56] Stewart, 2008, p. 16.

[57] Senebier, 1782, Vol. 1, pp. 1, 2-3; Magiels, 2010, pp. 248-262; Beale & Beale, 2011, pp. 326-330.

methodology, but with some sui generis adaptations, which suggested that he was not overly concerned with Ingenhousz's scrupulous precision.[58]

Senebier offered a more definitive opinion on the Fontana-Ingenhousz eudiometer in his work of 1783, *Recherches sur l'influence de la lumière solaire pour métamorphoser l'air fixe en air pur pour la végétation,* which consisted of four papers with a preface addressed to naturalists, particularly to those studying the plant kingdom. Senebier believed that chemistry could advance plant physiology as it had done in the animal and mineral kingdoms, since little more could be derived from strictly anatomical and physiological studies on plants. His advocacy for a chemical approach to experimental physics was such that all the research work involving aspects of physics and natural history would acquire a chemical nature if conducted thoroughly.[59] The third paper in the *Recherches* concerned new experiments and described the action of different reagents on different air samples. Senebier devoted a section to experiments that according to him revealed the cause of inaccuracy and even the uselessness of the nitrous air eudiometers.

These experiments originated in previous observations on the diminution in volume detected over time in gaseous samples collected over water in cylindrical recipients. Since this was the type of phenomenon that affected eudiometrical observations, Senebier conducted a series of experiments in the dark to prevent air from being released by the action of sunlight. He particularly observed contractions in volume over time (from two hours to days or months) of four individual airs (fixed, inflammable, nitrous and common air) and of mixtures of the same (binary, ternary and of all four airs). In the case of nitrous air alone he found that the ability of the residual nitrous air to perform contractions in volume when mixed with common air decayed over time.

In his conclusions, Senebier recalled previous objections to Ingenhousz's eudiometrical procedure and how his efforts to improve his own eudiometer had not had the desired effects. He lamented the difficulties in replicating Ingenhousz's test, since while one experimenter was expected to obtain similar results with these eudiometers, two experimenters would not be able to achieve the same unless they possessed the same dexterity when performing the test. Not only were different experimenters required to have the same comparable skills, but also a comparable swiftness in the

---

[58] Senebier, 1782, Vol.1, pp. 275-276.

[59] Senebier, 1783, pp. i-xv.

execution of the test, since the reduction in volume increased over time. These were the main reasons for Senebier's report on the inaccuracy and uselessness of the nitrous air eudiometers.[60] However, Senebier's objections to nitrous air eudiometers, and to Fontana-Ingenhousz's in particular, became somewhat patronizing in his work of 1788, *Expériences sur l'action de la lumière solaire dans la végétation*, in which he regarded his own eudiometrical device as an abridged version of Fontana-Ingenhousz's 'excellent' eudiometer, which he had used in situations where he had been unable to handle his larger original eudiometer. In one form or another, he stated that he had been using and still used that eudiometer.[61]

Henry Cavendish. The nitrous air test and the Fontana-Ingenhousz eudiometer under scrutiny

In January 1783, the natural philosopher Henry Cavendish (1731-1810) published the paper *An Account of a New Eudiometer* in which he presented a new eudiometrical device based on the nitrous air test. Cavendish had studied at Peterhouse College in Cambridge but left the University in 1753 without completing his degree. Back home in London, he was invited to attend the dinner meetings by members of the Royal Society as a regular guest. These gatherings gave him the opportunity to become immersed in diverse areas of scientific knowledge, and as a result of this peculiar association he was elected Fellow of the Society in May 1760. Cavendish's elevated financial status enabled him to rent a country estate on Clapham Common in the south of London. A building on the estate was adapted for experimental research, and it was here where Cavendish conducted his most significant investigations.[62]

Cavendish's first unpublished studies were in the field of chemistry, probably during 1764, on arsenic compounds. His skills in quantitative work were fully evident in these first researches in which he worked with unusually small amounts of substances. His other chemical research works, probably during the same year, were on the cream of tartar (potassium hydrogen tartrate) crusted in wine barrels.[63] In the meantime, Cavendish was gaining experience in the field of pneumatic chemistry. His first publication on this subject was in 1776 and consisted of three papers

---

[60] Ibid., pp. 297-312.

[61] Ibid., 1788, p. 113.

[62] Seitz, 2005, pp. 179-180.

grouped under the common title of *Three Papers, Containing Experiments on Factitious Air.* [64] These airs consisted of the inflammable air from metals and acids, the fixed air from alkalis, acids or calcinations, and the air mixtures from fermentations, distillations and putrefactions. These studies concerned a new range of methods devised by Hales for collecting, transferring and measuring gases and for isolating and characterizing different kinds of air.[65] At the same time, Cavendish evinced great neatness in the experimental methodology and concern for precision in the processing of experimental data.

In 1767, Cavendish went on to publish a second paper in the field of pneumatic chemistry, this time on the analysis of a mineral water (*Experiments on Rathbone Place-Water*). In the late 1760s, he focused his academic interests on electrical phenomena, only to return once more to the chemistry of gases during the period 1778-1786 , due apparently to the contributions made by other natural philosophers - notably Priestley – throughout the previous decade. This shift in his interests crystallized in the presentation of a new device based on the nitrous air test.

Cavendish's reasons for improving the Fontana-Ingenhousz eudiometer

Fontana-Ingenhousz's eudiometer was for Cavendish the most accurate of any of those known hitherto. He found in this instrument many ingenious means for overcoming small inbuilt errors, which constituted a series of tacit skills for its operational procedure. In particular, Cavendish underscored those technical improvements that enabled a greater and more complete contraction in volume. The importance of immediately shaking the air mixture in Ingenhouz's procedure led Cavendish to determine whether the contraction in volume would be even more complete and regular if either of the two kinds of air was bubbled slowly into the other while the vessel was continuously shaken. This paved the way for Cavendish to develop a new eudiometer or, rather, a new eudiometrical device.[66]

---

[63] Jungnickel & McCormmach, 1999, pp. 196, 201.

[64] Robert Boyle coined the term 'factitious air' to refer to any air embodied in any material in an inelastic state and released by laboratory operations.

[65] Cavendish, 1766, pp. 142-143.

[66] Cavendish, 1783, pp. 106-107. This device was not the apparatus later known as the 'Cavendish eudiometer'. This latter was nothing but a variation of Volta's gun that Cavendish used for his experiments on the condensation of water. He never

Cavendish's intention was not so much to invent a new eudiometer as to improve on those that already existed. Thus he designed an analytical device in the sense of a set of interacting laboratory instruments that did not constitute an apparatus with its individual, permanently or semi-permanently assembled components (Figure 3.5). The first remarkable feature of this new device was that it was originally conceived for a gravimetric trial. As in the case of Fontana's early instruments, presented in the *Descrizione* of 1775, Cavendish designed a device for a gravimetric determination of the purity of air.[67] However, there is no doubt that both pragmatic and epistemological reasons lay behind this gravimetric approach. Firstly, any inaccuracy associated with volumetric measurements was irrelevant in a gravimetric test.[68] Secondly, from an epistemological point of view, Cavendish considered that measuring weights was more accurate than measuring volumes. He was intellectually much closer to the Newtonian conception of matter consisting of massive particles quantifiable by the use of a balance, rather than to the Cartesian conception, which regards the essence and nature of matter is its extension and therefore quantifiable by volumetric measurements. Cavendish must have had his eudiometrical device ready in the first half of 1781, since in the second half of that year he used it to determine the purity of different air samples.

Figure 3.5 Cavendish's nitrous air eudiometer.

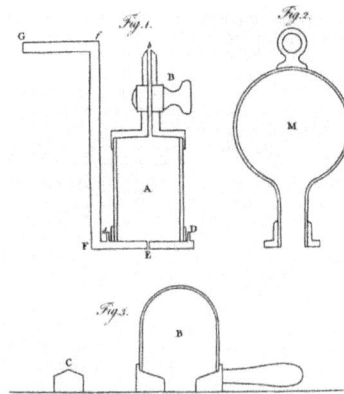

From *Philosophical Transactions* (1783), Vol. 73, plate 2, p. 134.

---

referred to it as an "eudiometer" (Jungnickel & McCormmach, 1999, p. 358, note). For the vicissitudes of this 'Cavendish eudiometer' see Farrar (1963).

[67] In 1779, H.B. Saussure had already designed a gravimetric eudiometrical procedure on the basis of a portable nitrous air eudiometer. This case is discussed in chapter 4.

[68] Ibid., pp. 107-108.

His device consisted of a cylindrical glass vessel (*Fig. 1*, A) that was taken as the standard measure. This vessel (A) had brass caps at both top and bottom. A brass tap (B) was fitted to the upper cap. The bottom cap was open but was made to fit closely into the brass socket (Dd). In addition to this vessel, there were three glass bottles as in (*Fig.2*, M), with the corresponding measure (*Fig. 3*, B) adapted to each bottle. For instance, a first bottle for testing common air (or worse) held three measures, each one corresponding to 1 ¼ of the standard measure. A quantity of nitrous air was introduced into one of the bottles (M) by means of its corresponding measure and then set on the knob (*Fig. 3*, C).[69] An appropriate quantity of respirable air was then let into the vessel (A) and also set on the knob (C). To test a sample of common air, a 1 ¼ measure of nitrous air and one measure of common air were used. Although one measure of nitrous air would have been sufficient to produce a complete contraction in volume, Cavendish chose to use a 1 ¼ measure for fear that the nitrous air might have been impure.

The contraction in volume when testing common air was determined by weighing in the following manner: the vessel (A) was fixed in the brass socket (Dd) that had a small hole (E) in the bottom. The socket was fastened to the board of the pneumatic trough by the brass support (FfG) so that the top (b) of the tap (B) was about half an inch under water. The bottle (M) was weighed by suspending it from one end of a balance and then placed with its mouth over the tap (B). On opening this tap, the air in the vessel (A) bubbled slowly into the bottle (M) in such a way that it could not escape faster than the water entering by the small hole (E) to replace it. The bottle (M) with its mouth still over the tap (B) was continuously shaken from side to side. Lastly, the bottle (M) was removed and weighed again to ascertain the increase in weight, which could be converted into measures, one measure being equivalent to 282 grains. This value, added to one measure, approximated very closely to the true diminution in volume.[70]

---

[69] There were two alternative procedural methods; the first was to add respirable air to the nitrous air and the second vice-versa.

[70] Ibid., pp. 108-112.

The problematic endpoint of the nitrous air test

The contraction in volume of the mixture of common and nitrous air was a key factor in all the different nitrous air eudiometers. In fact, Cavendish conducted a complete study of the factors that apparently influenced the determination of the endpoint of the eudiometrical test. He examined aspects such as how to dose the two airs; the ideal ratio of the air mixture; the rhythm at which the bubbles appeared; the size of the bubbles; the diffusion time of one air into the other; the speed at which the air mixture was shaken, and the kind of water to be used. Some of his remarks constituted true tacit knowledge that had remained unperceived in the description of other eudiometrical procedures and which testify to Cavendish's extraordinary experimental skills and the training required to perform the test.

One important point was whether the respirable air should be bubbled into the nitrous air or vice-versa. His initial conclusion was that the amount of nitrous air required for a complete contraction of the air mixture was greater in the first method than in the second. His second conclusion was that the second method was more accurate than the first, but the first was more reliable. It was for this reason that Cavendish chose to use the first method in preference to the second.[71]

With regard to the contraction in volume, he observed that it was more regular if one air was bubbled slowly into the other while continuously shaking the bottle containing the latter. However, this contraction in volume was not dependent on the size of the air bubbles. In addition, the contraction increased with the diffusion time of one air into the other. Cavendish set this time at 50 seconds. On examination of how the air mixture should be shaken, he found that the contraction in volume was significantly less without agitation, slightly less with gentle rather than energetic shaking, and that it remained notably steady with gentle shaking.

Cavendish was the first to conduct a thorough study on a matter that had received little attention from makers of nitrous air eudiometers extant at that time. In fact, the source of water used in the test was becoming a serious problem regarding the generation of significant variations in the contraction in volume of the air mixture, which appeared to be the main cause of uncertainty when determining the purity of the air. After a series of different trials, the greatest contraction in volume was achieved by using distilled water followed by rainwater. Consequently, in order to overcome these drawbacks, Cavendish recommended always using the

---

[71] Ibid., pp. 118-122.

same source of water, preferably distilled water. Furthermore, the capacity of water to absorb nitrous air varied with time and temperature, a factor which also affected the contraction in volume of the air mixture. Cavendish even went as far as to formulate an algorithm to partially counteract these effects.[72]

Testing Fontana-Ingenhousz's eudiometer and the need for a scale of air purity

Cavendish's other major contribution in his eudiometrical device was to test the Fontana-Ingenhousz eudiometer rigorously, something that had never been done before. He carried out a comparative study of Fontana-Ingenhousz's method with his own procedure, which implied the conversion of a volumetric test into a gravimetric test, so that the results of both procedures could be compared, otherwise it would have been difficult to determine which of the two procedures was better, since the apparently greater accuracy of Cavendish's procedure could have been attributed to the higher accuracy of a gravimetric procedure over a volumetric one. In order to proceed with this comparative study, he made some modifications in the abridged Ingenhousz procedure, such as first adding the nitrous air; weighing the eudiometrical tube under water; then mixing both airs in equal parts, and finally weighing the eudiometrical tube again under water. The contraction in volume was determined by the difference between the two weights.[73]

Cavendish concluded that his eudiometrical device provided more accurate results than the Fontana-Ingenhousz eudiometer and that it also required fewer manipulative skills by the experimenter. He devoted the second half of 1781 to testing air samples from one particular place for two months, and then from different parts of London and from around country. He found no differences attributable to different weather conditions, while the variations observed at different locations could be equated with experimental errors.[74]

---

[72] Ibid., pp. 116-118. In 1803, Berthollet reviewed this version of Cavendish's eudiometrical device and found that, as in the case of Fontana-Ingenhousz's eudiometer, if all these factors were not taken scrupulously into account, it was impossible to obtain comparable results. For this reason, he considered Cavendish's device to be unsafe (Berthollet, 1803b, pp. 506-507).

[73] Cavendish, 1783, p. 122.

[74] Ibid., pp. 123-124, 126-128. According to Wilson (1851, p. 41), the completion of these tests also led Cavendish to conclude that the amount of pure air in the common air was 10/48 (20.8%).

Cavendish not only followed Fontana's *Descrizione* of 1775 in proposing a gravimetric nitrous air test, but he also adopted Fontana's idea of establishing a scale for the purity (Fontana's salubrity) of air. Cavendish believed that such a scale was needed to standardize a test that provided different determinations, depending on each eudiometer, which were impossible to compare. Fontana had suggested a scale from zero to one hundred with two fixed points of reference: the minimum value set by a lethal air sample and the maximum by a sample of an air that was as salubrious as possible. Cavendish recommended as fixed points common air and perfectly phlogisticated air (i.e. lethal air).[75] Thus, standards below that of common air were to be established by mixing common and perfectly phlogisticated air in different proportions. Moreover, to prepare standards above of that common air, it was first necessary to procure a sample of good dephlogisticated air and find its standard by testing what proportion of phlogisticated air it should be mixed with in order to be as good as common air. Mixtures of this dephlogisticated air with different proportions of phlogisticated air were tested.[76]

The preparation of these standards enabled Cavendish to compare the results obtained by using his own device with those obtained with the gravimetric version of the abridged Ingenhousz procedure. Apart from of the aims pursued by Cavendish, the study provided analytical results obtained by a self-trained experimenter (Cavendish) to execute the experimental protocol designed by Fontana and Ingenhousz.[77] The statistical analysis of three series of experimental results confirms again that the Fontana-Ingenhousz nitrous air test was reasonably standardized.[78]

---

[75] Cavendish proposed to obtain this perfectly phlogisticated air by shaking common air with a solution of *sulphur of liver* or, alternatively, by placing it in contact with a wet paste of iron filings and sulphur (see chapter 5, notes 4 and 12). Both procedures consumed respirable dephlogisticated air (oxygen) leaving a residue of irrespirable and lethal phlogisticated air (mainly nitrogen). The values of both fixed points were 1.00 for common air and 0.00 for perfectly phlogisticated air. Perfectly dephlogisticated air (oxygen) would have a maximum value of 4.8.

[76] Cavendish, 1783, pp. 129-131.

[77] For Golinski (1992, p. 125), this paper by Cavendish constituted a further attempt to standardize the practice of eudiometrists, which required a considerable level of expertise. Boantza (2013b, p. 402) has rightly remarked the advanced level of proficiency and expertise required to operate with Cavendish's eudiometer.

[78] This statistical analysis shows high values for the coefficient of determination between the variation in volume of the gaseous mixture and the volume of nitrous

The nitrous air test in the hands and mind of Lavoisier

In 1774, when the first volume of *Experiments and Observations* was published, in which Priestley announced his nitrous air test, Lavoisier was already engaged on the study of everything that the British natural philosophers had discovered on the air. Arguably, Lavoisier began using this test with confidence when, on March 31st, 1775, he performed one of his first experiments on respiration by examining the air exhaled by a bird. Lavoisier's laboratory notes show how he used one measure of nitrous air for every two measures of the air sample in this examination.[79] In the hands of Lavoisier, the nitrous air test was in consonance with his research on respiration, which was inextricable from its general conceptual network. Lavoisier expected that respiration to have the same effect on the air as that produced by the calcination of mercury.[80]

However, Lavoisier conceptually appropriated Priestley's nitrous air test when he dealt with the issue of the salubrity of the air in hospitals and other crowded places such as theatres. This was reflected in the paper he deposited at the Academy of Sciences on May 10th, 1777, about the relevance of the effects of respiration on the atmosphere and on public health.[81] To this end, Lavoisier used the nitrous air test to determine the salubrity of hospital dormitories and theatres by collecting air samples from the ground floor to higher levels of these establishments. He was convinced by the experiments carried out by the Duke of Chaulnes to the effect that the components of atmospheric air tended to be stratified according to their densities.[82] Lavoisier believed that in crowded buildings fixed air remained fixed to the ground, whereas the irrespirable mephitic

---

air added to a constant volume of respirable air. A quadratic regression with a coefficient of determination (R squared) not lower than 0.98 has been estimated.

[79] Holmes, 1985, p. 47.

[80] Holmes, 1985, pp. 61, 89.

[81] This paper, entitled *Observations sur les altérations qui arrivent à l'air et sur les moyens de ramener l'air vicié à l'état d'air respirable*, was published as *Expériences et observations sur les fluides élastiques en général et sur l'air de l'atmosphère en particulier*. According to Édouard Grimaux, a "little different" version of the manuscript was read to the Academy (LO, Vol. 5, pp. 271-281). A fuller version of this paper was read at a meeting of the Académie Royale de Médecine of February 15th, 1785, with the title *Mémoire sur les altérations qui arrivent à l'air dans un grand nombre de circonstances* (Holmes, 1985, pp. 238-241; LO, Vol. 2, pp. 676-687).

[82] Marie Joseph Louis d'Albert d'Ailly, Duc de Chaulnes.

air rose to the ceiling.[83] Although praising Priestley's nitrous air test for estimating the salubrity of air, he reinterpreted the contraction in volume of the air mixture in different terms. For Lavoisier, the reduction in volume observed when respirable air came in contact with nitrous air made the test a touchstone of detection as well as for determining the proportion of respirable air in a sample of any air. According to the observed contraction in volume, it was possible to determine the part of an air sample that was respirable (true) air and that the remainder was irrespirable.[84] The historian Frederic L. Holmes interpreted Lavoisier's approach to the nitrous air test in the sense that the latter - unlike Priestley - was not as interested in establishing a scale of goodness or salubrity of different airs as much as in characterizing them by their pure air ratio, i.e. for their respirability in relation to common air.[85] Thus, Lavoisier's conceptual framework for the nature of the air would have led him to appropriate and reinterpret Priestley's nitrous air test. In fact, this test outlived the demise of the phlogiston conceptual framework in which it was conceived and proved to be equally operational in Lavoisier's new theoretical framework.[86]

After the first series of experiments on animal respiration – conducted between April and October 1776 - which led Lavoisier to conclude that respiration was nothing more than a slow burning of charcoal (*matière charbonneuse*) with the consumption of vital air and the release of fixed air, his interest in respiration became part of his agenda in January 1783. This was related to his first experiments with the calorimeter, conducted jointly with Pierre-Simon Laplace (1749-1827). Since that time, the heat involved in the process of respiration had become a key factor in this research. One of the problems that Lavoisier had to face was how to determine the amount of the vital air consumed by guinea pigs placed

---

[83] Lavoisier's commitment to the eudiometrical determination of the salubrity of closed rooms has been interpreted as a revival of Landriani's eudiometrical approach to medicine (Beretta, 2000, pp. 60-62).

[84] LO, Vol. 5, pp. 278-279.

[85] Nevertheless, Lavoisier was far from abandoning the idea that eudiometrical tests were effective in determining the salubrity of air (Beretta, 2000, pp. 63-64).

[86] Holmes, 1985, pp. 95-96.

inside the calorimeter.[87] However, in order to obtain a direct measurement of this vital air, Lavoisier and Laplace reverted to the classic experiment of allowing an animal (guinea pig and parrot) to breath for as long as possible inside a bell jar placed over mercury. In the initial phase of this experiment, they used the Fontana-Ingenhousz nitrous air eudiometer to determine the proportion of impurities in a sample of vital or dephlogisticated air.[88]

Another proof that fueled Lavoisier's revived interest in respiration during this same period was his careful examination of the different procedures when using the nitrous air test, which he expounded in the *Mémoire sur la combinaison de l'air nitreux avec les airs respirables et sur les conséquences qu'on en peut tirer, relativament à leur degré de salubrité*, presented on December 20th of the same year, 1783.[89] It is worth saying in advance that in his approach to the nitrous air test, Lavoisier tried to remain somewhat removed from Priestley's hygienist conception. The nitrous air test was to be regarded only as a means of determining the amount of the vital air contained in a given quantity of common in order to find its degree of salubrity.

Lavoisier then presented a comparison of the procedures used by Priestley and Ingenhousz, but with some inaccuracies. He remarked that Priestley had mixed both airs in equal proportion, even though that since 1779 it had been known that Priestley had executed a second addition of nitrous air, thereby doubling the dose of nitrous air to common air. He was also inaccurate in his commentary on Ingenhousz's preferences for Priestley's methodology, since the former had scrupulously followed that employed by Fontana. Lavoisier wanted to conduct the test with the rigour required to make it reproducible. To this end, he addressed it on three flanks: the search for the proportion at the point of saturation between

---

[87] Lavoisier and Laplace first determined the amount of fixed of air absorbed by a solution of caustic alkali. Secondly, they assumed that the amount of the absorbed vital air was equal to the amount of the released fixed air. They were eventually able to estimate quantitatively the amount of vital air consumed in the respiration of the guinea pig from approximate values of the densities of vital and fixed air (Ibid., p. 168).

[88] Ibid., 1985, p. 170.

[89] Lavoisier used this test in other experiments concerning the salubrity or respirability of the air. This is reflected in the annotation in his laboratory records of October 15th, 1783, and 9 October 9th, 1784. In the first case, comparing different trials using the nitrous air with the action of phosphorus. In the second case, assessing the quantity of oxygen absorbed in the detonation of nitre with charcoal (Berthelot, 1890, pp. 294, 298-299)

nitrous and vital air; the quality of the nitrous air, and the development of a formula for an arithmetical solution of the test.

Firstly, he used samples of pure nitrous and vital air to determine the proportion of both airs required to obtain their reciprocal saturation. Although Lavoisier found that this proportion was 40 parts of vital air to 66-69 parts of nitrous air, he actually used the ratio 69/40 (1.725), pointing out that a difference of three units (69 - 60) amounted only to an error of one hundredth in the determination of the vital air.[90] As for the quality of nitrous air, he proposed two procedures to obtain it. The first consisted of reacting sugar with nitric acid. According to Lavoisier, this procedure provided a purer nitrous air but did not assure a constant quality of air. This procedure was therefore only recommended for research experiments.[91] The other procedure for obtaining nitrous air was based on the already known reaction between nitric acid mercury.[92] Although it was not as pure as that recommended by Lavoisier, the nitrous air obtained by this procedure had a more consistent quality, and indeed Lavoisier preferred it for ordinary laboratory experiments. Lavoisier mathematically formalized the nitrous air test according to two assumptions: first, the addition of an excess of nitrous air to saturate all the vital air of the air sample, and secondly, adding less nitrous air than necessary to saturate the vital air. For both assumptions:

---

[90] LO, Vol. 2, pp. 503-508.

[91] In fact, Lavoisier stumbled upon this procedure as a result of the experiments carried out from July 22nd, 1777, with the aim of extending his theory of acidity. Lavoisier wanted to widen this theory, which attributed the acidity of some substances such as nitric, sulphuric or phosphoric acids, to the fact that the purer part of common air went into its composition. To add a new acid to his list, Lavoisier began a series of experiments to obtain a recently discovered new acid – acid of sugar (oxalic acid) - by reacting sugar with nitric acid. Lavoisier modified the procedure described by Bergman in order to collect the air released, but these first experiments were not entirely successful and Lavoisier did not resume this task until February, 1779. These new experiments were more successful, not only in relation to the production of the acid of sugar, but also for obtaining a high purity nitrous air (Holmes, 1985, pp. 110-112, 138-141):

$C_{12}H_{22}O_{11}$ (s) $+ 12\ HNO_3$ (aq) $= 6\ H_2C_2O_4$ (aq) $+ 12\ NO$ (g) $+ 11\ H_2O$ (l)

[92] In 1776, Lavoisier believed that mercury took a portion of pure air out of the nitrous acid to form its own calx, whereas the acid was transformed into nitrous air. Furthermore, if both airs combined again, a fuming nitrous acid was recovered (LO, Vol. 2, p. 137)

a: parts of the air sample.

b: parts of impure nitrous air added (containing a mephitic air as impurity).

c: parts of residual air after the contraction in volume.

μ: conversion factor that related the amount of nitrous and vital air necessary for their reciprocal saturation (i.e. the proportion at the point of saturation). In practice, the used value was 69/40 (1.725).

First assumption: determination of the vital air content.

x being parts of vital air in the air sample, there would be **a** - **x** parts of irrespirable air. The **x** parts of vital air had to be saturated with **μ· x** parts of nitrous air. Once completed, the contraction in volume of the residual air would contain **a** - **x** parts of irrespirable air, and **b** - **μ· x** parts of exceeding nitrous air containing mephitic air as an impurity. Therefore, **c** = (**a** – **x**) + (**b** - **μ· x**). Finally, Lavoisier developed the following formula for calculating the parts of vital air in the air sample:[93]

$$\frac{a+b-c}{1+\mu}$$

The procedural execution of the test was similar to that of Fontana-Ingenhousz, with the difference that Lavoisier first introduced a measure of the nitrous air in the eudiometrical tube to which measures of the air sample were sequentially added. In that paper, Lavoisier presented the results of the determination of the purity of a vital air obtained from the red oxide of mercury, and the vital air content of a sample of atmospheric air (Appendix III). Although the trials showed that vital air represented roughly a quarter part in volume of the atmospheric air, Lavoisier did not reject the values obtained by Scheele in Stockholm (27.5%) and by himself in 1777 in Paris.[94] The reason for this was that Lavoisier considered that the composition of the atmospheric air depended on circumstances such as the geographical area or the season of the year.[95] By means of such trials, Lavoisier was hopeful that the perfection of eudiometers would lead to an improvement in the accuracy of the results, especially after becoming aware of Cavendish's works in this direction.

---

[93] In practice $\frac{a+b-c}{2.725} = \frac{\Delta V}{2.725} = 0.367$ (ΔV), ΔV being the contraction in volume.

[94] Lavoisier, 1780, p. 367.

[95] LO, Vol. 2, p. 507.

Second assumption: determination of the purity of the nitrous air.

y being parts of pure nitrous air in the nitrous air sample, there would be **b – y** parts of mephitic air as an impurity. The **y** parts of pure nitrous air had to be combined with **y/μ** parts of vital air. Once completed the contraction in volume, the residual air volume would contain, on the one hand, **a - y/μ** parts composed by exceeding vital air and irrespirable air from the original common air sample, and on the other hand, the residue also contained the **b - y** parts of mephitic air from the nitrous air sample. Therefore, **c = (a - y/μ) + (b – y)**. Finally, the parts of pure nitrous air absorbed in the mixture could be calculated by the following formula.[96]

$$\mu \frac{a + b - c}{1 + \mu}$$

Lavoisier applied this second formula to determine the quality of the nitrous air used in the first experiment. This new trial indicates that it was Lavoisier's intention to show that the versatility of the nitrous air test, enabling it to be reverted it by adding measures of nitrous air to a pure vital air sample (Appendix III). One notable procedural feature of the nitrous air test is the mathematical algorithm proposed by Lavoisier to perform these calculations. To calculate the amount of vital air and the purity of the nitrous air, he proposed the following:[97]

'[...] take the sum of both airs, subtract the residue, find the logarithm of the remaining number, that is, subtract from this logarithm, for the vital air, the constant logarithm 0.4353665, and for the nitrous air, 0.198577,[98] the logarithm obtained is that of the requested number.'

Why did Lavoisier want to give mathematical formality to the nitrous air test? Why did this formality include a logarithmic calculation if the end result could simply be calculated with a sum followed by a subtraction and a multiplication? Firstly, the mathematical formulation of the test lent a certain elegance to the laboratory procedure. Moreover, providing a mathematical algorithm involving the (unnecessary) use of a logarithmic

---

[96] In practice $\frac{69(a+b-c)}{109} = 0.633$ (ΔV), ΔV being the contraction in volume.

[97] Ibid., pp. 505-50.

'[...] prendre la somme des deux airs, d'en retrancher le résidu, de chercher le logarithme du nombre restant, enfin, de retrancher de ce logarithme, pour l'air vital, le logarithme constant, 0.4353665, et, pour l'air nitreux, 0.198577, le logarithme qu'on obtient est celui du nombre cherché.'

[98] $0.4353665 = \log 2.725$ and $0.198577 = \log 109/69$

calculation allowed Lavoisier to introduce numbers with up to seven decimal places - well above the accuracy of the experimental data – in order to persuade the reader of the correctness of the procedure. All this suggests the staging of rhetorical resources reflecting the ideal of mathematization that Lavoisier intended for chemistry.

Thus, Lavoisier was still using the nitrous air test with full confidence in 1783, following a procedural variant of Ingenhousz's method to determine both the amount of vital air in common air samples and the purity of nitrous air samples.[99] In time, however, Lavoisier opted for other eudiometrical tests such as those based on alkaline sulphides or phosphorus, while distancing himself from the nitrous air test. From 1789 onwards Lavoisier found compelling reasons to abandon this test. Alkaline sulphides, phosphorus or the wet paste of sulphur with iron filings were easily affordable eudiometrical means with a comparable degree of purity, whereas the quality of nitrous gas was too dependent on innumerable circumstances. In addition, nitrous gas was unable to absorb all the vital air mixed with azotic gas and presented different degrees of oxygenation. Eventually, other non-gaseous eudiometrical means could be used in excess without affecting the corresponding test result, whereas an excessive amount of nitrous gas mixed with the residue modified the true contraction in volume, and therefore falsified the end result.[100]

Fontana's nitrous air test revisited

The publication in 1782 of Senebier's work *Mémoires physico-chimiques sur l'influence de la lumière pour modifier les êtres des trois regnes de la nature et surtout ceux du regne végétal* triggered a controversy, not only between him and Ingenhousz over the priority on the discovery of a dephlogisticated air (oxygen) released by the plant leaves under the influence of sunlight but also over Ingenhousz's observation of the *influxus nocturnum*. This phenomenon (plant respiration) consisted of the absorption of dephlogisticated air and the release of phlogisticated (fixed air) air by plants in the dark. Senebier rejected this phenomenon, and Ingenhousz was determined to confirm it. From May to November 1783, he devoted himself to repeating and expanding his experiments. The final results were published in 1784 in two papers in the *Observations sur la physique*. Ingenhousz had finally decided to publish in a French periodical instead of in the *Philosophical Transactions* of the Royal

---

[99] Ibid., p. 506.

[100] Ibid., pp. 720-721.

Society.[101] He reproached Senebier for failing to follow his procedure more scrupulously, which had led to uncertainty in the results. Ingenhousz made an even more cogent criticism when he stated that he had been unable to form an exact idea of Senebier's eudiometrical procedure. He made all these remarks in a footnote, where he also gave a few details about a new and abridged version of the nitrous air test for dephlogisticated airs. Ingenhousz had already presented a previously abridged version of this test for common air samples in his *Experiments upon Vegetables*.[102]

This simplified eudiometrical procedure was first published in 1781 in the journal of the Bataafsch Genootschap voor Proefondervindelijke Wijsbegeerte (Batavian Society for Experimental Philosophy of Rotterdam) in a paper on the therapeutic utility of dephlogisticated air.[103] The original version (in French) of this paper was not published until July 1785, although in the previous May this abridged procedure had already been explained, together with the presentation of Fontana-Ingenhousz's standard eudiometer in the *Observations sur la physique*. Ingenhousz's simplified procedure was mainly designed to save time in the testing of samples of dephlogisticated air. For an experimenter with some previous training, testing a sample of common air would take no more than a few seconds, while a sample of dephlogisticated air that required the addition of three or four consecutive measures of nitrous air might take three to four minutes.

In this abridged procedure, one measure of dephlogisticated air was transferred into a large glass vessel. The accuracy of this new procedure depended on the diameter of this glass vessel, which was determined by a process of trial and error using recipients of different sizes.[104] Four

---

[101] Senebier, 1782, Vol. 1, pp. 1, 2-3; Magiels, 2010, pp. 248-262; Beale & Beale, 2011, pp. 326-330.

[102] Ingenhousz, 1784, pp. 342-343.

[103] *Verhanderlingen over de gedéphlogisteerde lucht, en de manier hoe men dezelve kan bekomen en de ademhaaling doen dienen* (Dissertation on the dephlogisticated air and the way to obtain it and make it serviceable for breathing), *Verhandelingen van het Bataafsch Genootschap der proefondervindelyke wysbegeerte te Rotterdam*, 1781, VI, 107-159. The German version was published in 1782: *Vermischte Geschriften, physich-medizinische Inhalts*, Vienna, Kraus.

[104] Ingenhousz had already investigated the effect of the diameter of the recipient in which the mixture of nitrous and common air occurred on the determination of the goodness of an air sample (Ingenhousz, 1779, pp. 283-287).

measures of nitrous air were then immediately transferred into the same vessel. Finally, the mixture was moved to the eudiometrical tube, and the length of the air column could be read directly on the brass scale. This procedure involved the use of an amount of nitrous air exceeding the saturation of dephlogisticated air. This excess of nitrous air remained in the tube, and therefore the length of the air column was longer than usual, but the contraction in volume of the air mixture was still indicative of the goodness of the air. In principle, the abridged test was not as reliable as the standard test, but dephlogisticated airs did not demand such a high degree of accuracy as atmospheric air.[105]

The paper in the *Observations sur la physique* also contained a brief description of Fontana-Ingenhousz's eudiometer with frequent references to the 1780 French translation of the *Experiments upon Vegetables* of 1779, including information about the best makers and suppliers of Fontana's eudiometers.[106] The observation on the influence of the kind water used in the test constituted a remarkable procedural novelty. As mentioned above, Cavendish had recommended always using the same source of water, preferably distilled water, to obtain the greatest contraction in volume. When addressing this issue, Ingenhousz retrieved some observations made by his friend, doctor, editor and town official in Delft, Jacob van Breda (1743-1818) between 1780 and 1782.[107] These observations had already been published in the German editions of *Vermischte Geschriften, physich-Medizinische Inhalts* in 1782 and 1784, to which Ingenhousz referred.[108] Van Breda had first concluded that the eudiometrical results were not dependent on the source of water (well water, pure distilled water or rainwater), and secondly that it was only necessary to fill the eudiometrical tube with pure distilled water or rainwater to obtain the same results as those obtained when the tube and the pneumatic trough were filled with the same kind of water. On the whole, it was advisable to fill the tube with distilled water (since pure rainwater was not always

---

[105] Ibid., 1785a, pp. 345-348, 357.

[106] Clindworth, mechanic of the King of England in Gottingen (2 gold louis and a half). Ampichel, near the church of Saint-Etienne in Vienna. Megnier & Sikes in Paris. (Ibid., p. 342, footnote).

[107] Van Breda was responsible for the Dutch translation of *Experiments upon Vegetables,* published in 1780.

[108] These observations by van Breda appeared in a letter to Ingenhousz (translated into French) and included in the second volume of *Nouvelles expériences et observations sur divers objets de physique* (Ibid., 1785b, pp. 137-167).

available) immediately before starting the test, since it did not matter what kind of water was used in the pneumatic trough. According to Ingenhousz, van Breda's conclusions eliminated any uncertainty from the reliability of the test.[109]

The nitrous air test in the research for a therapeutic use of dephlogisticated air

The origin of the paper on the therapeutic use of dephlogisticated air mentioned above (*Dissertation on the Dephlogisticated Air and the Way to Obtain it and Make it Serviceable for Breathing*), published in Dutch in 1781, lies in the context of John Pringle's recommendation to examine the quality of the marine air in order to assess its therapeutic potential for people suffering from breathing disorders. Ingenhousz was echoing Pringle's words in his address of November 30th, 1773, when Priestley was presented with the Copley Medal. Pringle had quoted Priestley's theses on the two major natural resources for the restoration of the stale air produced by combustion and breathing; the plant kingdom and large masses of water (oceans, seas, lakes and rivers), which were able to absorb the harmful parts of air.[110]

In November 1780, Ingenhousz had already constructed a device for dispensing dephlogisticated air through the respiratory tract and had begun to draw up his report on this research. When Ingenhousz realized the poor circulation his research work published in the journal of the Batavian Society for Experimental Philosophy, he decided to publish the original French version. This was not possible until July 1785 with the work *Nouvelles expériences et observations sur divers objets de physique*,[111] even though the manuscript had been completed in December of the previous year. Although the main subject of this work was concerned with electricity, it included the corresponding French version of the paper published in 1781 under the title of *Sur l'air dephlogistiqué, la manière de l'obtenir, et d'en faire usage pour la guérison des maladies*, but considerably extended.[112]

---

[109] Ibid., 1785a, pp. 351-353.

[110] Priestley, 1772, pp. 199-200; Pringle, pp. 1783, 33-34.

[111] The work had been previously described as *Mélanges de physique et de médecine*, in a translation of the title of the German edition of 1782 (*Vermischte Geschriften, Physich-Medizinische Inhalts*).

[112] Ingenhousz, 1785b, pp. 192-288.

Ingenhousz was convinced of the utility of dephlogisticated air for curing diseases thanks to its purifying action on the lungs infected by fixed air. In this regard, the report describes the device (Figure 3.6) that he designed to produce and dispense dephlogisticated air (*Figs. I* and *II*), its therapeutic use and the different ways to obtain that air. In addition to this essentially operational part, Ingenhousz made a careful study of the quality of the dephlogisticated air obtained in different ways, and the degradation of respirable air as it was breathed. This study shows that Ingenhousz continued to rely on his nitrous air procedure and that he was using his eudiometer again in further research of a chemical-physiological nature.

Figure 3.6 Ingenhousz's apparatus for dispensing and breathing oxygen regularly.

From Jan Ingenhousz, *Nouvelles expériences et observations sur divers objets de physique* (Paris, 1785), plate III.

In this work, he examined the purity of the dephlogisticated air obtained by the thermal decomposition of nitre.[113] To that end, he collected six consecutive samples of that air as it was produced. These samples were then eudiometrically tested by adding four consecutive measures of nitrous air to one measure of each sample. In certain respects it was a kinetic study

---

[113] Basically, it was the nitre used for the manufacture of gunpowder (potassium nitrate), although mineral nitre (sodium nitrate) was also used. The decomposition occurs according this reaction:

KNO_3 (s) = KNO_2 (s) + $1/2\ O_2$ (g)

The purity of the dephlogisticated air released depended largely on the purity of nitre and on achieving a steady rate of heating in order to keep nitre in a state of fusion.

of the decomposition of nitre. His general conclusion was that the air of the first sample, obtained at the beginning of the decomposition, was the best. Regarding the degradation undergone by respirable air passing through the lungs, Ingenhousz drew all the air – of known purity – into a bladder and expelled it through the same bladder. This operation of inhalation-expulsion was repeated from five to eight times. The purity of an air sample after each operation was eudiometrically tested, and the final result showed a significant and progressive loss of purity of the air breathed.[114]

To encourage the use of his eudiometer by those who considered it overly complicated, Ingenhousz promoted the abridged procedure on the grounds that it enabled any air sample to be analyzed with a considerable saving of time. In addition to a description of the abridged procedure, he included a theoretical exposition on the processes of exchange of phlogiston in the air mixture. After presenting a series of twelve experiments, he concluded that the contraction in volume was much faster if the mixture was contained in a vessel wider than the eudiometrical tube.[115]

Replying to criticisms of Fontana-Ingenhousz's nitrous air test. The last calls for recognition of his eudiometer

Since the publication in 1779 of *Experiments upon Vegetables*, somewhat virulent criticisms had been levelled against the Fontana-Ingenhousz eudiometer. It may be said that the paper published in the *Observations sur la physique* in May 1785 presented Ingenhousz with the first opportunity to publicly refute some of those criticisms. It is not surprising that this paper began with a series of reflections on the expected ethical reaction of men of science when assessing the findings of their colleagues. For Ingenhousz, the appropriate reaction would first to have personally examined the experiments underpinning new discoveries before judging them, and secondly, to argue publicly any objections based on experimental data that clearly proved the errors in key experiments.[116] Ingenhousz continued fiercely to defend his eudiometer in the face of criticism, affirming that insufficient verifiable evidence had been found against it, only rather vague statements lacking any well-detailed accounts of the existence of anomalies. He vindicated the instrument as being the most suitable for determining the purity of the air. He also claimed the

[114] Ibid., pp. 220-232.

[115] Ibid., pp. 272-288.

[116] Ibid., 1785a p. 339.

merit of having made it public and of acknowledging more than anyone the contributions of others who, in attempting to modify it had only impaired it, making the instrument a more complicated and cumbersome device.[117] All these refutations indicated that Ingenhousz believed that the criticisms had come from those who either did not own a Fontana-Ingenhousz eudiometer or did not know how to use one.

Ingenhousz referred to three recurrent objections; namely, the uncertainty of determining the contraction in volume of the air mixture; that the eudiometer required special operational skills, and the irrelevance of the strength of the nitrous air. In particular, the first objection concerned the assertion that there was no absolute contraction in volume, since this could last for some time (hours or days). For Ingenhousz this did not amount to a real objection, because if the goodness of an air sample was determined at a given time, it did not matter that a few hours later this same goodness could still not be determined. Furthermore, it was mistaken to believe that only the possible maximum contraction in volume was indicative of the goodness of air. What really mattered was the accuracy of the instrument, which depended on its reliability.[118] The determination of the contraction in volume was, however, a more serious objection to the Fontana-Ingenhousz eudiometer, because the need for special operative skills made the replication of experimental results difficult. In this case, Ingenhousz was adamant that he had yet to see an untrained physicist who after practising for a short period of time was unable to handle the eudiometer well enough to carry out concordant tests.[119]

By 1785 or early 1786, Ingenhousz had already completed the manuscript of the first volume of the second French edition of his *Expériences sur les végétaux*. Although this volume was ready for publication in January 1786, it did not appear until 1787. During this time, Ingenhousz was able to finish writing the second volume of the work, which was published in October, 1789, while simultaneously in the same year he also compiled enough material to publish the second volume of *Nouvelles expériences et observations sur divers objets de physique*, the second installment of which enabled him to update his eudiometrical

---

[117] Ibid., pp. 341-342.

[118] This would be similar to the case of an acid-base indicator, in which once the color had changed, the new color changed its tone with time. What matters is the determination at the time of the initial colour shift.

[119] Ibid., pp. 343-344.

procedure, respond again to some criticisms about his researches, and consolidate the recognition of his eudiometer. He announced some changes in the names and sizes of certain parts of the eudiometer, extending and updating the list of makers and suppliers of eudiometers.[120] In contrast to the 1779 edition, the new edition included the description of an apparatus for generating nitrous air (Figure 3.7).[121]

Figure 3.7 Ingenhousz's apparatus for generating nitrous air.

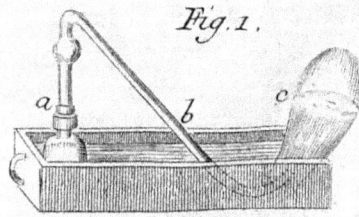

From Jan Ingenhousz, *Expériences sur les végétaux*. (Paris, 1787), Vol.1, plate p. 465.

The extensive experience that Ingenhousz had gained in the use of his eudiometer had made him an expert in the matter and had therefore equipped him with the authority to pronounce on those tacit practices and knowledge that only a continuous use of the instrument could provide. Some of the procedural changes introduced in the second edition of the *Expériences sur les végétaux* had already been made known, such as filling the tube with distilled water immediately before starting the test to ensure maximum contraction in volume of the air mixture. Other incorporations were actual novelties, such as leaving the tube and the measure in the water of the pneumatic trough to ensure that both acquired the same temperature as that of the water, thereby preventing the formation of air bubbles beneath the measure brass slider, because when emptying the measure they might have slipped into the tube, thus changing the total amount of air and eventually leaving the tube upright in the water for 1-2 minutes once the air mixture had been completed.[122]

---

[120] Magellan's eudiometer, Fleet Street in London; Fontana's eudiometer, Benjamin Martin in London (Museo Galileo); Sikes in London and Paris; Megnier in Paris and Madrid; George Adams Jr. In London (Museo Galileo); J. Reghter in Delft; Clindworth, mechanic of the King of England in Gottingen. (Ingenhousz, 1787, pp. ci-cii).

[121] Ibid., pp. 198-200, 216-218, 224-225.

[122] Ibid., pp. 206-207, 214.

Ingenhousz continued to refute some of the criticism of his eudiometrical test. In particular, he again dismissed Priestley's objection regarding the irrelevance of the strength of the nitrous air. Ingenhousz had already argued that the strength of the nitrous air was of little importance, provided that it saturated the common air. However, at that time Ingenhousz was able to back up his arguments with concrete experimental data. He tested a common air sample with two kinds of nitrous air of different quality (i.e. strong and weak). A total amount of five measures (two of common air and three of nitrous air) were used to reach the endpoint of the test with the strong nitrous air. The endpoint with the weak nitrous air needed a total amount of six measures (two of common air and four of nitrous air). Nevertheless, the contraction in volume in the first case was 1.94 measures and in the second case 1.93 measures. In the eyes of Ingenhousz, this negligible difference considerably weakened concerns about the nitrous air quality on the result of the test.[123]

Ingenhousz took advantage of the second volume of the *Expériences sur les végétaux* (1789) to counter Senebier's complaints arising from the accounts of his research work in *Mémoires physico-chymiques* (1782) and in *Expériences sur l'action de la lumière solaire dans la végétation* (1788). The preface to the second volume is practically a page-by-page refutation of Senebier's objections. Ingenhousz was extremely critical of Senebier's eudiometrical skills and regarded his eudiometrical device a bad imitation of his own. He was of the opinion that it was unworthy of any confidence whatsoever and that its misuse had been a source of errors. Senebier's inconsistent position on the nitrous air eudiometers led Ingenhousz to place no trust in his eudiometric trials.[124]

In the *Observations sur la physique*, his paper of 1785, he stated that what mattered in an instrument was its accuracy, which depended on its reliability. In fact, in his work of 1779 Ingenhousz had emphasized the need to follow the same eudiometrical procedure scrupulously and systematically, giving warning of a number of potential errors and their corresponding quantitative impact on the test results. In the second edition of the *Expériences sur les végétaux*, Ingenhousz departed from that list of errors and instead introduced the procedural novelties mentioned

---

[123] Ibid., pp. 222-223.

[124] Ibid., 1789, pp. xii-xiv.

above, insisting that the accuracy of the test depended to a much greater extent on the uniformity (standardization) of all operations.[125]

To illustrate this uniformity, Ingenhousz was especially careful about the time taken by certain operations, such as the immersion of an air measure in water, the final reading of the length of the residual air column or the time between the addition of nitrous air and the shaking of the tube.[126] Standardizing the time spent performing certain operations was certainly more complicated than standardizing instrumental manipulations of the test. For Ingenhousz, simplicity and uniformity, as well as versatility, were the characteristics that made his eudiometer the best, and the only one that was superior to Priestley's original eudiometrical apparatus, which he had never actually handled himself. Ingenhousz believed that an instrument became more awkward, more accident prone, more difficult to construct and more expensive as it became more complicated. Thus, in addition to its simplicity, the Fontana-Ingenhousz eudiometer had the great advantage that the same measure, the same tube and the same scale, which were all necessary parts of the instrument, were perfectly suitable for testing common air as well as relatively phlogisticated airs.[127]

This need for simplification had led Ingenhousz to design a eudiometrical abridged procedure as accurate as the standard one, provided that the nitrous and the common air mixed uniformly. This was because the reliability of the results depended on the uniformity of diffusion between both airs, which in turn depended on the surface of their first encounter. Finally, Ingenhousz arrived at a device that was able to provide a better diffusion between both airs. This consisted of a glass globe with an inner diameter of approximately 3 inches and with a hole larger than that of the funnel of the pneumatic trough. The inner diameter of this funnel also had to be larger than that of the eudiometrical tube to allow the air to rise without interruption. With this design, the globe could not be shaken while mixing both airs (one measure of vital with four measures of nitrous air),

---

[125] Ibid., 1787, p. 203. Ingenhousz warned that the accuracy of eudiometers could not at that time be greater than that of thermometers and barometers. He considered that expanding the scale of the eudiometer with a nonius (as in astronomical quadrants) would lead to unwanted errors, the reason being that variations inherent to nature of the eudiometrical test would be an unfounded attribution to a supposed variability in the goodness of the examined air (Ibid., pp. 231-232).

[126] Ibid., p. 209.

[127] Ibid., pp.228-229.

because then the diffusion process would accelerate and the final result would not be comparable to that obtained with the standard procedure. Finally, the residual air mixture was transferred to the eudiometrical tube.[128]

When Ingenhousz was writing the second volume of the *Expériences sur les végétaux*, other eudiometrical devices such as the sulphur of liver, the wet mixture of iron filings with sulphur and Volta's eudiometer had already appeared (see chapters 2 and 5). Without entering into great detail about these alternative eudiometrical devices, Ingenhousz recognized that the first two might have been suitable for determining the purity of respirable air, but he was not convinced that they could replace the nitrous air test.[129] Consequently, he recommended his eudiometer over the others for its simplicity, accuracy and effectiveness in his last statement of recognition:[130]

'My familiarity with its use [Fontana-Ingenhousz's eudiometer], and on comparing its use with other eudiometers, make me still prefer it to all those that have been published so far. One may imagine a large number of these kinds of instruments, possibly also certain in their effects. I myself have not been far from imagining this. But it seems to me that for the simplicity, celerity and accuracy of such trials one could hardly imagine anything more perfect than M. Fontana's instrument.' [131]

The late nitrous air eudiometer of Adam Wilhelm Hauch

Adam Wilhelm Hauch (1755-1838) abandoned a successful military career in the Danish army in 1786 for his great passion, natural philosophy. It is quite likely that Peter Christian Abildgaard, his friend and professor at the School of Veterinary Surgery influenced him to take this decision. By the time that Hauch joined the Royal Horse Guards in 1775, Abiildgaard already held an appointment at the Royal Stables. At the end of the 1780s, to extend his scientific knowledge, Hauch attended Christian Gottlieb Kratzenstein's lectures on experimental physics at the University, and

---

[128] Ibid., 1789, pp.346-348.

[129] Ibid., pp. 357-358.

[130] Ibid., p. 360.

[131] 'L'habitude de m'en servir, et d'en comparer l'usage avec celui des autres eudiomètres, me le fait encore préférer à tous ceux qui ont été publiés jusqu'à présent. On pourrait imaginer un grand nombre de ces sortes d'instruments, peut-être également certains dans leurs effets. Je n'ai pas été moi-même éloigné d'en imaginer aussi. Mai il m'a paru que pour la simplicité, la célérité et la précision des essais de ce genre, on ne saurait guère imaginer quelque chose la plus parfait, que l'instrument de M. Fontana.'

Nicolai Tychsen's lectures on chemistry at the Academy of Surgery, both in Copenhagen.[132] During the years 1788-1789, he embarked on a training tour abroad that enabled him to establish personal contacts with prominent figures in science such as van Marum, Banks, Priestley, Cavendish and Lavoisier.

One of Hauch's main interests during his European tour was to acquire scientific instruments with which he established an excellent physical cabinet for performing demonstrations sessions for the nobility, university professors, students and military officers after his return to Copenhagen.[133] This set him on course for a highly regarded scientific and institutional career that secured him good relations with the nobility and royalty. These relationships ensured him a financial well-being that probably explains how he was able to devote himself to the construction of frontline research instruments.[134]

One of the instruments in this collection was a new nitrous air eudiometer that Hauch presented officially on May 25th, 1792, at the Royal Danish Academy of Sciences and Letters. Hauch's stated aim in constructing the instrument was to determine the purity of common air, especially in houses and workplaces. Healthy air was as essential as the nutritional value of food for preventing premature deaths. Hauch's alternative use of the terms "dephlogisticated air" and "oxygen" reveals the changing times affecting chemistry as a consequence of the two rival theories.[135] Hauch began to spread the new theory of oxygen in both his private lectures and in meetings of the Royal Danish Academy of Sciences and Letters, where he reproduced the key experiments from Lavoisier's *Traité* using replicas of essential instruments such as gasometers and round-bottom flasks for combustions. In 1794 he published his textbook (*Begyndelsesgrunde til Naturlaeren*) that was based on his chemistry demonstrations and in which he presented chemistry in terms of Lavoisier's antiphlogistic theory.[136] Hauch held his eudiometer in such regard as to dedicate an article exclusively to it

---

[132] At that time the Kingdom of Denmark included, apart from Greenland, Iceland, Norway and the duchies of Schleswig and Holstein.

[133] This collection of instruments was built up in the period 1790-1815 and was initially kept at Hauch's official residence in Copenhagen. King Frederik VI purchased the collection in 1815 and donated it to the Sorø Akademi in 1827, where it has remained ever since.

[134] Jacobsen, 2000, pp. 77-78, 90.

[135] Hauch, 1793.

[136] Levere, 1999; Jacobsen, 2000, pp. 78-90.

(*Beskrivelse af nye Luftprøver eller Eudiometer*), published in 1793 in the *Nye Samling af det Kongelige Danske Videnskabernes Selskabs skrifter*.[137] The eudiometer had an ingenious design but a complicated operational procedure involving a set of instructions that appeared difficult to execute correctly (Appendix IV).[138]

Hauch's eudiometer was similar in design to those belonging to the first generation of Fontana's volumetric eudiometers, especially to the fifth and seventh. Like them, an equal proportion of both airs was mixed irrespective of their saturation point. One of the advantages of this instrument was the small amount of water required for conducting the test compared with the other nitrous air eudiometers. Hauch introduced a pair of innovative features in relation to the other eudiometers: the shutters akin to the sliders of Fontana-Ingenhousz's measure and the valve for balancing internal with atmospheric pressure. This automation mechanism made the work of the experimenter easier, reducing the need for intervention and therefore the occurrence of errors due to the operator interfacing with the device.

Summary

The progress of the nitrous air test and eudiometer was mainly reflected in their use in research experiments and in the search to achieve a standard test procedure. The nitrous air eudiometer proved to be more than an instrument for ascertaining the salubrity of the air, since it started to be used as an apparatus for monitoring experiments on plant physiology, animal respiration and the medicinal use of dephlogisticated air.

Lavoisier had already been using the nitrous air test as early as 1775 in his first experiments on animal respiration, which were resumed from 1783 onwards for testing the purity of vital air with the Fontana-Ingenhousz eudiometer. Although Lavoisier did not completely abandon the nitrous air test for determining the salubrity of the air in crowded, closed spaces, he was more concerned with characterizing airs by their ratio between respirable and non-respirable air. The nitrous air test survived the transition from phlogiston to the oxygen-centred chemistry. In a case similar to that of Lavoisier, Ingenhousz was more interested in his researches on plant physiology rather than in measuring the salubrity of air.

---

[137] New Collection of the Royal Danish Academy of Sciences and Arts.

[138] Hauch, 1793.

Plant physiology was the battleground for the Fontana-Ingenhousz nitrous air eudiometer. While Ingenhousz was redesigning the instrument and his eudiometrical procedure in the summer of 1779, the intensive experimental work he was carrying out was mostly focused on studying the influence of sunlight on plant release of dephlogisticated air (photosynthesis). In the winter of 1782, he employed his nitrous air test once again when demonstrating that dephlogisticated air was freed from plants and not produced, as Priestley maintained, from the air dissolved in water. The latter had also made extensive use of his nitrous air test procedure to justify his standpoint. The priority given to the discovery of the release of dephlogisticated air by the plant leaves under the influence of sunlight not only sparked a controversy between Priestley and Ingenhousz but also between the latter and Senebier. Furthermore, the dispute between Ingenhousz and Senebier was further fueled by Ingenhousz's discovery of the *influxus nocturnum* (plant respiration). Between May and November 1783, Ingenhousz repeated his experiments to confirm this phenomenon, while developing an abridged eudiometrical procedure for testing highly dephlogisticated air samples. Prior to that, in November 1780, Ingenhousz had constructed a device to dispense dephlogisticated air for curing people with respiratory complaints, a form of medical treatment he was sure that his nitrous air test was able to control.

The nitrous air test was unable to detect the air impurities that tainted common air, making it unhealthy to breathe. Furthermore, the test raised serious doubts about its procedural standardization. Fontana, Ingenhousz and Magellan contributed to the redesign of the instrument by integrating accessory appliances to improve its performance: a magnifying glass to take more accurate volume readings; a brass ring to avoid parallax errors in these readings; the measure device for collecting replicable amounts of any air sample (later used by Volta), and the stabilizer to keep the eudiometrical tube in a vertical position. However, the real challenge of the nitrous air test for eudiometrists was the reliability of its experimental procedure. Ingenhousz was probably the one who made the most efforts to establish the trustworthiness of the nitrous air test and the eudiometer. He designed abridged procedures for different purposes, such as to reduce the execution time of the test when dealing with a large number of tests, in 1779. Because the shortened procedure was not suitable for highly dephlogisticated air samples, by 1781 he had developed a new abridged version for this kind of air, and in 1789 he presented the latest version of his eudiometrical procedure for improving the accuracy of the test.

The test reliability depended on many intrinsic and extrinsic factors, which were not always easy to define and control, but those that proved to have the most significant effect were: the establishment of the test endpoint; the dosing of the air samples; the time spent on particular operations; the water source used, and the quality of the nitrous air. Establishing a criterion for detecting the test endpoint was a persistent handicap, because this final point was associated to the reaction ratio at the point of saturation, which despite Lavoisier's efforts was far from being known, and also to the dosing of the air samples. The original and most commonly employed combination ratio of nitrous and common air was 1:1, but the habitual use of the test gave rise to the consideration of other ratios. Ingenhousez, for instance, habitually used a ratio of 3:2 when dosing the nitrous air in the common air sample.

The question of the test endpoint concerned at which point in time the contraction in volume should be observed. One option (Landriani and White) was to proceed to read the remaining volume in the eudiometrical tube immediately after mixing both airs in a determinate ratio. Another option consisted in taking the reading after a period of time that may be unspecified (Priestley): 'after a proper or sufficient time', or uncertain (Géradin and Magellan): 'after the greatest or complete contraction in volume'. Ingenhousz's method was more precise; the final reading was taken one and half a minute after the mixing of both airs while in the meantime performing certain operations. There were also variants to these options; in 1779 , Priestley suggested waiting for two minutes after adding a new dose of nitrous air. In the rigorous study conducted by Cavendish in 1781 on the factors influencing the determination of the test endpoint, he concluded that the contraction in volume increased as the diffusion between both airs progressed. In this study Cavendish also paid attention to the influence of the source of water used in the test, with the suggestion of always using the same source of water, preferably distilled water. During this same period of time, the physician van Breda, a friend of Ingenhousz's, reached the same conclusion.

From 1779 onwards Ingenhousz had been introducing procedural novelties into his nitrous air method with the aim of standardizing it. In this respect, he was very much involved in controlling the time spent on certain key procedural operations. Together with the above-mentioned time criterion for taking readings of the final residual volume, he also set the timing for other operations such as keeping the air measure under the water and shaking the eudiometrical tube after mixing both airs. The variable quality (i.e. purity) of the nitrous air as the reagent of the test had always been one of its inseparable weak points until Fontana and

Ingenhousz stated its irrelevance in 1779. This position pitted them against Priestley, who argued that a variable quality of the reagent implied uncertainty in the test results. By 1787, Ingenhousz had been able to validate empirically that the purity of the reagent was of little significance as long as it could saturate the common air.

Standardizing the nitrous air test to make it more reliable involved working out an operating protocol that demanded skilful experimenters with the appropriate training. Fontana had already manifested at an early stage the difficulties in the standardization of the test procedure, and Ingenhousz was well aware of the need for eudiometrists to be capable of following methodically the same test procedure to ensure its reliability. Ingenhousz eventually managed to devise a relatively simple instrument and to work out a procedural protocol that enabled eudiometrists to test common as well as dephlogisticated air samples. The statistical analysis carried out on Ingenhousz's test procedure using his experimental results confirms that the test was reasonably standardized. The problem was not that skilful experimenters such as Ingenhousz and Cavendish were able to execute that protocol successfully, but rather that it could be achieved by novices in eudiometry. Although it was significant that Fontana-Ingenhousz's eudiometer was highly praised by an expert and skilful experimenter such as Cavendish, it was nonetheless true that Priestley criticized it authoritatively on the basis of the irrelevance that Fontana and Ingenhousz had attached to the quality of nitrous air as a key reagent of the test. Ingenhousz responded to criticisms by claiming that the objections should have been based, first and foremost, on evidence-based experimental data. He was aware that the underlying problem of the test procedure was its reliability, which was closely related to the replication of the test results. For him this was a matter of training apprentices, but the lack of effectiveness in the transmission of the tacit knowledge involved in the test procedure became a serious setback for achieving that apprenticeship. From 1778 onwards, the nitrous air eudiometers had to coexist, not only with Volta's eudiometer but also with two different kinds of eudiometers, one based on the absorbent activity of a wet paste made with iron filings and sulphur, and the other grounded in the diminution of volume undergone by an air sample due to the combustion of a piece of phosphorus. These competing eudiometers and the aforementioned difficulties surrounding the nitrous air eudiometer overshadowed its utility. Nevertheless, the nitrous air test was far from being dismissed, and it would mutate into a new version at the beginning of the nineteenth century.

4.

Portable eudiometry. The cradle of the phosphorous eudiometers

In 1779, the year in which Volta interrupted his eudiometrical publications, a new family of eudiometrical instruments saw the light of day. On October 16th of that year, at the Royal Academy of Sciences in Berlin, the German chemist Franz-Carl Achard presented a new type of nitrous air eudiometer and the first member of the new generation of phosphorous eudiometers.[1] The paper describing both eudiometers was published in 1780 in the *Nouveaux mémoires de l'Académie Royale des Sciences et Belles-Lettres de Bruxelles*. The same paper was also published in the *Observations sur la physique* in 1784, this time with a more detailed graphical depiction of both instruments. A common feature of these Achard eudiometers was that both were portable. Although Volta had been the first to develop a portable inflammable air eudiometer between 1778 and 1779, during the period 1779-1780 Ingenhousz and Saussure also realized the need for portable eudiometers. Chemical portability was associated with the idea that certain chemical operations could be completed more promptly if performed in situ rather than in a chemical workshop at home. The use of portable chemical laboratories or facilities goes back to the seventeenth century when they were used in the field of mineralogy and metallurgy in order to conduct tests in mining regions distant from the local workshops. The notion of portability began to be appreciated by peripatetic naturalists and mineralogists towards the end of the eighteenth century, and the late eighteenth and early nineteenth centuries saw the proliferation of the chemical chests, or *nécessaires chimiques,* containing reagents, glassware, blowers and other chemical equipment.[2]

---

[1] Müller, 2002, p. 40.

[2] Gee, 1989, pp. 40-41; Homburg, 1999, pp. 2-5.

The early portable nitrous air eudiometers of Ingenhousz, Saussure and Achard

Ingenhousz

After visiting John Pringle in Bath in October 1779, Ingenhousz returned to London to prepare for his journey to Belgium and Paris. He took advantage of these stopovers in his journey for two reasons. First, to fulfil Pringle's request that he examine the quality (salubrity) of the air around coasts to assess the therapeutic value of sea air for patients suffering from consumptive diseases. Ingenhousz voiced Pringle's words again in his Royal Society address of November 30th, 1773, when Priestley was awarded the Copley Medal. The second reason for his journey was to set up a portable version of his nitrous air eudiometer acquired from the respected instrument maker Benjamin Martin of Fleet Street. During November and December 1779, Ingenhousz examined the air from coastal locations such as Gravesend, in the Thames estuary, offshore and other places such as Ostend, Bruges, Ghent, Brussels, Antwerp, Breda, Moordyke, Rotterdam, Delpht and The Hague. The research was inconclusive because of inconsistencies found in the experimental results. Ingenhousz therefore limited his comments to expressions of confidence about sending chronically ill patients to coastal places (without marshes or wetlands) because the sea air was apparently healthier for animal life. The inconsistencies pointed out by Ingenhousz were largely due to the experimental conditions, such as taking measurements on board a ship in motion, and also to operational circumstances specific to the new portable and travelling apparatus. Portable eudiometers would also help Ingenhousz in his project to establish a network of control stations in London, and perhaps all over the country, for testing the salubrity of the air in order to determine the influence of trees on the quality of the air in large open spaces.[3]

This portable eudiometer was packed in a case approximately 10 inches long, 5 inches wide and 3.5 inches high. The glass tube was 16 inches long, slightly shorter than the eudiometrical tube described in Ingenhousz's *Experiments upon Vegetables* (18-20 inches), and was divided into two sections that could be assembled by means of brass screws adapted to the divided extremities. Ingenhousz provides no further information, and apart from the partition of the tube the instrument was therefore very similar to the one described in his book. Presumably, the case also

---

[3] Ingenhousz, 1780a, pp. 362, 375.

contained other items such as a measure, vials, funnels and chemicals for producing nitrous air. A wooden bucket, with a brass funnel fixed in its rim, replaced the pneumatic through. It appeared from the tests carried out during the long journey (between November 1779 and January 1780) that the air at sea and the immediate environs was in general fitter for animal life than the air on the land, although it appeared to be subject to the same inconstancy in its degree of purity as the air on land.[4] Ingenhousz finally arrived at his destination on December 29th, 1779.

Saussure

At the end of this same year, 1779, the first volume of *Voyages dans les Alpes* by the naturalist Horace-Bénedict de Saussure (1740-1799) was published. This work included another version of a portable nitrous air eudiometer. Saussure was influenced by two other naturalists; his uncle, Albrecht von Haller, and Charles Bonnet, from whom Saussure acquired his passion for botany. It was thanks to Haller's strong support that in 1762 Saussure was appointed to the chair of philosophy at the Academy of Geneva. From 1764 onwards, he began to concentrate on geology, but without abandoning his work on botany completely. In 1768 he moved to Paris, where he met Buffon and held discussions with a number of French geologists. While in Paris, he attended the courses given by Petit, Rouelle and Jussieu, and entered into contact with members of enlightened society such as Guyton de Morveau. In 1776, he undertook extensive alpine explorations that he later described in the four volumes of the aforementioned work *Voyages dans les Alpes*, published between 1779 and 1796. By 1779 he had already traversed the Alps fourteen times along eight different routes. Saussure regarded the great mountain ranges, with their summits rising to the highest zones of the atmosphere, as nature's laboratory in which all physical phenomena made themselves felt on an impressive scale.[5] The collection of data on these trips required scientific instruments that were easy to manage and transport. Some of these instruments were already portable, such as thermometers, barometers or compasses, but others such as the electrometer and the magnetometer were ordered from expert instrument makers in England.

Saussure explained the functioning of his portable nitrous air eudiometer at the end of the tenth chapter in the first volume of the *Voyages dans les Alpes*, which was devoted to the observations made at the

---

[4] Ibid., pp. 356-357, 375.

[5] Saussure, 1779, p. viii.

summit of Mont Buet. The preface to the first volume is dated December 28th, 1779, which shows that Saussure was using his portable eudiometer before Ingenhousz employed his own portable instrument. It was Saussure's intention to use the eudiometer to determine the purity of the air in the mountains. His idea of purity implied the absence of putrid or phlogisticated materials, at some remove from the idea of assimilating the purity of the air with its salubrity. The nitrous air eudiometers known to that time seemed too fragile and bulky to be carried across mountain peaks, as well as being inappropriate for on-the-spot analysis because preparing the air mixture was too time-consuming. It was for this reason that Saussure invented a portable eudiometrical device that fitted into a small light box that was easy to transport. According to him, up to eight tests could be made in an hour with higher accuracy than that obtained with standard nitrous air eudiometers.[6]

> The experimental device consisted of a cylindrical glass tube with a diameter equal to its height and a capacity of about 5½ ounces of water.[7] The tube was closed at one end and open at the other with a ground glass stopper. To execute the test, the tube was first filled with water and weighed. It was then stood upright with the open end submerged in a bucket with water. Two measures of common and one of nitrous air were passed through the water with the aid of a funnel. The measure consisted of a small vial with a capacity equivalent to the third part of the tube so that the latter could contain up to three measures. The air mixture turned orange, while water rose in the tube and a diminution in volume of the air mixture occurred. Then the tube was closed and shaken when completely submerged in water. Afterwards, the tube was opened while maintaining it in an upright position with the open end submerged in water. A little more water was observed to enter the tube. After repeating this operation three times, the contraction in volume was completed. The tube was then closed, removed from the water and completely dried. It was weighed again, the difference in weight being indicative of the purity of the air sample.

Saussure analyzed two underlying causes for any inaccuracy that may have arisen from the change of location for the execution of the test. One

---

[6] Ibid., pp. 513-514.

[7] This was equivalent to 170 cm[3] of water, which means that the diameter and height of the container would have been approximately 7.5 cm.

cause was that the modification that nitrous air may have undergone during transportation, and the other concerned any alterations in the reactivity of the air samples due to possible differences in their densities in the valleys and on the mountain peaks. Saussure performed duplicate tests in order to forestall these sources of inaccuracy; that is, at the summit of a mountain he was analysing by using the same batch of nitrous air; on air samples collected previously in the valley, and on fresh air samples from the summit. Then, before descending, he collected air samples at the summit and once in the valley and conducted an analysis of these air samples as well as new air samples collected in the valley. After analysing air samples coming from five peaks and valleys, it turned out that the purity of the air diminished with height. Nevertheless, Saussure did not generalize these results, and for two reasons; firstly, because they were based on few samples, and secondly, because Volta's accepted hypothesis about the continuous production of inflammable air in nature, which because it was less dense than atmospheric air would found be in a higher proportion at the summits than in the valleys.[8] This inflammable air mixed with common air would make the latter less respirable and less sensitive to the nitrous air test.[9] This was a conjecture similar to Ingenhousz's conclusion about the interference of inflammable air in the nitrous air test.

The eudiometrical test proposed by Saussure was the second attempt at a gravimetric determination of the purity of the air after the four versions of Fontana's first generation of eudiometers. The idea, however, was the same; that is, to ascertain the salubrity of the air by equating it with the weight of a liquid (mercury in Fontana's instruments and water in the case of Saussure's portable eudiometer) that replaced the lost volume of the air mixture. In Saussure's test, the parts of common air were twice that of nitrous air. Obviously, this single dosage of both airs was not for the purpose of determining the test endpoint on the basis of a reciprocal saturation. Rather, this endpoint was closer to the saturation point of water by the residual air mixture. Saussure would subsequently replace the nitrous air eudiometer with the alkaline sulphide eudiometrical test and Giobert's phosphorus eudiometer. He considered that the nitrous air test

---

[8] A measure of atmospheric air from the summit, being less dense, contained less oxygen than a measure of air from the valley, which was denser.

[9] Ibid., pp. 516-518.

was inaccurate for his research into the influence of water, air and soil on vegetation, and the changes that the latter caused in the atmosphere.[10]

When Henry Cavendish submitted Fontana's eudiometer to scrutiny in 1783, he also paid attention to Saussure's gravimetric eudiometrical proposal.[11] Cavendish believed that Saussure's eudiometrical procedure was prone to some inaccuracies; for example, that the air contained in the tube could be compressed when it was enclosed, which would make the tube heavier. However, if the tube was closed slowly this error became smaller. Cavendish also found Saussure's procedure less expeditious than his own, since it involved weighing the device when submerged in water, which incurred a loss of time when drying moisture from the tube before weighing it. However, he accepted that the less spectacular nature of Saussure's eudiometer rendered it more portable, as Saussure had intended. However, Cavendish advocated the adaptation of Saussure's procedure in accordance with his own warning that a standard balance was inappropriate for weighing recipients submerged in water.[12]

Achard

In his early twenties, Franz-Carl Achard (1753–1821) received official approval as Marggraf's collaborator, and in June 1776, he took up his seat in the Berlin Academy. In 1782, he replaced Marggraf as Director of the Class of Physics at the Academy. Apart from a number of achievements in applied chemistry, such as his process for fabricating lye from common salt and litharge, he was best known for his method of extracting sugar from beets. By 1779, Achard had tackled questions concerning the exchange of airs between persons, one instance of which was asphyxia. In this same field, he undertook interesting and original researches, which were not exempt from risk, on the introduction of air into tissues under the skin of animals to cause a smooth bulging of the skin (subcutaneous emphysema).[13] Achard was interested in studying the therapeutic or lethal

---

[10] Ibid., 1803, pp. iv-v.

[11] See chapter 3; Henry Cavendish. The nitrous air test and Fontana-Ingenhousz's eudiometer under scrutiny.

[12] Cavendish, 1783, p.110, note.

[13] The faking of subcutaneous emphysemas was not an unusual practice in daily life at that time. Thus, beggars displayed false skin protuberances in order to inspire pity. Butchers believed that with this trick they could impart a better appearance to meat. Country farmers used the same method to simulate the fattening of cattle

effect of dephlogisticated and fixed air when insufflated, as well as the degree of phlogistication reached by common air in close contact with the tissues of different animals, a procedure that required the use of a eudiometer.[14]

In this context, Achard believed that to best of his knowledge the nitrous air eudiometers with which he was familiar (Landriani's and Magellan's eudiometers) were not portable enough to be used anywhere. His idea was to construct a portable instrument that could be used everywhere to enable him to determine quickly and easily the salubrity of any location. The entire eudiometrical equipment, with all its parts, could be enclosed in a case 5 inches long by 4 inches wide and 3 inches high and could be carried around without discomfort (Appendix V).[15] This portable nitrous air eudiometer by Achard was designed to determine easily the salubrity of an air sample, but its operating procedure did not appear to be simpler than Ingenhousz's and Saussure's portable instruments. Actually, Achard never reported any experimental results with this eudiometer.

The portable phosphorous eudiometers of Achard, Reboul and Giobert

Achard

Like other eudiometrists, Achard also complained that a common fault of the nitrous air eudiometers was the difficulty in obtaining a reagent (nitrous air) of uniform quality, which made the test unreliable unless the same batch of nitrous air was used for a successive series of tests. In order to overcome this drawback, Achard embarked on the design of a portable, non-nitrous air-based eudiometer that was grounded instead on the combustion of phosphorus.[16] The idea was to determine the diminution

---

and they also adopted this process with cows because they believed that it would increase milk production.

[14] Müller, 2002, pp. 45-47.

[15] Achard, 1780, pp. 93-95; 1784, pp. 35-36.

[16] Achard used the expression *phosphore d'urine* (urinous phosphorus) to refer to the white allotropic form of phosphorus. The physician and chemist Hennig Brand of Hamburg, who sold the secret of his discovery to a certain Johann Daniel Krafft, first discovered this waxy white substance around 1668-1669. Krafft travelled to England to demonstrate the new substance at the Court of Charles II. It was thus that Robert Boyle first learned of it in 1677. Like Boyle, the new substance fascinated the German chemist Johann Kunckel. Both Kunckel and Boyle, independently, discovered that phosphorus could be obtained by distilling evaporated urine. During the eighteenth century phosphorus – called Kunckel's phosphorus or English phosphorus by the

in volume that a common air sample underwent when saturated by the
phlogiston released from a piece of phosphorus during its inflammation.[17]
Since Achard's objective was the construction of a eudiometer with an
acceptable degree of portability, he designed a case 4 inches high, 6 inches
wide and 6 inches long that housed not only the eudiometer but also three
little bottles containing water, spirit of wine and some pieces of
phosphorus.

The first version of this device had an auxiliary cylindrical tube
(a'b'c'd') (Figure 4.1, *Fig. 3*) of approximately 2 lines in diameter and 6
inches in length. This tube was open at its upper end (a'b'), with a
brass tap (K) in the lower end (c'd'), and was attached to a small
board (f'g'h'i') scaled in lines. The main part of the eudiometer was
shaped like two communicating vessels (*Fig. 2*). Before packing the
apparatus in its case, it was necessary to fill it with water up to a
certain level and to close both branches with glass stoppers. By
blowing air across the mouth (op) of the ball-shaped branch (mopn),
of about 1 inch in diameter, the portion of air in the upper part of the
vessel (bdef) of the other branch was pushed out, while the water
rose to replace the air. This vessel and the ball-shaped branch were
closed once the vessel was full of water. To take a sample of the local
air, the stopper of the vessel (bdef) was removed and the local air
entered into the vessel to fill the space (bstd). The glass vessel (bdef),
with a large diameter of 3 inches and a small one of 2 inches, had an
upper neck (abcd) that could be completely sealed with a glass
stopper transfixed by a silver wire (qr) with a welded bowl in the
upper extremity (q) outside the vessel. The whole device was
attached to a board (WYXZ), together with a thermometer (n'o') and
a small pendulum (HI) to keep the system in a vertical position.

To conduct the test, a piece of phosphorus was placed on a small
dish at the lower extremity of the wire (r) before the vessel was
closed, so that the piece of phosphorus was surrounded by the air
sample. The ignition of the piece of phosphorus occurred when heat

---

French chemists – was prepared by Kunckel's and Boyle's method, and later improved
by Hellot and Marggraf.

The oxidation of phosphorous requires an excess of oxygen and produces
diphosphorous pentaoxide (P_4O_{10}). This whitish oxide is very soluble in water, initially
forming metaphosphoric acids [$(HPO_3)_n$] which hydrolyze to form orthophosphoric
acid (H_3PO_4) and also, probably, pyrophosphoric acid ($H_4P_2O_7$).

[17] Ibid., pp. 96-100; Ibid.,1784, pp. 37-40.

was conducted by the silver wire from the burning of an amount of spirit of wine inside the upper bowl. According to Achard, phlogiston was released from phosphorous during combustion, thereby saturating the air sample, while at the same time certain fixed air flowed from the air and was absorbed by the water, which is what caused the reduction in volume.

After leaving the apparatus to cool, the tap (K) of the cylindrical tube (a'b'c'd') filled with water was inserted into the mouth of the left branch (opmn). The tap was then opened to let the water inside flow down until it was equal to the water level in the tube (mnKL) and in the vessel (bdef). The reading of the amount of water that had left the tube (a'b'c'd') indicated the reduction in volume due to the combustion of phosphorous, as well as the phlogiston content and the salubrity of the original air sample.

Figure 4.1 Achard's portable phosphorous eudiometer.

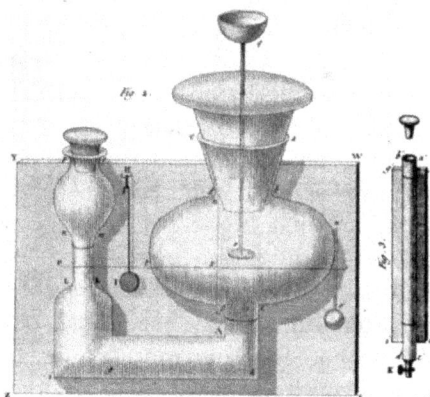

From *Observations sur la physique, sur l'histoire naturelle et sur les arts* (1784), plate 2, p. 86[18].

There is one feature in the design of this and other subsequent eudiometers that deserves to be analyzed due to its association with both theoretical frameworks and operational as well as pragmatic considerations. This feature concerns the use of water or mercury as the medium with which the test could be performed. Firstly, it is worth mentioning that the fact that mercury was heavier and more expensive than water, while water was much more affordable and portable; these were factors that favoured the use of water instead of mercury. Furthermore, Achard decided to use

---

[18] See Appendix V, note 8 .

water instead of mercury for his first phosphorous eudiometer on theoretical grounds, since he believed that water facilitated the diminution in volume of the air sample because it was able to absorb the fixed air dropped from the atmospheric air, while phlogiston was liberated by phosphorus during its burning. It appears that this eudiometer was not much used in practice because of its problematic construction, strange design and cumbersome handling and, in addition, the complete combustion of phosphorus was also difficult to achieve.[19] However, Magellan stated that had received an order to construct two of these Achard nitrous air and phosphorus eudiometers.[20]

Later, in 1784, Achard said that his attempt to find the causes of the errors he had observed in different types of eudiometers had led him to develop a device for determining the degree of phlogistication of a common air sample with the greatest accuracy, and for comparing the relative levels of salubrity of distinct airs. As he remarked about his first eudiometer, the main common fault of the nitrous air eudiometers was the difficulty in obtaining nitrous air of uniform quality.[21] Volta's eudiometer suffered from the same and possibly worse lack of reliability as the air nitrous test, due to the variable quality of the inflammable air needed for this eudiometrical test. At that time, Volta's eudiometer was already open to scrutiny and therefore the object of criticism. Achard pointed out that the drawback with Volta's eudiometer was that it was only useful for determining certain degrees of salubrity. The reason for this was that if the air sample was highly phlogisticated, then it would not detonate when mixed with any proportion of inflammable air because it could not support more phlogiston. Additionally, Achard was of the opinion that Volta's eudiometer was not sufficiently portable for use in all possible locations. Scheele's test, which was based on the absorbent capacity of the wet mixture of sulphur with iron filings, was rejected outright by Achard due to its many inconveniences.[22]

To overcome these obstacles, Achard devoted his efforts to a simpler version of his first phosphorous eudiometer with the novelty that

---

[19] Reboul, 1788, p. 379. Achard reported having used the nitrous air test to check the effect on the salubrity of the air of diverse kind of fumes produced when burning different materials such as the leaves, flowers and roots of a variety of plants (Achard, 1783).

[20] Magellan, 1780.

[21] Achard explicitly mentions Priestley's, Magellan's and Fontana's eudiometers.

[22] Achard, 1786a, pp. 30-33.

combustion was performed with mercury instead of water.[23] This modification was introduced as a result of Achard's finding that the inflammation of phosphorus caused irregular contractions in the volume of the air sample that made the test results irreproducible. Achard attributed this anomaly to the phlogiston, which in certain circumstances could unite with water to form an aerial fluid that might be the cause the observed irregularities. Furthermore, Achard regarded mercury as the only fluid that could come into contact with air both during and after the phosphorus inflammation without causing additional changes. Phosphorus was considered a suitable eudiometrical material because it was extremely inflammable and contained no volatile principle other than phlogiston; it completely saturated the air sample with this principle without producing other unwanted side-effects.[24]

There is no available graphic design supporting the description of Achard's second phosphorous eudiometer. It consisted of two glass tubes of the same length (18 inches) but with different diameters and both open at one end only. The narrowest tube (4 inches in diameter) had a scale attached that was divided into 12 parts. The volume of each part was equal to the volume of a small measure made with a glass tube of the same diameter. The narrowest tube was filled with mercury, closed by placing a finger over the open end, and inverted with the open end submerged in a vessel also containing mercury. Then a number of measures of the air to be tested were run in under the bottom of the tube. Immediately afterwards, a piece of phosphorous was also placed underneath the bottom of the tube rising up to the surface of the mercury. At that point, the length of the air column was recorded.

The first step of the test consisted of igniting the piece of phosphorus by bringing the flame of a candle close to the glass tube at the level of the surface of the mercury. The burning of phosphorous yielded some white vapours that immediately condensed. After leaving the tube to cool, it was immersed (preventing the entrance of air from without) inside the wider tube that was partially filled with mercury. The inner tube was then moved vertically until the mercury level in both tubes was equal. Finally, the reading of the length of the

[23] Scheele had also used phosphorous as a eudiometrical material in 1779. This phosphorous eudiometer may have been made by the artisan Gio Balta Schiavetto (Müller, 2002, p. 114).

[24] Achard, 1786a, pp. 33, 36-38.

remaining air column was compared with its initial value. The phlogiston content and the salubrity of the air sample were given by the ratio of both values.[25]

It seems unlikely that Achard had complete confidence in his phosphorus eudiometrical devices. In January 1786, he read another paper presenting the results of a substantial research project to determine the salubrity of the air at certain times of day, in different locations and seasons of the year. No phosphorous-based device was used in this research. Achard preferred to use the widely approved Fontana-Ingenhousz eudiometer and his own simplified version of Volta's eudiometer.[26] Despite the lack of agreement between the results obtained by these two eudiometers, Achard tended to give preference only to the results from the nitrous air eudiometer. He believed that the diminution in volume observed in the Volta's eudiometer came not only from the aerial phlogiston content but also from changes produced by other aerial components that on the contrary did not interfere with the nitrous air test. Nevertheless, the final results were unexpected as well as being inconclusive. It was assumed that air phlogistication was due to animal respiration, putrefaction and combustion. The more densely populated places were therefore expected to be the less salubrious, while the places with vegetation were the most salubrious. Likewise, air was expected to be healthier in winter than in summer in inhabited places - cold weather would prevent rot – while the opposite would be true in uninhabited places. Nevertheless, the results provided by the nitrous air test showed the contrary.[27]

---

[25] Ibid., pp. 38-39. In contrast to Achard's first phosphorous eudiometer, the design of this second phosphorous eudiometer bears a closer resemblance to Hales' device, presented in his *Vegetable Staticks*, for determining the quantity of air interchanged by a burning material or a living animal while breathing. Phosphorous was one of these burning materials, as shown in the experiment LIV (Hales, 1727, pp. 97,162; fig. 35; plate 16).

[26] Achard, 1786b, pp. 44-48. Achard met Volta and his eudiometer in 1784, during Volta's visit to Berlin. Achard is thought to have made a sketch of the instrument that later was lost. For this reason he wrote a letter to Volta on July 3rd, 1785, asking him for a detailed drawing of the eudiometer in order to make it himself, and thus be able to conduct the aforementioned experiments (VO, Vol. 2, pp. 306-307, 315). Later, on January 12th, 1786, Achard requested the Class of Physics of the Berlin Academy of Sciences to purchase a Volta eudiometer (Müller, 2002, p. 115)

[27] Achard, 1786b, pp. 52-53, 56-57.

Reboul

The naturalist and chemist Henri-Paul-Irénée Reboul (1763-1839) had already expressed his dissatisfaction with Achard's first phosphorous eudiometer when in July 1788, he read a paper at the Académie des Sciences of Toulouse describing a new type of phosphorus eudiometer that resembled Achard's second phosphorus eudiometer. Reboul was educated in the Oratorian Colleges of Pézenas, Lyon and Paris, and in 1782 began his studies in law in Toulouse. Motivated by his friend Casimir du Puymaurin, son of the syndic of the region of Languedoc and a passionate devotee of chemistry, he joined a savant society of Toulouse, and with Puymaurin's aid he dedicated himself wholeheartedly to the study of chemistry. In 1792 he was appointed member of the Toulouse Académie des Sciences. Reboul's eudiometrical researches were associated with his quest to acquire detailed knowledge about the Pyrenees. In the manner of Horace Bénédict de Saussure's explorations in the Alps, Reboul and his colleagues planned climbs in the mountains of the Pyrenees to collect data such as the height of the mountains, mineralogical specimens, geological observations and, of course, the salubrity of the air.[28]

> Reboul's eudiometer (Figure 4.2) consisted of a glass tube graduated in 100 parts, each 6 inches long and 2.5-3 lines of interior diameter with one ball-shaped extremity (B) and the other open (*Fig. 1*). The capacity of this ball was between twice and three times of that of the tube. A cylindrical iron component (AC) was sealed with mastic in the open end of the tube.
>
> The piece of phosphorus was placed inside the glass ball, and the mouth (A) of the tube was closed airtight with the square-head nut (D). A spanner (E) was of help to screw the nut. The piece of phosphorus was ignited by bringing the candle flame close to the glass ball. The air expanded by the effect of the heat and was eventually absorbed by the burning phosphorus. The absorption of air was completed by bringing the candle close to the glass ball three or four times.
>
> After that, the tube was inverted with its mouth submerged inside the square aperture, nearly full of mercury, drilled in a cylindrical wooden case (*Fig. 2*). Then, as the nut (D) was unscrewed, the tube being held by the indentations (i,i) of the spanner, mercury

---

[28] Grao, 2013, pp. 19-23. Reboul used his eudiometer in his ascents to the French Pyrenean summits of the Pic d'Anie (1786) and the Pic du Midi de Bigorre (1786 and 1787)

rose inside the tube to replace the air consumed. The tube was moved vertically until the mercury level both inside and outside the tube was equal. The length of the column of mercury inside the tube indicated the amount of vital air absorbed.[29]

Figure 4.2 Reboul's portable phosphorous eudiometer.

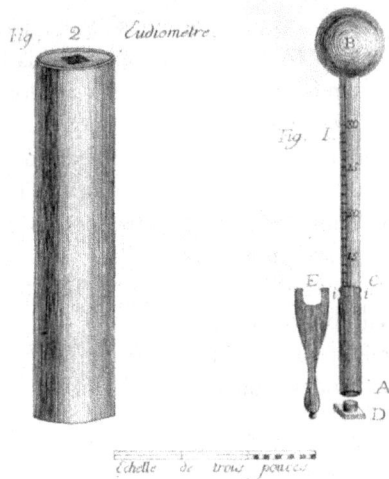

From *Mémoires de l'académie royale des sciences, inscriptions et belles-lettres de Toulouse* (1788), plate 15, p. 443.

It is worth remarking that this eudiometer by Reboul provided meter readings in percentages of the proportion of vital air in the atmosphere. However, Reboul did not explicitly state the need for setting up a percentage scale as a means of standardizing the measurements, which provided a universal representation of air composition.[30] Actually, Reboul acknowledged that his instrument fell short of meeting all requirements since improvements were needed to make it more portable and easier to operate.[31]

It is also worth noting that Reboul's eudiometer resembled Achard's second device in several respects, such as the use of mercury instead of

---

[29] Reboul,1788, pp. 381-383.

[30] Capuano & Calvachi, 2010, p. 324.

[31] In any case, Humboldt later used this Reboul eudiometer for a comparative study of the phosphorus and nitrous air eudiometers (Humboldt, 1798e, pp. 150-152; 1799c, pp. 63-80). He also used this eudiometer during his expedition to explore the lands of the Pacific coast of South America between 1799 and 1804.

water and the technique for igniting the phosphorus. Of particular note in the design of Reboul's eudiometer is the immersion of the glass tube in the aperture of the wooden case partly full of mercury, since this is a feature similar to Achard's two tubes, one inside the other. All things considered, Reboul managed to reassemble the parts of Achard's device into a compact portable instrument.

The description of Reboul's eudiometer constitutes a significant example of the departure from the phlogiston system in chemistry at the end of the 1780s. Reboul defined his eudiometer as an instrument for measuring the quantity of vital air in a sample of common air. There is no reference to phlogiston in this definition or in the explanation of how the instrument functions. Reboul based this function on the separation of the vital air when being absorbed by a piece of burning phosphorus. Although combustion was yet to be interpreted in terms of the coming theory of oxygen, it is clear that atmospheric air was destined to play a leading role in this process.[32] In actual fact, the main reason for the rejection of Achard's first phosphorous eudiometer was theory-laden. According to Reboul, fixed air was no longer precipitated from the common air during the combustion of phosphorous.[33] Reboul's eudiometer would stand as proof of an apparently unproblematic accommodation of an instrument to a new and imminent theoretical framework.

Reboul's description of his eudiometer was republished four years later in the *Annales de chimie*. At that time, he had embraced the new chemical nomenclature and wanted to record it in a footnote in which the term "azotic gas" replaced the old term of "phlogisticated air". However, he still preferred to use the term "vital air", even though the term "oxygen" was already in use.[34] No new feature or development figured in the 1792 description of the phosphorous eudiometer, but since Reboul had had the opportunity to use it, he proposed a modification to simplify the instrument further, but at the expense of making it less portable. The change consisted in the ball blown at the end of a very short tube of a larger diameter, and consequently removing its graduation. This was no minor modification, since it entailed determining the amount of

---

[32] Reboul corresponded with Lavoisier and was a guest at his home in Paris for some months during 1785.

[33] Reboul, 1788, p. 379, note.

[34] Ibid., 1792, p. 39.

decomposed vital air by weighing the amount of mercury introduced into the eudiometer after combustion and cooling.[35]

Giobert

The same year, 1784, in which Achard presented the latest version of his phosphorous eudiometer, the Italian chemist Giovanni Antonio Giobert (1761-1834) had already begun to use his own phosphorus eudiometer that he presented before the Società Físico-Medica of Turin in 1785. This Medical-Physical Society was an endeavour to set up a private academic organization of physicians and surgeons in Piedmont at the end of the eighteenth century. The meetings of the society were held in the home of its secretary, Gian Batista Anforni, and one Michele Vincenzo Malacarne was the chairman. The Society ceased its activities in 1790 when Malacarne moved to Pavia. The Society's areas of interest were miscellaneous: general and applied chemistry, physics, physiology, zoology and mineralogy, and it also issued its own scientific publications.[36] Giobert played an important role in the acceptance of the new chemistry in Italy, and in 1792 the Accademia Reale di Scienze, Belle Lettere ed Arti of Mantua (Royal Academy of Sciences, Humanities and Arts) awarded him a prize for his paper on the controversy surrounding the simple or compound nature of water according to the theories of phlogiston and oxygen.[37] Giobert was an enthusiastic advocate of the principles of the French Revolution, and his endorsement of Lavoisier's chemistry (the first one in Italy to be motivated by experimental considerations) should be understood in the context of his political commitment to the ideals of the Enlightenment.[38] From 1795 onwards, Giobert gave up chemical research in order to devote himself to the applications of chemistry in agriculture.

The description of Giobert's phosphorus eudiometer was not published until 1793, eight years after its academic presentation, by which time he was already engaged in the chemistry of oxygen with an analysis of the sulphurous waters of the thermal baths in the town of Vaudier.[39] The aim of this study was to conduct analyses of the air from this site in order to

---

[35] Ibid., p. 46, note.

[36] Maffiodo, 1996, p. 55.

[37] Abbri, 1996, pp. 4-5.

[38] Beretta, 1989, pp. 141-142.

[39] "Vaudier" is the Occitan name for the present "Valdieri". It is a municipality of the Italian region of Piedmont, situated in the valley of the river Gesso ("Gesse" in Occitan) in the southwest of Turin.

compare the results with those from different locations in and around the city of Turin, where he lived. The need for such analyses was associated with the interest in the curative powers of the sulphurous baths in Vaudier. One detail to note in this case is that Giobert's eudiometrical researches were no longer be limited to determining the proportion of oxygen and nitrogen in the air, but also of other gases such as carbon dioxide and hydrogen. Before describing his eudiometer, Gioberti made a brief evaluation of other eudiometrical tests, paying particular attention to Volta's eudiometer and the difficulty of obtaining enough pure hydrogen. Similarly, he ruled out the alkaline sulphide test on the grounds that the absorption of other gases, different from vital air, provided a content of this air higher than the true value.[40]

Giobert's eudiometer was quite simple and had a novel design that made the combustion of phosphorus easier over water instead of mercury. The reasons for this replacement were the reduction in the cost of the instrument and the improvement of its portability. But since phosphorus is denser than water but less dense than mercury, the shape of the eudiometric tube could no longer be straight and vertical but bent at a right angle (Figure 4.3, left). The illustration corresponds to the description that Lazzaro Spallanzani gave of Giobert's eudiometer in his work *Chimico Esame* and which served him as a model for his own eudiometer.[41]

The tube (abc) was 18 inches long with a diameter of half an inch. The vertical part (ab) represented two-thirds of its full length.[42] The lower end (a) of the tube was open and the opposite end (c) closed. At 3-4 inches above the lower end there was a signal (r) made of a welded coloured glass. The space between this signal, which was marked as "0", and the top end was divided into 100 parts. The vertical part of the tube was attached to a wooden board with two strings so that the tube could turn on the board. In this way, the horizontal part of the tube was separated from the board, whereas the vertical part remained attached to it.[43]

---

[40] Giobert, 1793, pp. 59-63.

[41] Spallanzani, 1796a, *Fig. I*.

[42] To ensure a uniform inner diameter in the curvature of the tube, it was filled with well-dried, fine-grained sand before warming to bend it.

[43] Spallanzani reported that Giobert had recommended marking the graduated scale on the wooden board. However, the figure of Giobert's eudiometer in the *Chimico Esame* showed the scale marked on the tube (Spallanzani, 1796a, p. 6). In the modern replica of Giobert's eudiometer exhibited at the Centro Studi Lazzaro Spallanzani in Scandiano, the graduated scale is attached to a wooden board (Figure 4.3, right).

The test was executed by placing a piece of phosphorus at the bottom of the bent part of the tube. The tube was then filled with water, ensuring that the piece of phosphorus remained at the bottom. After that, the tube was partially emptied of water to allow the entrance of an air sample, and the open end (a) of the tube was submerged in a vessel containing water. The next step consisted of sucking air through a thin curved tube introduced into the eudiometer until the water level reached the lower mark (r). The tube was rotated about its vertical part, and the horizontal part (bc) was heated by bringing the flame of a burning paper close to it. First, air expanded and the water level dropped, then after cooling, water rose inside the tube replacing the consumed vital air. The reading of the level of water in the graduated scale gave the percentage of vital air in the air sample.

The residual air was mainly nitrogen, although it was often mixed with carbonic acid gas that had to be removed and quantified. To this end, after capping the lower end under the water, the tube was removed, and its open end was submerged again in another vessel containing limewater. The carbonic acid gas would be absorbed by the limewater and, accordingly, water would rise above the level reached by the plain water (after the preceding absorption of vital air). This increase in the level of water would indicate the amount of carbonic acid gas in the air sample.

Figure 4.3 Giobert's portable phosphorous eudiometer.

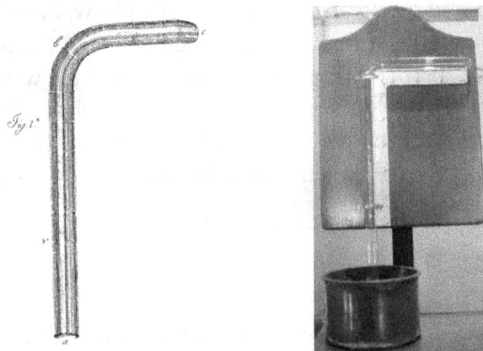

(Left) From Lazzaro Spallanzani, *Chimico Esame degli esperimenti del Sig. Göttling* (Modena, 1796). (Right) Modern replica of Giobert's eudiometer, at the Centro Studi Lazzaro Spallanzani, Scandiano. Courtesy of Centro Studi Lazzaro Spallanzani, photography by the author.

Although Giobert remarked on the simplicity of the test, he also indicated some of its shortcomings. Firstly, the vital air in the air sample could not be completely absorbed by phosphorus in a first combustion. It was necessary to heat the piece of phosphorus several times, and when after heating the tube no glimpse of light was observed in the darkness, the test was deemed to be finished. It was also necessary to use an excess of phosphorus. An amount of 6 - 8 g of phosphorous was sufficient for the first combustion, and 2-3 g were required for the complementary combustions. Furthermore, the closed end of the tube containing the piece of phosphorus had to be heated gently. Otherwise the glass might crack and break. In addition, if ignition occurred too quickly it could rarefy the air inside the eudiometer to excess and facilitate the output of air from the open end of the tube submerged in water. For that reason, the segment of the tube from the mark (r) to the open end (a) acted as a safety margin to prevent the test from being spoiled. The temperature had to be controlled during the test, either to obtain results at the same initial temperature or, if that was not possible, to correct them according to Lavoisier's tables.[44] Giobert found a higher proportion of vital air in Vaudier (24% - 33%) than in Turin (24% - 28%), and a nearly constant presence of carbon dioxide gas in Turin, while this gas was virtually absent in Vaudier. According to Giobert, these results confirmed the better respirability of the air in Vaudier and the healing virtues of its sulphurous water baths.[45]

The original idea of using a bent tube in the eudiometer might not be attributable to Giobert himself. Traces of this novelty can be found in a paper written by the French chemist Guyton de Morveau in 1795, where he described his own eudiometer in which he used solid potassium sulphide as the oxygen absorbent. Guyton reported that his colleague François Chaussier had used a somewhat different instrument for his eudiometrical experiments with phosphorus. To understand this overlap, it is essential to examine some experimental aspects of Guyton's contribution to the composition of atmospheric air.

A plausible source of inspiration for Giobert's eudiometer

Louis-Bernard Guyton of Morveau (1737-1816) was a renowned lawyer and a member of the Dijon Académie des Sciences, Arts et Belles-Lettres. He was a self-educated chemist who did not become a full-time science researcher until he was 30 years old. Guyton maintained good relations

---

[44] Giobert referred to Lavosier's method for calculating the corrections relative to the degrees of pressure and temperature (Lavoisier, 1789, Vol. 2, pp. 378-383)

[45] Giobert, 1793, pp. 64-75.

with Lavoisier, who highly appreciated his work. Although Guyton had advocated the theory of phlogiston for interpreting some anomalies in the calcination of metals,[46] Lavoisier acknowledged his experiments in that regard as the most comprehensive and accurate conducted to that time. The personal encounters between both chemists began in the spring of 1775 when Guyton visited Paris for the first time and Lavoisier welcomed him to his laboratory. In 1779, Guyton visited Lavoisier again, and during this second stay he was elected a member of the Société Royale de Médecine of Paris on a recommendation by Lavoisier. His third visit to Lavoisier, which lasted from December 1786 until June 1787, turned out to be the most profitable. Both chemists worked together, and by the time Guyton ended his stay in Paris he had adopted Lavoisier's new chemistry. This change of mind would be reflected in the article *Air* in the *Encyclopédie Méthodique*. Guyton was the author of the first volume, which was published in two parts. The first part, issued in May 1786, had been drawn up on the basis of the chemistry of phlogiston. The second part came out in November 1789, after Guyton's last encounter with Lavoisier, and had already been written on the chemistry of oxygen. It begins with a *Second advertissement* consisting of forty pages and is followed by the article *Air*.[47]

Guyton devoted the second section of the article – about three quarters in all - to describing as many as fifty experiments for the purpose of determining the constituent parts of atmospheric air. Phosphorus was the main element in seven of these experiments, the first of which (Exp. XII) consisted of the known combustion of phosphorous over water, where ignition was achieved by concentrating solar radiation through a magnifying glass. Since on occasion neither a suitable magnifying glass nor a sufficient intensity of solar radiation were available, Guyton proposed a new experimental setup (Exp. XIII) which in practice went beyond the mere replacement of the magnifying glass as a means for igniting phosphorus.[48]

---

[46] Guyton articulated these arguments in the essay *Dissertation sur le phlogistique*, published in his work *Digressions chimiques* of 1772. On Guyton's experiments and their reception, see Guerlac (1961, pp. 125-145)

[47] Bouchard, 1938, pp. 127-128, 133-134; Bret, 2017, pp. 10, 22-23.

[48] Guyton, 1789, pp.705-706.

The experimental device consisted of a small inverted retort (*récipient cornu*)[49] with a neck about 8-10 inches long and a bulb (A) 14-15 lines in diameter (Figure 4.4, left). A circular mark (C) was engraved just beneath the curvature, with the retort in a vertical position and the open end (d) pointing down. If such a small retort was unavailable, Guyton suggested using a curved glass tube with only one open end. The bent part of the tube, with the closed end, was a third of the full length and corresponded to approximately one-quarter of a circle with a radius of 3 inches. The inverted retort was filled with water in a pneumatic trough, and a number of measures of the air sample were introduced into it in order to fill the space up to the (C) mark. In this case, the measure was a glass tube whose size corresponded to a part of the capacity of the bulb up to the (C) mark.

The open end of the retort was closed by placing a finger over it and submerged just beneath the surface of the water in the vessel (B). This vessel was filled with water to about ¾ of its capacity. A piece of phosphorous was then introduced through the open end, which was closed again with a finger. After that, the retort was removed and tilted to move the piece of phosphorus to the other end. The retort was again put in an inverted vertical position and tilted slightly to drain any water from the neck without losing the piece of phosphorus. Once the inverted retort was back in the vessel, a lit candle was placed under the bulb. The piece of phosphorus was ignited, and dense white fumes filled the upper part of the retort. First, the water level in the neck of the retort dropped because of heat, but then rose again due to the reduction in the amount of air.

---

[49] This small retort (*récipient cornu*) was one piece of the apparatus in Guyton's portable chemical laboratory (*nécessaire chymique*), which he presented in 1783. Actually, Guyton had learnt about the use of this kind of retort in Bergman's laboratory (Smeaton, 1966, pp. 86-87)

Figure 4.4 Guyton's experimental device with the inverted retort.

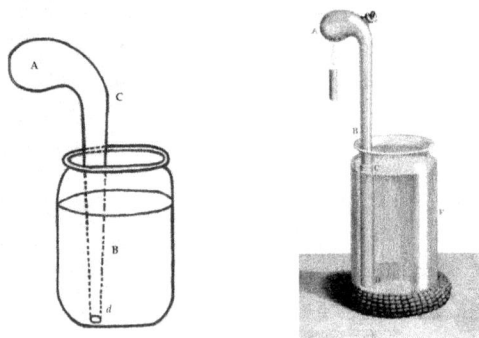

(Left) From *Recueil des planches du dictionnaire de chimie et métallurgie, premier partie, chimie* (Paris, 1813), plate 26. (Right) Early version of Guyton's potassium sulphide eudiometer. From *The Philosophical Magazine* (1799), plate 5.

Guyton did not redesign this experiment as a eudiometrical test. The remaining five experiments were variations on the first two - such as making the burning of phosphorous over limewater or mercury - and only the penultimate experiment (Exp. XVII) served Guyton to describe the transfer of the residual air into a eudiometrical tube, but only to verify quantitatively the diminution in volume. At the end of his description of all these experiments Guyton added some remarks, one of which referred to the eudiometrical test with phosphorus. He stated that with the burning of phosphorus in common air he had been unable to achieve a reduction in volume exactly proportional to the amount of oxygen, which had been obtained by other eudiometrical means such as the nitrous air.[50]

Six years later Guyton described his own eudiometer in which he used solid potassium sulphide as the eudiometrical means.[51] In a footnote he mentioned that the instrument depicted in the article *Air* (the inverted retort) provided all the advantages needed for reuse by changing the piece of phosphorus for one of potassium sulphide. Moreover, he pointed out that his colleague François Chaussier had already used a somewhat different

---

[50] Guyton, 1789, p. 709. See chapter 5.

[51] See chapter 5. Reprising the inverted retort apparatus with solid potassium sulphide.

instrument for his eudiometrical experiments with phosphorus.[52] According to Guyton, Chaussier's apparatus was equipped with an inlet tube in the bulb that was closed after causing the water to rise up inside the neck to two-thirds of its height. However, he also noted that although the inlet tube made the apparatus very effective for testing atmospheric air, this was not the same for other gases, which could only be introduced by transference.[53] The illustration of an early version of Guyton's eudiometer (Figure 4.4, right) published in *The Philosophical Magazine* in 1799, which shows an inlet tube (e), gives an example of what Chaussier's eudiometer might have been like.

A contemporary description of Guyton's laboratory by Arthur Young, an English writer on agriculture and economics, as a result of his visit to Dijon on August 1st, 1789, recounts that Guyton was at that time engaged in a series of eudiometrical experiments using the Fontana-Ingenhousz and Volta's eudiometers, which had probably been made by Nicolas Fortin.[54] Young reported that Guyton believed that the results of these experiments depended on specific procedural factors, such as keeping nitrous air in bottles sealed with cork stoppers and placed upside down. Young particularly stressed that the eudiometrical test with phosphorus using an inverted retort was a simple and elegant method for determining the proportion of vital air in common air. Thus, it seems clear that by 1789 Guyton had already been using that phosphorous eudiometer,[55] and it is therefore likely that the experiment of the burning of phosphorus in the bulb of the inverted retort, as well as the reintroduction of water as a means

---

[52] François Chaussier (1746-1828) practiced medicine as a physician and gradually gained a reputation as a teacher of anatomy, chemistry and *materia medica* at the Dijon Académie des Sciences, Arts et Belles-Lettres. As regards the teaching of chemistry, he was, together with Hugues Maret, the secretary of the Academy, one of the two assistant professors of the chemistry course created by the États de Bourgogne with Guyton as professor. Guyton was actively involved in the development of this course, organizing the laboratory of the Academy in order to give practical classes. The course began on April 28th, 1776, and was repeated each spring for two and half months with three lectures a week until 1789. On the death of Maret in 1786, Chaussier became the second assistant professor of the course and in 1790 replaced Guyton as professor. In 1796, he became professor of anatomy and physiology at the École de Santé in Paris. He also replaced Berthollet in his chemistry lectures at the École Polytechnique during the latter's expedition to Italy between 1796 and 1797. After this time, he devoted himself entirely to his duties as a physician.

[53] Guyton, 1795, p. 167.

[54] Bret, 2017, p. 13.

[55] Young, 1792, p. 153.

of confining that combustion, were in fact Chaussier's contributions. Since phosphorus is denser than water but less dense than mercury, the eudiometric device could not have been a vertical tube. Chaussier's inverted retort with its bulb almost at a right angle to its neck remedied that shortcoming.

Giobert's eudiometer was an early simpler version of Guyton's alternative to the inverted retort. It is worth recalling that Guyton had proposed the use of a curved glass tube instead of an inverted retort, and that the ratio between the lengths of the two branches of Giobert's eudiometer was the same as that pointed out by Guyton for this alternative curved glass tube. It therefore seems plausible that when Giobert presented his eudiometer to the Medical-Physical Society of Turin in 1785, the use of a tube bent at a right angle for eudiometrical purposes was already a known practice. As mentioned above, the design of this phosphorous eudiometer inspired its reuse with potassium sulphide as the eudiometrical means. In any event, this episode requires an exploration of the development of eudiometers based on the alkaline sulphides.

Summary

Portable eudiometers were conceived for determining the respirability of atmospheric air in locations far removed from the usual workshops or laboratories, as well as in different climatological and meteorological conditions. The characteristics required for these portable instruments included ease of handling and of conveyance, virtues that were especially useful for naturalists like Saussure and Reboul. This meant that it had to be possible for these types of eudiometers to be packed in small cases similar to the chemical chests that already existed.

Volta developed his first and portable eudiometer from his earlier electric pistol, at a time when he was much more concerned about the nature of inflammable air than the respirability of common air. Ingenhousz was engaged in the making of a nitrous air portable eudiometer for his project of assessing the therapeutic value of the air of coastal locations, as well as creating a network of monitoring stations for determining the influence of vegetation on the salubrity of air. The uncertain quality of both nitrous and inflammable air and the insensitivity of the latter to highly dephlogisticated air samples were the main reasons that between 1778 and 1784 led Achard to contemplate phosphorous as an alternative eudiometrical means. Additionally, he considered that nitrous and inflammable air eudiometers were not sufficiently portable to be used in all types of locations. It was thus that the phosphorous eudiometer emerged from this context of portable eudiometry in order to remedy the inadequacies of the nitrous and inflammable air eudiometers.

The rapid ignitibility of phosphorous, and the assumed absence of any other volatile ingredient except phlogiston in its composition, qualified phosphorous as an appropriate candidate for these new eudiometrical means. Achard's alternative use of water and mercury as a medium upon which to conduct the combustion of phosphorous exemplifies the influence of a theoretical framework on the design of an experimental device. It appears that Achard was not satisfied with his portable phosphorous eudiometer since he preferred to use the standard Fontana-Ingenhousz and Volta's eudiometers for his further researches.

In the following years, Reboul and Giobert continued working on the development of new versions of the portable phosphorous eudiometer. Reboul actually converted Achard's second phosphorous device into a compact apparatus and, as in the case of the nitrous and inflammable air eudiometers, the description of his eudiometer shows that the phosphorous eudiometer was permeable to the transition from the phlogiston to the oxygen-centred chemistry. Reboul and Saussure proposed modifications in their respective phosphorous eudiometers in order to convert them into gravimetric instruments.

Giobert, for his part, ideated a minimalist design of the phosphorous eudiometer that enabled the use of water again instead of mercury, thereby making the entire instrument more affordable and more portable. Giobert's idea of using a glass tube bent at a right angle as the core of his eudiometer turned out to be both simple and successful. Nevertheless, the credit for the originality of this development should rightly be shared between Guyton de Morveau and his colleague Chaussier in Dijon, given their contributions to the experiments on the combustion of phosphorous. The phosphorous eudiometer would undergo further significant developments in the coming years.

5.

The alkaline sulphides
based eudiometers

Scheele's experiments on air in the background of alternative eudiometrical tests

During the last decade of the eighteenth century, the nitrous air eudiometer had to defend its status against other new arrivals on the scene. In addition to Volta's inflammable air eudiometer, other eudiometrical means such as phosphorus, alkaline and calcium sulphides, and the wet paste of sulphur with iron filings were all jostling in competition. The two latter ones stemmed from the experimental work of Scheele. In 1777, the Swedish Pomeranian pharmacist and chemist Carl-Wilhelm Scheele (1742–1786)[1] published his first and only monograph *Chemische abhandlungen von der luft und dem feur*. Prior to this publication, Scheele had already established that common air was a mixture of two kinds of elastic fluids: 'empyreal or fire-air' (oxygen), which supported combustion and calcination, and 'foul or corrupted-air' (nitrogen), which did not. Additionally, air was a respirable fluid mixed with another elastic fluid, aerial acid or fixed air, and differed from it in many respects.

Shortly after obtaining his pharmacy journeyman's certificate in 1765, Scheele moved to a pharmacy in Malmö. At that time, he was already an able laboratory chemist who had read all the chemistry books he could find. In 1768, he moved to Stockholm to take up employment at a pharmacy, and two years later he found a new appointment as a laboratory assistant at the Upland's Arms pharmacy in Uppsala. The publication in 1774 of his paper *On the Brown-stone or Magnesia Nigra and its Properties* marked him out as a skilful chemist and an excellent experimentalist. Furthermore, the fact that

---

[1] Although Scheele was born in Stralsund, a town in present-day Germany, which at the time was part of the Swedish province of Pomerania, he spoke and wrote in German for the whole of his life. This should be understood in light of the fact that many Swedish towns had large German ethnic communities and that German ethnicity was common in the pharmaceutical profession (Fors, 2008, p. 34)

the paper was published in the *Transactions of the Royal Swedish Academy of Sciences* implied his acceptance by the Swedish community of chemists.[2] In 1775 he gained the post of pharmacist in the town of Köpping, where he stayed for the rest of his life.

The content of his 1777 monograph was based on research dating back to the beginning of his career, and although it was published in 1777, the manuscript itself had already been written in the autumn of 1775. It was translated into English in 1780 with the title *Chemical Observations and Experiments on Air and Fire*[3], and into French in 1781 as the *Traité chimique de l'air et du feu*. Just to prove that air was composed of those two elastic fluids, Scheele set up an experimental device that he described in a series of four experiments in his monograph. This device became a reference for future eudiometrical tests based on the alkaline sulphides.

> The basic experiment consisted of dissolving one ounce of alkaline liver of sulphur in eight ounces of water. Four ounces of this solution were poured into an empty bottle whose capacity was of twenty-four ounces and then well-corked. After turning the bottle, its neck was immersed in a small vessel containing water and kept in this position a fortnight. The solution partly lost its red colour, and some sulphur precipitated from it during this time.[4] After this, the bottle was placed in the same position in a larger vessel also containing water. While keeping the mouth and neck of the bottle underwater, the bottle was uncorked, and water immediately rushed into it. The bottle was corked again and removed from the water. The remaining solution weighed ten ounces, and when the four-ounce solution of liver of sulphur was subtracted from it, there remained six ounces. Consequently, the experiment proved that, in a fortnight, six parts of

---

[2] Ibid., p. 44.

[3] The historian of science William Smeaton (1986, p. 30, note 5) regarded this translation as inaccurate. A better translation of this book was included in the new English translation of all Scheele's papers in *The Collected Papers of Carl Wilhelm Scheele*, London, G. Bell, 1931, reprint, New York, Klaus Reprint Co., 1971.

[4] Alkaline liver of sulphur was prepared by boiling sulphur with a saturated solution of caustic alkali (soda, potash or lime). This liver of sulphur was an orangey solution containing mainly alkaline trisulphide, tetrasulphide and thiosulfate, which reacted with oxygen according to:

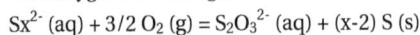

Sx^{2-} (aq) $+ 3/2 \ O_2$ (g) $= S_2O_3^{2-}$ (aq) $+ (x-2) \ S$ (s)

air out of twenty had been lost. In a second experiment, after keeping the bottle corked for one week, no more than four parts were lost.[5]

Scheele explained the phenomenon as a decomposition of liver of sulphur according to the laws of double affinity. Sulphur was believed to be a compound of vitriolic acid and phlogiston. During the experiment, the alkali attracted the vitriolic acid, whereas air (i.e. fire-air) attracted phlogiston.[6]

In 1779, Scheele described a more physiological approach to atmospheric air in his paper *On the Quantity of Pure Air which is Daily Present in our Atmosphere*, published in the *Transactions of the Royal Swedish Academy of Sciences*.[7] This paper was translated into French and published in the journal *Observations sur la physique*, in the *Supplement au Traité chimique de l'air et du feu* and in the second volume of the *Mémoires de chymie de C. W. Scheele*.[8] The English translation of that paper was first published in 1786 in *The Chemical Essays of Charles-William Scheele*.[9] Scheele confirmed that corrupted-air constituted the greatest part of atmospheric air and was dangerous for living animals and vegetables. On the other hand, the other kind of elastic fluid in the atmosphere, pure or fire-air, was beneficial for health, supported respiration and constituted the smallest part of the whole atmosphere.

---

[5] Scheele would have found in these experiments that common air contained roughly between 20% and 30% of vital air. The French chemist Louis Joseph Troost (1825-1911), who became a lecturer at the École Normale Supérieure (from 1868) and professor of chemistry at the Faculté des Sciences (from 1874) in Paris, reported to have replicated these experiments when serving as Chair of chemistry at the Lycée Bonaparte in 1866. He found that common air contained nearly 20% of oxygen (Troost, 1866, pp. 30-31).

[6] Scheele, 1780, pp. 7-8, 14; 1781, pp. 52-54, 60.

[7] Scheele, 1779.

[8] Ibid., 1782, pp. 79-82; 1785a, pp. 179-189; 1785b, pp. 1-12. All Scheele's papers were available in French in the two volumes of this work. French translations of Scheele's papers published in the *Observations sur la physique* and in the *Mémoires de chymie de C. W. Scheele*, were made by Madame Claudine Picardet. She was a member of a small team of translators from Swedish, German, English and Italian recruited by Guyton de Morveau in Dijon (Smeaton, 1992, p. 131). French translations from Swedish were first signed by Magnien as «M. M. de Dijon» or «M. Mgn. de Dijon». The signature of Madame Picardet «Mme. P***» appeared eleven months later.

[9] Scheele, 1901, pp. 190-194. Scheele's papers were not accessible to English readers until the publication of this work, which was re-issued in 1901. New translations of all Scheele's papers and his book are in *The Collected Papers of Carl Wilhelm Scheele*, London, G. Bell, 1931; reprint, New York, Klaus Reprint Co., 1971

Scheele's uncertainty about the existence of a constant proportion of both airs in the atmosphere led him to work out a new eudiometrical procedure, which he used throughout 1778 to determine the proportion of fire-air in common air.

The eudiometrical test of the wet paste made with iron filings and sulphur

This new eudiometrical test was based on the absorption of the atmospheric fire-air by a wet paste made with iron filings and sulphur. Scheele had already been experimenting with this kind of mixture in the pursuit of other aims. In the only letter that Scheele addressed to Lavoisier, dated on September 30th, 1774, he gave an account of his failed trial for producing common air from fixed air by a mixture of iron filings, sulphur and water, attributing the lack of success to the combination of fixed air with iron.[10] An experiment with the same mixture was also carried out by Scheele with the intention of demonstrating that heat was composed of phlogiston and fire-air, and ascribing the heat caused by this mixture to the union of the phlogiston in iron with the fire-air.[11] Such a mixture would therefore absorb fire-air from a given quantity of common air in a closed vessel, and the quantity of the remaining corrupted-air would indirectly indicate how much pure air was contained in the original sample of common air.[12]

Scheele's experimental setup (Figure 5.1) consisted of a pedestal of lead (B) placed in the middle of a vessel (A) with a fixed glass rod. A flat piece of wood was fastened on the top of the rod to support a small vessel (C). This device was covered with an

---

[10] Boklund, 1968, pp. 52-53. The importance of this letter is the experiment that Scheele asked Lavoisier to carry out to produce oxygen with the aid of Tschirnhauser's burning glass.

[11] Scheele, 1780, p. 64; 1781, p. 116.

[12] Scheele, 1782, p. 80. Sulphur and iron combine to give iron sulphides, the properties of which vary according to the proportions of their components. A paste of iron filings, sulphur and water blackens and may burst into a flame. This phenomenon was the key factor in a workshop simulation of the so-called underground fires, from the ancient theories about the formation of volcanoes and earthquakes. In the seventeenth century, Nicolas Lemery presented the first conjectures concerning the origin of those fires, which related to the production of fire by making a wet paste of iron filings and sulphur. The experience, also known as the 'artificial volcano of Lemery', was frequently repeated after that time. Non-stoichiometric iron sulphides produced in the wet paste combine with atmospheric oxygen giving oxide-hydroxides of iron [$FeO(OH) \cdot nH_2O$] and sulphur.

inverted glass cylinder (D) with a capacity of 34 ounces of water, which had a strip of paper (E) pasted on the outside. This strip was equal in length to 1/3 of the capacity of the cylinder[13] (11 ounces of water)[14] and was divided into equal parts marked with lines and ciphers, so that each line showed one-thirty-third part of the space in the cylinder. The piece of paper was varnished over to prevent the effect of water upon it. The vessel (C) was filled with a mixture of two parts iron filings and one part fine powdered sulphur, moistened with a little water. After inverting the glass cylinder over the vessel with its stand, the vessel (A) was filled with water,[15] and the barometer and thermometer readings were observed. The water began slowly to ascend into the cylindrical glass and stopped after eight hours.

Figure 5.1 Scheele's experimental device using a wet paste of iron filings and sulphur.

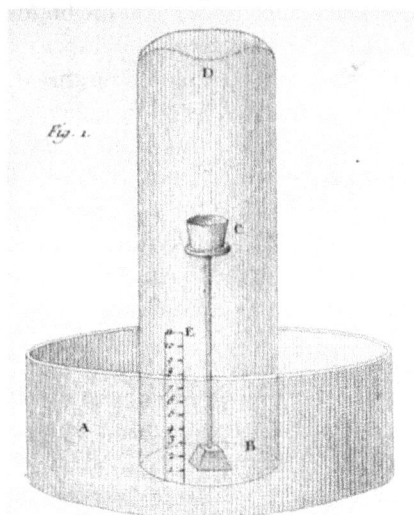

From *Observations sur la physique, sur l'histoire naturelle et sur les arts* (1782), Vol. 19, plate 2, fig. 1.

[13] Scheele was already aware, from the experiments of others as well as his own, that pure air never constituted more than one-third of the whole atmosphere.

[14] The small vessel (C), with its mixture of iron and sulphur, and its stand, occupied the space of 1 once of water. There remained, therefore, room for 33 ounces.

[15] Scheele made use of spirit of wine instead of water when the cold was so intense as to freeze it.

Scheele repeated the test four times a week on average during 1778. The mean value of the level of water in the cylinder was nº 9 on the strip of paper, which represented 9/33 parts of the whole cylinder. For Scheele, the height reached by the water seemed to depend on the rise and fall of temperature and atmospheric pressure. He finally concluded that atmospheric air always contained, though with some slight difference, nearly 9/33 parts of pure or fire-air (nearly 27%).[16]

Stephen Hales' influence on the design of the nitrous air test, explicitly recognized by Priestley, might also be extended to Scheele's eudiometrical test described above. Nine of the experiments described in the sixth chapter of Hales' *Vegetable Staticks* used the apparatus illustrated in Figure 5.2. This device was designed to estimate the quantity of air absorbed or released by a burning material or the breath of a living animal, and it was quite similar to Scheele's experimental contrivance.

> The device consisted of a pedestal set at the bottom of a wide vessel (xx) full of water. The subject of the experiment (inflammable materials or living animals) was placed on this stand. The pedestal was covered with an inverted glass cylinder (zzaa) with a flared mouth (rr) suspended from the top with a string so that the mouth was positioned 3-4 inches under water. Then with the aid of a siphon air was sucked out of the tube until the water level rose to a certain position (zz). When the materials on the pedestal released air, the water level dropped from (zz) to (aa). However, when the materials absorbed any part of the air, then the water level rose from (aa) to (zz).[17]

The decisive difference between this device by Hales' device and Scheele's contrivance was that the glass cylinder of the latter was graduated. Many of the experiments reported by Hales consisted of burning materials such as candles, matches of sulphur, phosphorus or iron filings with sulphur; detonating substances such as nitre (potassium nitrate) and, additionally, breathing animals such as rats and cats. The most common means of ignition were the burning glasses used to converge sunlight on the material once it was inside the tube or, alternatively, starting the ignition with the material outside the tube. Experiments CIV and CVI deserve special attention because they report on the combustion of a mixture of iron filings with sulphur poured on a hot

---

[16] Ibid., 1782, pp. 80-82.

[17] Hales, 1727, p. 93.

iron placed on the pedestal under the inverted glass cylinder.[18] The similarity of these experiments regarding the materials used – iron filings and sulphur - make the influence of Hales' device on Scheele's eudiometer even more likely.

Figure 5.2 Hales' experimental device.

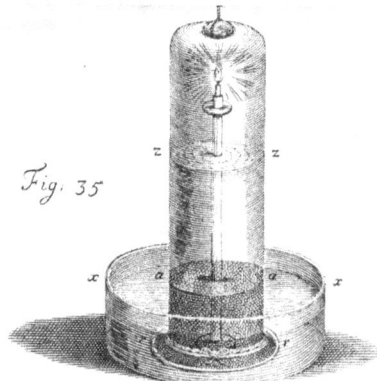

From Stephen Hales, *Vegetable Staticks* (London, 1727), plate 16, fig. 35.

The letter that Scheele addressed to Lavoisier, dated September 30th, 1774, may throw additional light on this issue. Actually, this was a letter of appreciation occasioned by Lavoisier's gift to Scheele of a copy of his *Opuscules physiques et chimiques,* published in the same year, 1774. Although Scheele had been experimenting with various kinds of air some years before, he said in his reply that he had been looking forward for a long time to reading a report on all the experiments carried out in England, France and Germany. Therefore, Scheele might have first acquired a comprehensive knowledge of Hales and Priestley's pneumatic experiments from Lavoisier's book.[19] Lavoisier quoted Hales a number of times in his book; in particular, his experiments with the wet paste of iron filings and sulphur.[20] Nevertheless, Scheele only quoted Hales once in his *Chemical Observations and Experiments on Air and Fire* regarding the nature of fixed air, and did not quote him at all in his paper *On the Quantity of Pure Air which is Daily Present in our Atmosphere.*[21] Scheele's first contribution to the composition of atmospheric air – the alkaline

---

[18] Ibid., pp. 130-133.

[19] Boklund, 1968, pp. 52-53, 366-367.

[20] Lavoisier, 1774, pp. 114, 132.

[21] Scheele 1777, p. 4; 1780, p. 4; 1781, p. 49.

sulphides test – proved to have more adepts and followers than the above-described test based on the moistened paste of iron filings with sulphur.

Lavoisier. Enabling Scheele's alkaline sulphide solutions

Lavoisier retrieved Scheele's discovery of the alkaline sulphides as a suitable eudiometrical means in a paper written between 1789 and 1792, which did not enter into circulation until 1805.[22] Although he found it more suitable to use the paste of sulphur with iron filings as a eudiometrical means, he believed that this test did not warrant complete confidence.[23] Lavoisier finally opted for the phosphorus test for routine determinations and recommended combining this test with the potassium sulphide test for those analyses requiring a more scrupulous accuracy. In any case, he warned of the importance of keeping the temperature and pressure of the air sample constant.[24]

> The reagent, i.e. potassium sulphide, should be prepared before performing the test. A mixture of equal parts of sulphur and potash was placed in a crucible and heated until it melted. After cooling, the mixture was crushed and dissolved in three or four times its weight in water. A turbid solution was formed and left to stand before being decanted to obtain a liquor of potassium sulphide. This liquor had to be kept in partially vacuum-sealed jars to prevent oxygen from somehow joining with sulphur to form sulphuric acid. A number of glass bottles were then filled with the liquor of potassium sulphide and placed face down on the shelf of a marble pneumatic trough. The first stage of the test was to introduce a precisely measured amount of the air to be examined, and secondly to close the bottles

---

[22] LO, Vol. 2, pp. 715-723; LM, Vol. 2, pp. 154-168. This was a paper partially devoted to the use of potassium sulphide as a eudiometrical means. Lavoisier reported in this paper that "vitriolic acid" had been renamed as "sulphuric acid" in the new nomenclature (LO, Vol. 2, p. 716; LM, Vol. 2, p. 156). Since the new chemical nomenclature entered into use in April, 1787, the paper may already have been written by 1788.

[23] Lavoisier believed that the sulphuric acid produced reacted with iron to form hydrogen, which combined with the remaining nitrogen to form ammonia (LO, Vol. 2, p. 721; LM, Vol. 2, p. 165).

[24] To improve gas manipulation, Guyton de Morveau designed an experimental device for the eudiometrical test based on the alkaline sulphides. He presented it to Lavoisier, Monge, Fourcroy, Vandermonde and Berthollet during their stay in Dijon in October, 1787 (Bret, 2017, p. 14)

immediately to prevent the loss of liquor. The next step consisted in moving the bottles to a bowl containing enough water so that the neck of each bottle remained underwater. After 15 to 20 hours, a bottle was submerged upside down in a bucket of water and opened in order to transfer the remaining air into a graduated tube. In this way the residue of pure azotic gas (nitrogen) was determined, since vital air had been absorbed.[25]

Lavoisier believed that the potassium sulphide test had no shortcomings, except for the fact that it was very time-consuming (between fifteen and twenty days) and left something to be desired in terms of cleanliness. Moreover, he encouraged others to use alkaline sulphide and phosphorous eudiometrical tests to determine the quality of the air in different places, seasons and circumstances. This was a project that he envisaged as a continuation of the research on the salubrity of the air in theatres and hospital dormitories.[26]

Reprising the inverted retort apparatus with solid potassium sulphide

In 1795, the French chemist Louis-Bernard Guyton de Morveau proposed a eudiometer based on heating solid potassium sulphide to accelerate oxygen absorption.[27] Guyton's eudiometer was a remake of one of the

---

[25] For calculating the amount of gasses, Lavoisier recommended that they should be reduced to a mean barometrical pressure of 28 inches of mercury and to a standard temperature of 10 degrees. The reasons for taking both references were as follows: the average height of a column of mercury in equilibrium with the weight of a column of air from the highest part of the atmosphere to the surface of the earth was about twenty-eight French inches in Paris and in other quarters of the city. For the temperature, 10 degrees on the Réaumur scale (12.5 °C) was the average between the heat of summer and the cold of winter. Moreover, the temperature was taken in subterraneous locations, which were easy to access in all seasons of the year (Lavoisier, 1789, Vol.2, pp. 371-372, 378-379). The Réaumur scale was a temperature scale in which the freezing and boiling points of water were set to 0 and 80 degrees, respectively. The scale was named after René-Antoine Ferchault de Réaumur.

[26] LO, Vol. 2, p. 721, 723; LM, Vol. 2, p. 165, 168. See chapter 3; The nitrous air test in the hands and mind of Lavoisier.

[27] Guyton quoted neither Scheele's nor Lavoisier's use of alkaline sulphides for eudiometrical purposes. Nitrous gas, hydrogen, phosphorous and the mixture of sulphur with iron filings were the only substances that, to the best of his knowledge, had been used as eudiometrical means (Guyton, 1795, p. 166). However, in 1787 Guyton had reused the inverted retort (*récipient cornu*) to devise and operate a

experimental devices depicted in his article *Air* of the *Encyclopédie Méthodique* for identifying the constituent parts of atmospheric air.[28] He reused the inverted retort – the central component of that device – and replaced the piece of phosphorous with another of potassium sulphide (Figure 5.3). Guyton also provided a more detailed description of the experimental device and procedures at this time. This is an instance that illustrates an example of the transfer of experimental designs between two different kinds of eudiometers.

Figure 5.3 Guyton's potassium sulphide eudiometer.

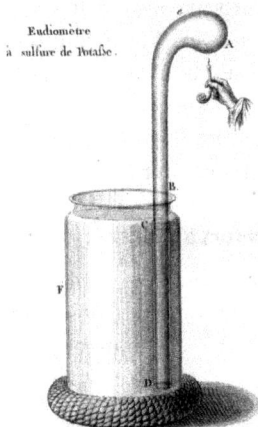

From *Journal de l'École Polytechnique* (1795), plate 5.

The inverted retort (AB) had a long neck and a total capacity of 12-15 cl. It had to be sufficiently curved so that when the neck was held in a vertical position the bottom of the bulb (A) could retain whatever material was introduced. After placing two or three pea-sized pieces of potassium sulphide within the bulb, the retort was filled with water. Then it was tilted to evacuate the remaining air from the bulb. Afterwards, the retort was closed with a finger and moved over a pneumatic trough to take an air sample for examination. To ensure that the potassium sulphide remained in the bulb without any residue of water, the retort was tilted again in both directions

---

eudiometer based on the liver of sulphur (Bret, 2017, p. 14). Additionally, Guyton's quote from Berthollet included a reference to the use of alkaline sulphides as a eudiometrical reactant (Berthollet, 1795, p. 74).

[28] Guyton, 1789, pp. 705-706. See chapter 4; A plausible source of inspiration for Giobert's eudiometer.

alternately. The open end of the retort was then connected to a glass tube (CD) below the water surface of the cylindrical glass vessel (F). The device was made airtight by means of a ground-glass joint (at C) to prevent loss of air. The tube of 20-25 cm in length was open at both ends. To hold the retort in his vertical position, a wooden cover with a notch was adapted to the mouth of the vessel.

The test was initiated by placing a lit candle under the bulb. Initially, heat dilated the air to the point at which the water level dropped almost to the bottom of the tube (CD). This tube was specifically designed to receive the dilated air and prevent it from escaping. When the sample of potassium sulphide began to melt, the water level rose quickly both in the tube and in the neck of the retort. With pure vital air the absorption was complete. In this case, to prevent the retort from breaking due to cooling too rapidly, it was necessary to slow the rise of water either by retiring the candle a little or by tilting the retort. After cooling, the residual air was brought to atmospheric pressure by lowering the tube. The open end of the tube was then closed with a shutter. A graduated recipient was used to determine the amount of water that had entered into the retort, which represented the volume of absorbed air. Alternatively, a varnished scaled strip of paper attached to the neck of the retort could be used for this purpose.

Although this Guyton's eudiometer was successfully tested on September 30th, 1795, at the École Polytechnique, where he was one of the chemistry professors, by 1805 it could have been used only rarely.[29] Actually, by 1800 Berthollet had already formed an unfavourable opinion of this eudiometer, and in fact did not mention it at all in his *Essai de statique chimique* of 1803.[30] Berthollet devoted an entire chapter of this work to assessing the different kinds of eudiometers and quoted an abridged version of a paper written by the Catalan naturalist Antoni de Martí-Franquès, published in the *Journal de physique, de chimie et d'histoire naturelle* of 1801. Martí suggested using a solution of "calcareous liver of sulphur" as an alternative eudiometrical means.[31]

---

[29] Smeaton, 2000, p. 222.

[30] Berthollet, 1800a, p. 78; 1800b, p. 288.

[31] Ibid., 1803b, Vol. 1, p. 512. Berthollet erroneously referred to Martí as Macarty, a mistake that occasionally led to Martí being attributed English citizenship (Libes, 1813, Vol. 4, p. 263)

Antoni de Martí Franquès. Improving the sulphide test

Antoni de Martí Franquès (1750–1832) was born in Altafulla - a town in the province of Tarragona on the southern coast of Catalonia - into a local family of nobles who possessed land and held industrial interests. He was basically self-taught, and besides studying Latin and philosophy, he also studied several European languages such as French, English, German, Greek and Italian. Throughout his life he built up a remarkable library that included the most important contemporary European scientific publications, as well as a cabinet where he conducted his experimental activities. He lived in his hometown until 1798, when he moved to the city of Tarragona, which is where he died in August 1832, but it was in Altafulla where he carried out much of his experimental work. However, he spent periods of time in Barcelona where entered into contact with scientific institutions, and in 1786 he was elected a member of the Real Academia de Ciencias y Artes de Barcelona (Royal Academy of Sciences and Arts of Barcelona).

Martí also spent much of his time studying diverse science subjects. On the one hand, he devoted considerable effort to acquiring knowledge about the natural environment, accumulating geological and botanical collections, and also about meteorology. On the other hand, he conducted exhaustive experimental work on the study of common air and the reproduction of plants, as well as on other topics such as plant physiology and spontaneous generation. Evidence exists of five papers read by Martí at academic institutions in Barcelona, two of which were devoted to the study of atmospheric air that Martí had begun in June 1786, one of which, *Memoir on the Quantity of Vital Air in the Atmosphere, and the Different Methods of Measuring it,* remained unpublished until 1795 in the *Memorial literario, instructivo y curioso de la Corte de Madrid,* and had a significant impact abroad.[32] Abridged translations of this paper into

---

[32] A transcription and a facsimile edition of this paper were published in Grau & Bonet (2011, pp. 47-70; 75-120). The manuscript of the other paper (*On Some Productions Resulting from the Combination of Several Aerial Substances*), read on January 24th, 1787, at the Royal Academy of Sciences and Arts of Barcelona, was lost, but a copy was eventually found in the 20th century in the Municipal Archive of Tarragona (Quintana,1935, p. 147). The manuscript of the third paper (*On the Vital Air from Plants and Particularly from the American Agave*), read over two sessions (October 10th, 1787 – February 27th, 1788) at the Royal Academy of Sciences and Arts of Barcelona, was definitively lost and only a short summary by the author has been preserved. Martí read a fourth paper (*Experiments and Observations on the Sexes and Fecundation of Plants*) on March 28th, 1791, at the Royal Academy of Practical Medicine of Barcelona. The manuscript of Martí's last paper (*The Results of Mixing*

French, English and German were published respectively in the *Journal de physique, de chimie et d'histoire naturelle*, the *Philosophical Magazine* and the *Annalen der Physik*.[33]

The paper, read on May 12th, 1790, at the Royal Academy of Sciences and Arts of Barcelona, was the fruit of Martí's eudiometrical experiments carried out in mid-1787 and was already written in terms of antiphlogistic chemistry. Martí was very well acquainted with the different eudiometrical tests known to that date (nitrous air, alkaline sulphide,[34] paste of sulphur and iron filings, inflammable air and phosphorus). The subject of his essay was the examination of the suitability and accuracy of these tests, with the intention of presenting his own device for the analysis of atmospheric air. Observations considered trustworthy by Martí appeared to have proved that atmospheric air contained not more than 30 % (Scheele) and not less than 20 % (Cavendish) of vital air. Martí was committed to narrowing this margin of uncertainty.

He concluded that the substance employed for determining the purity of the air should be neither gaseous nor in a state of combustion. For these reasons, he gave no recommendation for the nitrous and the inflammable air tests or for the phosphorous test.[35] Thus it appears that he had more trust in the tests conducted with the wet paste of sulphur with iron filings and those with the alkaline sulphides.[36] However, since the results of the latter tests were always found to be higher than those of the former, Martí favoured the alkaline sulphide test. This decision was supported by Lavoisier's observation of the formation of sulphuric acid in the wet paste of iron and sulphur (see note 23), and Priestley's finding of the release of inflammable air in the same mixture. Martí regarded the sulphide trial as the best for ascertaining the quantity of vital air contained in any gaseous

---

some Aerial Substances), read on June 20th, 1792, at the Royal Academy of Sciences and Arts of Barcelona, was also lost and no summary of it was recorded.

[33] Martí, 1801a; 1801lb; 1805. The English version came from the French translation and not from the original paper. The publication in the *Annalen der Physik* was a review rather than an abridged translation of Martí's paper.

[34] Martí referred to this sulphide with different names: "liver of sulphur", "sulphurous liver", "hepatic liquor" or simply "liver", before presenting his own eudiometrical test. The French version used only the term "sulfure" and the English one the terms "liver of sulphur" and "sulphuret". I have decide to continue using the term "alkaline sulphide" to maintain the terminological consistency of the text.

[35] Martí, 1795, p. 274; 1801a, pp. 175-176; 1801b, p. 252.

[36] Martí, 1795, p. 348; 1801a, p. 176; 1801b, p. 253.

fluid, since it left the mephitic air (nitrogen), and the other kinds of air that did not combine with it, without risk of any other gaseous substance being produced or lost, except for the vital air, which alone was absorbed by the alkaline sulphide.[37] The drawback was that complete vital air absorption required an exposure time of at least three days. In order to shorten this time, he replaced the alkaline sulphide with calcium sulphide (liquid calcareous liver of sulphur) to achieve a prompt absorption of vital air of between three and five minutes.[38] It has been suggested that this slight but decisive change in the composition of the eudiometrical means should be placed in the context of Martí's interests in agriculture.[39] Martí was a landowner who oversaw different types of agricultural production, such as vineyards, olive groves and fruit and vegetables. At that time, lime-sulphur (the calcareous liver of sulphur)[40] was a chemical whose use was already well-known as a pesticide for treating crop diseases such as powdery and downy mildew, and it is possible that Martí may have been presented with the opportunity of replacing the alkaline sulphide by the calcareous one.

Replacing one sulphide with another was not the only novelty that Martí introduced into the eudiometrical test. In his paper on the vital air from plants, he had already remarked on the observation of small variations in the determination of the proportion of vital air in the common air when using the nitrous air test. Since then, he had continued his experiments on the same subject in order to ascertain whether such variations might not arise from operational circumstances rather than from the nature of the

---

[37] Martí, 1795, pp. 349-350; 1801a, p. 177; 1801b, p. 254.

[38] Martí used the expression "liquid calcareous liver of sulphur" only once, when presenting the new eudiometrical means. After that, he continued to refer to it with the usual terms used for alkaline sulphide. The French version employed the term "sulfure calcaire liquid" and the English translation the term "liquid sulphuret of lime". This calcareous liver of sulphur was prepared by boiling sulphur with a saturated solution of lime. This orange-red liquid mixture mainly contained calcium tetrasulphide and thiosulfate. The calcium thiosulfate easily decomposes into calcium sulphite and free sulphur. When sulphur is used in excess, a secondary reaction occurs in which it combines with the tetrasulphide to form pentasulphide (van Slyke et al., 1910, pp. 405-411; Tartar, 1914, pp. 495, 498). Polysulphides react with oxygen according to:

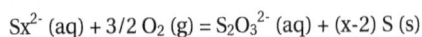

$$S_x^{2-} (aq) + 3/2 \, O_2 (g) = S_2O_3^{2-} (aq) + (x-2) \, S (s)$$

[39] Grau & Bonet, 2011, p. 32.

[40] In France, lime-sulphur was known as *chaux soufré* or *bouille soufrée* or *bouille nantaise*. The product was also used as a skin remedy.

atmospheric air.[41] Meanwhile, he was maturing the notion of the intrusive part played by nitrogen in the different eudiometrical tests. Echoing Van Breda's observations on the influence of water on the nitrous test, Martí suggested that the effect of water depended mainly on its degree of saturation with carbon dioxide, oxygen and nitrogen.[42] Concerning the inflammable air test, he referred to Cavendish's discovery that a quantity of nitrogen could unite with oxygen in the state of ignition, and with regard to Achard's phosphorous test he also suspected that not only oxygen was absorbed from the common air but also a part of its nitrogen.[43]

When Martí eventually decided to use a solution of calcium sulphide as the eudiometrical means, results from different experiments had already made it clear that this means was capable of containing a certain portion of nitrogen interposed between its particles. However, he noted that this absorption of nitrogen could not be specifically attributed to the liquid sulphide, since this acted like any other liquid when deprived of the amount of nitrogen that it could naturally absorb.[44]

Actually, Martí had been experimenting with the absorption (i.e. solubility) of different kinds of airs (oxygen, hydrogen and nitrogen) in water prior to his reading of the 1790 paper, although he made only slight mention of these experiments.[45] However, the physicist Jean-Baptiste Biot shed more light on them in a letter to Berthollet reproduced in the *Annales de chimie*.[46] He had met Martí when he was invited as a guest to his home in Tarragona in December 1806, in order to recover from a fever that he may have contracted in the convent of the Desert de les Palmes in Benicassim, where he went to replace Pierre Méchain after his death; this

---

[41] Martí, 1795, p. 264; 1801a, pp. 174-175; 1801b, p. 251.

[42] Ibid., 1795, p. 270; 1801a, p. 175; 1801b, p. 140.

[43] Ibid., 1795, pp. 274-275; 1801a, pp. 175-176; 1801b, p. 252.

[44] Ibid., 1795, p. 353.

[45] Ibid., 1795, p. 391; 1801a, p. 181; 1801b, p. 258.

[46] Biot, 1807a. This letter, dated December, 1806, should be seen in the context of Biot's attendance at the meetings held at Berthollet's country house, the Société d'Arcueil, from the summer of 1807, when he was already back to Paris, (Crosland, 1967, p. 122). The letter was also published in the *Journal of Natural Philosophy, Chemistry and the Arts* and in the *Annalen der physik* (Biot, 1807b; 1808).

surveyor and astronomer had been commissioned along with François Aragó to measure the arc of the Paris meridian.[47]

In the first part of his letter, Biot stated that Martí had made the observation that if water impregnated with nitrogen was placed in contact with hydrogen or oxygen, it absorbed both gases without relinquishing its nitrogen. Furthermore, an accurate analysis of atmospheric air could thus be made by the absorbent action of water alone. To do this, it was sufficient for the water to have been previously impregnated with nitrogen. Thus, water absorbed exactly 21 hundredths of the volume of the atmospheric air coming in contact with it, precisely as a sulphide would do. According to Biot, Martí had affirmed that water employed in large quantities to prevent the process from taking too much time, acted as an excellent eudiometer to which he had repeatedly had recourse in the past.[48] The calcium sulphide solution was more expeditious than water alone as a eudiometrical agent.

Martí eventually decided to use the solution of calcium sulphide impregnated with nitrogen as a eudiometrical means to prevent the absorption of the atmospheric nitrogen.[49] Calcium sulphide made the test quicker, and the impregnation of nitrogen made it more accurate. After several attempts, Martí managed to develop a simple eudiometrical procedure. None of Marti's eudiometers, some of which were brought from Paris and others built in Barcelona, have been preserved. Antoni Quintana-Martí,[50] the first scholar to study Martí and his work, replicated the instrument (Figure 5.4) by following Martí's own description.[51]

> The whole device consisted of a glass tube of 5 lines in diameter and 10 inches in length, capable of containing about an ounce of water. The tube was closed at one of its ends and graduated into 100 equal

---

[47] Quintana, 1985, p. 53.

[48] Water could be impregnated with nitrogen by shaking it in contact with atmospheric air, and leaving the two together in contact for some time. By these means water absorbed all the nitrogen it could contain. The oxygen that water took up from the atmospheric air at the same time did not prevent it from absorbing that of the air sample to be analyzed. (Biot, 1807a, pp. 275-276; 1807b, pp. 125-126; 1808, pp. 420-421.)

[49] Martí, 1795, pp. 353-354, 356; 1801a, p. 178; 1801b, p. 255.

[50] Quintana, 1985, pp. 58-59. (Quintana-Marí, 1935) is the seminal work on Martí-Franquès.

[51] Martí, 1795, pp. 357-359; 1801a, p. 179; 1801b, p. 256.

parts, each of 1 line. To take a sample of common air, the tube was first filled with water, keeping it in a vertical position with the orifice downwards and closed with a thumb. Then the tube was inclined a little at the surface of the water, and the finger was removed at intervals to allow the external air to enter into the tube, replacing the water, until it occupied all the 100 lines. After that, the orifice was closed again with a thumb, and the tube was immersed in a water trough to balance the temperature inside and outside the tube. After removing the tube, it was examined to check if the air surpassed the space of 100 lines or not and to remove or add the quantity of air necessary to make it stand exactly on a level where the divisions began. This air sample was then introduced in the usual manner into a flask containing from twice to four times its volume of liquid calcium sulphide, previously shaken with nitrogen in order to saturate it fully.[52] The flask was covered and shaken for five minutes, opened inside the water and covered and shaken again, after which the residue was transferred back into the graduated tube.[53] The space occupied by this residual air after the operation indicated the percentage of nitrogen in the air sample and, consequently, the missing parts indicated the percentage of oxygen. If the graduated tube ended in a neck with a ground stopper, it might be first filled with calcium sulphide rather than water. Proceeding thus, the operation would be quicker, without the necessity to employ water, or to introduce air into the flask and to transfer it.

---

[52] Martí did not mention how he obtained nitrogen. Cavendish had proposed that it could be obtained by common air with a solution of calcium sulphide or, alternatively, by bringing it into contact with a wet paste of iron filings and sulphur. Both procedures consumed oxygen and left a residue of nitrogen.

[53] Since the liquid sulphide was saturated with nitrogen, the former could no longer absorb the latter from the air sample. Martí argued that because of the liquid sulphide absorbed oxygen from the air sample, a vacuum occurred inside the flask. At the same time, the rarified air exerted less pressure on the liquid sulphide, the interposed nitrogen was released and joined with the remaining nitrogen in the air sample. However, after opening and covering the flask in the water and shaking it again, the nitrogen previously removed penetrated again into the interstices of the liquid sulphide. (Ibid., 1795, pp. 354, 356-357)

Figure 5.4 Modern replica of Martí's eudiometer by Antoni Quintana-Marí.

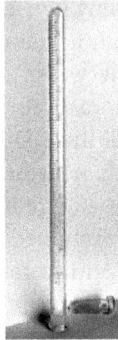

Courtesy of Marta and Antoni Quintana, heirs of Antoni Quintana-Marí, photography by the author.

Martí stated that he had repeated this procedure for a long time and on so many days that the uniformity of the results obtained clearly demonstrated the exactness of the method.[54] His firm conclusion was that in all seasons, in every month and at all hours, the air of his country taken in the open fields was always composed of between 21 and 22 parts of oxygen, and between 78 and 79 of nitrogen.[55] He was convinced that this small difference did not arise from the nature of the air, but rather from some negligence in the operation. He reported to have collected air in places where a great many persons were assembled, or near ponds of stagnant water, and always found this air to be as pure as the common air.[56]

As regards eudiometers as instruments for measuring the salubrity of the air, he was persuaded of their uselessness, since insalubrity could not arise from the disproportion between oxygen and nitrogen in the atmosphere. Although the three different kinds of air disengaged from stagnant water (nitrogen, inflammable air and carbon dioxide) were incapable of maintaining animal life, it appeared that they were released in very small quantities in comparison with the large amount of atmospheric air.[57] Since eudiometrical tests were unable to account for the dangerous effects produced in the neighbourhood of stagnant waters, Martí suggested that the cause might perhaps be found by analysing the water suspended in

---

[54] Ibid., 1795, pp. 357-359; 1801a, p. 179; 1801b, p. 256.

[55] Ibid., 1795, p. 392; 1801a, p. 131; 1801b, p. 258-259.

[56] Ibid., 1795, pp. 392, 395-396; 1801a, pp. 181-182; 1801b, p. 259.

[57] Ibid., 1795, pp. 397-398; 1801a, p. 182; 1801b, p. 259.

the atmosphere.[58] His aim had been to find a eudiometrical test capable of determining whether or not it was appropriate to attribute the accidental insalubrity of atmospheric air to its composition, rather than participating in a eudiometrical test competition.

The reception of Martí's eudiometrical test

Martí's eudiometrical trials had a significant impact on chemists interested in the composition of the atmospheric air. His eudiometer was appreciatively received in chemistry texts of the early nineteenth century, although his eudiometrical test was occasionally misinterpreted.[59] The second part of Biot's letter to Berthollet provides some instances of this misinterpretation by highlighting how Martí's abridged paper, published in the *Journal de physique* (as well as its English translation, published in *The Philosophical Magazine*), not only omitted some experiments but also some details about the experimental outcome.[60] The consequence of all this was that chemists such as Berthollet, Gay-Lussac, Fourcroy and the naturalist Humboldt, attributed to Martí inaccurate opinions and results.

For example, Gay-Lussac and Humboldt considered that Martí's eudiometer incurred more inaccuracies than Volta's instrument, because in their opinion Martí had determined the proportion of oxygen in the atmospheric air as being between 21% and 23%.[61] In reality, however, that was not the case; Martí had certainly found this margin of uncertainty, but only in the first trials, and his efforts to narrow this margin led him to discover the influence of the absorption of nitrogen as an important source of uncertainty. Once he had perfected his method, he succeeded in reducing this margin of uncertainty to between 21% and 22%.[62]

Biot also reproached Berthollet for having attributed to Martí the finding that the liquid sulphide was responsible for the absorption of nitrogen, when this was not the case, and Gay-Lussac and Humboldt spread this idea abroad even more widely.[63] For Biot, this misunderstanding came from an inadequate rendering of the abridged French translation of

---

[58] Ibid., 1795, p. 399; 1801a, pp. 182-183; 1801b, p. 260.

[59] Grau & Bonet, 2011, pp. 34-40.

[60] Biot, 1807a, p. 277; 1807b, p.126; 1808, p. 277.

[61] Humbolt & Gay-Lussac, 1805, p.134.

[62] Biot, 1807a, pp. 278-279; 1807b, p. 127; 1808, pp.423-424.

[63] Berthollet, 1803b, Vol. 1, p. 513; Humbolt & Gay-Lussac, 1805, p. 133.

Martí's paper.[64] In fact, Berthollet was unable to observe the absorption of nitrogen from the air when using solutions of alkaline sulphide, and he admitted not having used solutions of calcium sulphide as Martí had. Unfortunately, the misunderstandings surrounding Martí's eudiometrical test using calcium sulphide with the alkaline sulphide test were reproduced in Gay-Lussac and Humboldt's paper. Worse still, Fourcroy in his article *Eudiométrie* of the *Encyclopédie Méthodique* not only repeated these mistakes but also confused Martí's test with that of Scheele.[65]

Summary

In the last decade of the eighteenth century, sufficient convincing reasons already existed for shelving the nitrous air test and replacing it with the phosphorous or the alkaline sulphide tests. These two latter substances, together with the wet paste of sulphur with iron filings, were eudiometrical means with a comparable degree of purity and were more manageable than the nitrous air, which was seriously compromised by for many reasons. Furthermore, the experimental devices required for performing the phosphorous or the sulphides eudiometrical tests were much more affordable than the nitrous air and Volta's eudiometers. With the arrival of Giobert's phosphorous eudiometer and the sulphide eudiometrical devices, simplicity in experimental design was restored to eudiometry.

The prelude to the emergence of Guyton's eudiometer merits special attention. Guyton redesigned the inverted retort that he used in his experiments on the combustion of phosphorous, replacing phosphorous by solid potassium sulphide. However, according to Guyton himself,

---

[64] Biot, 1807a, p. 278.; Ibid., 1807b, p. 127; 1808, p.423.

[65] Fourcroy, 1805, pp. 280-281. Between 1820 and 1823, Martí's eudiometrical test was still being used by Eugène Julia de Fontenelle (1780–1842) in a study on the unhealthy emanations from marshes, latrines, stables and sewers. At that time, Julia de Fontenelle was professor of medical chemistry at the Paris Faculty of Medicine. He analysed air samples from marshes and ponds in different locations in the Roussillion and the Hérault in France, as well as from the mountains of Montjuic and Sant Jeroni de la Vall d'Hebron, and the cities of Girona, Figueres and Barcelona in Spain. The air of the latter city had a particular interest because of the yellow fever epidemic that had ravaged Barcelona in 1820. Julia de Fontenelle preferred to use Martí's eudiometer, which he described as 'Scheele's apparatus perfected by Martí', because he considered the calcium sulphide solution to be an excellent eudiometrical means once saturated with nitrogen (Julia de Fontenelle, 1823, pp. 52, 92).

Chaussier had previously used the same inverted retort for eudiometrical purposes with phosphorous. Giobert, for his part, eventually exploited the same design for his eudiometer. Thus, the development of the phosphorous and solid potassium sulphide eudiometers reached a crossroad that led to a transfer of experimental designs between both kinds of eudiometers. In this respect, it is worth remembering that Volta did not discard his first prototype of inflammable eudiometer for performing the nitrous air test either. Chemists were not necessarily involved in the creation of new devices, but rather were engaged in adapting and converting the available equipment in order to address new challenges.

All the different versions of the sulphide test were indebted to Scheele's pioneering experiments on the composition of atmospheric air. It was the problem of the excessive duration of the procedures using solutions of alkaline sulphides that led Martí to replace these solutions with solutions of calcium sulphide. In retrospect, the idea of impregnating the solution of calcium sulphide with mephitic air (nitrogen) explains the fine results obtained by Martí in his analysis of the atmospheric air. It was probably due to the deficiencies in the French translation of his original paper that his eudiometrical test was so subject to misinterpretation. Additionally, criticisms of Martí's eudiometrical procedure by prominent figures such as Gay-Lussac and Humboldt did little to help in its reception.

6.

The evolution of the
phosphorous eudiometer

The phosphorous eudiometers had got on the road as portable instruments at a time when the phlogiston-based chemistry was the prevailing theoretical framework for interpreting the combustion of phosphorous. The instrument was immune to its transit from the phlogiston to the oxygen-based chemistry. In parallel with the evolution of these portable eudiometers, a new laboratory bench phosphorous eudiometer was gaining momentum. Thus, at some point between 1789 and 1792, Lavoisier advised combining both the potassium sulphide and the phosphorous tests for analyses demanding accuracy and the phosphorus test alone for routine determinations.[1] Moreover, from 1790 onwards, when Lavoisier resumed his experiments on respiration with the assistance of Armand Séguin, the latter used a version of the phosphorous eudiometer resulting from Lavoisier's researches on the combustion of phosphorus.

The burning of phosphorous in common air had been a key point in Lavoisier's investigative pathway on the active role of air in calcinations and combustions, as well as in the formulation of his antiphlogistic chemical system. In this context, phosphorous performed the dual function of being a necessary contributor to discover 'the purest part of air' – the vital air, later called oxygen – as well as of a chemical agent useful to determine its composition. Therefore, the coming of a new phosphorous eudiometer was inextricably linked to Lavoisier's researches on animal respiration, as well as to his experiments aimed at the identification of oxygen as a chemical distinct part of the atmospheric air. These researches had originated in his ideas on the role of atmospheric air in the experiments carried out between November 1772 and October 1773, which served to put in evidence the gain of weight of some metals, such as

---

[1] LO, Vol. 2, p. 721; LM, Vol. 2, p. 165.

lead, once calcinated and of phosphorus and sulphur after burning.[2] All this forms a complex panorama that requires the following concise introduction to Lavoisier's first experiments on the combustion of phosphorous and on animal respiration.

The theoretical and experimental background of the laboratory bench phosphorous eudiometers

Lavoisier's experiments on the combustion of phosphorous

Lavoisier described his first experiments on the combustion of phosphorus in a closed recipient in a legendary sealed note (*pli cacheté*) that he deposited with the Perpetual Secretary of the Academy of Sciences on 1 November 1772. These experiments were carried out between early September and late October of 1772.[3] On 27 June 1773, Lavoisier resumed these experiments after the reading of the sealed note during the session of the Academy on 5 May of the same year. Lavoisier probably wanted to determine the proportion of phosphorus and air combined in the resulting phosphoric acid.[4] In this case the combustion of phosphorus was conducted inside a bell jar inverted over a water basin covered by a layer of oil. Lavoisier was persuaded that the combustion of phosphorus absorbed the fixed air contained in the common air and to keep fixed air from dissolving in the water he resorted to a layer of oil over it.[5] The piece of phosphorus was ignited with the aid of a magnifying glass to concentrate sun's rays on the phosphorus sample.

A surprising fact made him change this experimental design. Lavoisier thought that by bringing back fixed air to the air in which he had performed the combustion he would be able to restore the original

---

[2] Lavoisier already knew before August 1772 that the combustion of phosphorus involved the absorption of air and the production of an acid.

[3] In a first experiment, Lavoisier introduced a piece of phosphorus into a capsule placed on a tray of clay covered by a bell jar and ignited the piece of phosphorus with the hot tip of a knife. In a second experiment, the piece of phosphorus was deposited in a balloon flask. Once closed the flask, a candle flame near to the part of the flask in contact with the piece of phosphorus, first heated and then ignited the piece of phosphorus. See *Mémoire sur lacide du phosphore et sur ses combinaissons avec diferentes substances salines terreuses et metalliques*, reproduced in (Guerlac, 1961, pp. 224-227). This paper was drafted on October 20, 1772.

[4] Holmes, 1998, p. 82.

common air. Contrary to expectations, the mixture of the residual air after the combustion of phosphorous with fixed air extinguished a flame candle even more promptly than common air. Then, on June 30 Lavoisier repeated the combustion phosphorous, but in a jar inverted over mercury instead of water covered with a layer of oil. The most important difference with the previous experiment was that dry flakes of phosphoric acid stuck on the walls of the jar, which gave him the hope to measure the weight gained by phosphorus in the combustion. A relevant aspect of the experiment was to find that performing the combustion over mercury instead of water covered with oil was sufficiently advantageous to continue using it commonly.[6]

Lavoisier published the experiments conducted during the year 1773 in the second part of his *Opuscules physiques et chymiques*.[7] However, during the first week of August, before drafting the chapter on the combustion of phosphorus, he performed the last experiments intending to weight the dry phosphoric acid. He concluded that these experiments seemed to lead to think that atmospheric air, or some other elastic fluid contained in the air, combined with phosphorus during combustion.[8] Apparently, at that time, Lavoisier was not enough convinced that this elastic fluid was a substance essentially different from the common air.[9] In all these experiments Lavoisier achieved to ignite the piece of phosphorous with a magnifying glass or touching it with the hot tip or blade of a knife, or with a red-hot crooked iron wire that he managed to pass underneath the bell jar.

In the sealed note of November 1772, Lavoisier also expressed his belief that the gain of weight observed in the combustion of phosphorous and sulphur was too similar to that in the calcination of metals not to presume that both arose from the same cause.[10] During the next months he prepared his definitive paper on a new theory of the calcination of metals presented at the Easter meeting of the Academy in 1773. A decade after

---

[5] This was not an uncommon belief. It is worth remembering that Achard also believed that certain fixed air flowed from the atmospheric air during combustion.

[6] Holmes, 1998, pp. 84-86. To use water covered with a layer of oil had certain disadvantages such as to soil recipients leaving residual water droplets with emulsified oil that made their complete cleaning difficult (Daumas, 1955, pp. 117-118).

[7] Lavoisier, 1774, pp. 327-346.

[8] Ibid., p. 332.

[9] Holmes, 1998, pp. 121-124, 133, 136.

[10] Lavoisier, *The Sealed Note of November 1, 1772*, reproduced in (Guerlac, 1961, pp. 227-228).

discovering that air combined with metals during calcination, at a meeting held early in summer 1783, he announced that he and Laplace had been able to demonstrate that inflammable air and vital air, when burned together, formed water.[11] His discovery of 1772 on the role of air in calcination had finally led him to complete his unified theory of calcination, combustion, acidification and respiration.

Lavoisier experiments on animal respiration

Priestley's paper presented in January 1776 (*Observations on Respiration, and the Use of the Blood*) stimulated Lavoisier's interest in respiration.[12] Contrary to Priestly, Lavoisier intended to interpret the effects of the respiration on the air in terms of his forthcoming views about the composition of the atmosphere. From a draft of Lavoisier's first paper on respiration (*Expériences sur la décomposition de l'air dans le poulmon, et sur un des principaux effets de la respiration dans l'économie animale*) read to the Academy on 9 April 1777, some details are known of his experiments on this topic performed between April and October 1776. He experimented with birds (sparrows and robins) introduced under a glass jar over mercury and filled with common air. He analysed the residual air, after the death of the bird, concluding that the air had become completely unrespirable and that, in this respect, it was quite similar to which had remained after the calcination of mercury.[13]

In summer 1777 Lavoisier decided to present to his colleagues a synthesis of his views on combustion. This work became the *Mémoire sur la combustion en général* read to the Academy in November of the same year. Considering that respiration was an integral part of his system, he terminated the paper showing how his theory on combustion could serve to explain a part of the phenomenon of respiration. He now added to his precedent views on respiration that this phenomenon was combustion

---

[11] See chapter 2; The development of the inflammable air eudiometer.

[12] Philosophical Transactions of the Royal Society, 1776, 66, pp. 226-248.

[13] However, he perceived remarkable differences and similarities between the residual air after calcination and respiration. First, respiration acted only on a portion (pure air) of common air. Second, the calcination of a metal absorbed the same portion. Third, the bird perished when it had absorbed this portion converting it to fixed air, which precipitated limewater whereas the air of calcination caused no change in it. And fourth, the remaining unrespirable air (mophete) could be restored to respirable air by removing the fixed air and adding an amount of common air equal to that lost (Holmes, 1985, pp. 64-65, 89).

analogous to that of charcoal (*charbon*), converting pure to fixed air, and realising the heat (matter of fire) responsible of maintaining the constant temperature of warm-blooded animals.[14] Few days after the reading of this paper, Lavoisier read another paper titled *Considérations générales sur la nature des acides*, which included his first direct confrontation with the theory of phlogiston. He announced that he would hereafter refer to dephlogisticated air or eminently respirable air (the vital air) by the name of oxygen (*principe oxygine*). The gain in weight observed in a calcined substance was due to the fixation of oxygen not to the release of phlogiston and, furthermore, oxygen was also the principle of acidity and played a key part in animal respiration.

Between February and March 1783, Lavoisier and Laplace initiated their first experiments on respiration with a newly arrived instrument, the calorimeter. They placed a guinea pig in the wire basket inside the calorimeter and intended to measure the quantity of heat produced during a certain period of time. After these experiments, Lavoisier was more compelling stating that respiration was identical rather than analogous to the combustion of charcoal or some other substance containing *matière charboneuse*.[15] By 1785, Lavoisier had given up his calorimetric researches with Laplace on animal respiration, resolving that respiration was the source of animal heat. He should probably have considered sufficiently proven his theory of respiration so as not to pursue this line of research. Instead of this, he devoted much of his experimental activity between 1785 and 1790 to investigate the processes of fermentation and vegetation, and to disentangle the composition of plant substances.[16]

But in 1790 Lavoisier resumed his experiments on respiration. The objective of his new experimental agenda was to measure the rate of human respiration in various physiological conditions. Nevertheless, the first experiments were carried out with animals. Unlike his previous experiments during the years 1783-1784 with Laplace, Lavoisier now believed that the product of respiration was not only fixed air but also water. This change in mind raised, however, a major problem since the rate of respiration in terms of oxygen consumption, the common component of fixed air and water, could no longer be done from the loss in volume of the respired air once the fixed air had been removed. The solution to the problem was to use eudiometrical methods to determine

---

[14] Ibid., pp. 113-114, 121.

[15] Ibid., p. 169.

[16] Ibid., pp. 258-261.

the amount of oxygen in the air before and after being respired.[17] This was the beginning of a close collaboration between Lavoisier and Séguin that materialized in the four papers on respiration and perspiration read at the Académie des Sciences between 1790 and 1792.

Lavoisier began to read the *Première mémoire sur la respiration des animaux* on 13 and 17 November 1790 and ended it on 7 and 11 December 1790.[18] However, the paper was not published until 1793. In the meantime, Lavoisier completely rewrote the original paper for its publication.[19] In this revised version Lavoisier reappraised Séguin's scientific merits for the experimental design and the theoretical insights of their researches on respiration. Lavoisier changed Séguin's status of mere assistant in the original version to that of co-author in the published paper, admitting that he had developed a simple and convenient method to analyse respirable air without which it would not have been possible to have done such accurate experiments.[20] Therefore, the new phosphorous eudiometer was developed for monitoring those experiments during the year 1790, before the month of November.

Séguin and the fast combustion of phosphorous eudiometer

Armand Séguin (1767-1835) was one among other young students that worked as assistants in Lavoisier's laboratory at the Arsenal. His first steps in this laboratory were taken during the second tour of Lavoisier's experiments on the decomposition of water in 1785. Because of his experimental skills, Lavoisier recruited him as a full-time assistant. As a chemist, he is remembered for his experiments and papers on human respiration read at the Académie des Sciences between 1790 and 1792, as well as a demonstrator in Lavoisier's laboratory and as an instrument designer. In the Revolution he profited from methods of rationalising the process of tanning and became a successful financier during the Empire. His entrepreneurial activities moved him away from science.[21]

Séguin's first experiments on the physiology of respiration, carried out jointly with Lavoisier, began in 1789. His preliminary reflections on the different eudiometrical means highlight the doubts about their suitability

---

[17] Ibid., pp. 441-443.

[18] Séguin & Lavoisier, 1793.

[19] For a complete study of this rewriting see Holmes (1985, pp. 460-467).

[20] Séguin & Lavoisier, 1793, pp. 567-568.

[21] Mercier, 1976; Beretta, 2001b, pp. 331-334.

to determine the salubrity of the air. At that time it was already undisputed that the available eudiometrical tests served to know the relative amount of vital air in a respirable air, which was not far enough to determine its degree of salubrity. To this aim, it was necessary to know the miasma flowing in the air on which, however, very little was known. Despite ascertaining how far it was to have a real eudiometrical science, Séguin's commitment was continuing to make progress in devising a new version of the phosphorus eudiometer. This new eudiometrical method was presented in a paper read at the Académie des Sciences on 28 March 1791 and published in *Annales de chimie* as *Mémoire sur l'eudiomètre* and in Lavoisier's *Mémoires* as *Combustion du phosphore, employé comme moyen eudimétrique*. Séguin was looking for an assay that, unlike all the others, was able to determine the absolute volume of vital air in a gas mixture.[22]

As mentioned above, Séguin's first version of the phosphorus eudiometer was, in large part, the result of Lavoisier's precedent experiments with phosphorous. Nevertheless, these experiments were designed to quantitatively confirm that a portion of the atmospheric air – the oxygen gas - could be decomposed by phosphorus. They were not designed to determine the oxygen content of gaseous mixtures. Therefore, Lavoisier and Séguin had to readdress all those experimental devices for eudiometrical purposes.

The following description of Séguin's eudiometrical test has been worked out from different sources: Seguin's paper of 1791, Bouillon-Lagrange's description in his *Manuel d'un cours de chimie* of 1788-1789 and Fourcroy's article on eudiometry of 1805 in the *Encyclopédie Méthodique*. Unfortunately, these descriptions do not always provide all the necessary procedural details to understand and assess the test. Lavoisier's descriptions of his experiments on the fast combustion of phosphorus in his *Traité élémentaire de chimie* of 1789 have proved to be useful to cover these shortcomings.[23]

Lavoisier described these experiments twice in his *Traité*. Firstly, to show quantitatively the decomposition of oxygen by phosphorus with the

---

[22] Séguin, 1791, pp. 294-297; LM, Vol. 2; pp. 144-147. In a subsequent paper (*Observations sur l'eudiomètre à gaz nitreux*), Séguin adduced the well-known inconveniences of the nitrous air eudiometer (LM, Vol. 2, pp. 216-218)

[23] Hereafter in this chapter the common phenomenon of the quick ignition of phosphorous by means of a flame, a magnifying glass or a red-hot metal wire will be recognised as the "fast combustion" of phosphorous to distinguish it from the "slow combustion", i.e. chemiluminescence, of phosphorous. See note 54.

release of caloric from the base of the gas. The first part of the experiment was designed to determine the amounts of oxygen and phosphorus consumed in the combustion of the latter substance. Making an implicit use of the principle of the conservation of matter, Lavoisier deduced from these results the mass of the substance resulting from the combustion. The second part of the same experience was devoted to determining directly - with the express use of the balance - the masses of phosphorus and oxygen consumed as well as of the end product of combustion.[24] The experiment was described again in the second volume of the *Traité*, presenting the procedural details for the sake of the experiment reproducibility.[25] Both accounts of the experiment have been of great help to complete the depiction of the eudiometrical procedure explained by Séguin and Fourcroy. The first trials to implement the test, carried out jointly by Lavoisier and Séguin, were executed with a graduated bell jar filled with mercury in a pneumatic through (Figure 6.1, left).[26]

The bell jar was 3 inches in diameter and 5-6 inches high.[27] A sample of 12-15 cubical inches of respirable air was transferred to the jar by passing the neck of the bottle containing the air to be analysed underneath the jar. The level of the mercury in the jar was noted. After this, a piece of phosphorous was placed in a small flat shallow capsule (D) of iron, which was passed through the mercury into the jar. Then, a part of the respirable air in the jar was sucked out to raise the mercury to (EF).[28] Otherwise when the piece of phosphorous was ignited the expanded air would be partially forced out and it would

---

[24] Lavoisier, 1789, Vol. 1, pp. 57-66.

[25] Ibid., Vol. 2, pp. 482-486.

[26] 'In the first experiments we have made, **Lavoisier** and **I**, on respiration, we determine, with the aid of the following procedure the volume of vital air contained in our respirable fluids.' ['Dans les premières expériences que nous avons faites, **Lavoisier** et **moi**, sur la respiration, nous déterminons, à l'aide du procédé suivant, le volume de l'air vital que contenaient nos fluides respirables.'] (Séguin, 1791, p. 297; LM, Vol. 2, p.147). The highlighting of individual words is of the author.

[27] Lavoisier gave a thorough description of the manner of graduating bell jars (Lavoisier, 1789, Vol. 2, pp. 362-365).

[28] A procedure to draw out the air was by means of an air pump syringe adapted to a glass syphon (GHI) underneath the jar. Another procedure consisted in sucking out the air by breathing in through the same syphon adapted to the muscles of the mouth. To avoid the siphon become full of mercury, a small piece of paper (I) was twisted in its extremity (Ibid., p. 484)

no longer be possible to make an accurate calculation of the quantities before and after the assay.

The piece of phosphorus was ignited by means of a crooked red-hot iron wire (MN) and passed quickly through the mercury (Figure 6.1, right). To achieve complete combustion, the tip of the crooked wire was plunged into the capsule to pick up a bit of glowing phosphorus, which was moved around the top of the bell to increase contacts. The combustion was extremely rapid, with a bright flame and a considerable disengagement of light and heat. Because of the heat released, air was at first much dilated but soon after the mercury returned to its level with a considerable diminution in volume. At the same time, the whole internal side of the jar became covered with light white flakes of concrete phosphoric acid.

When phosphorus no longer burned, the iron wire was removed, and the device was let to cool down. The same operation was repeated a quarter of an hour later. However, if the first combustion was made carefully phosphorus no longer burned. Once vital air was fully decomposed, a little amount of caustic alkali was introduced into the jar to absorb the gaseous carbonic and phosphorous acids that could have been formed. Finally, the volume of the residual air was measured to calculate the volume of vital air absorbed.[29]

Figure 6.1 (Left) Lavoisier's experimental device for the combustion of phosphorous. (Right) Crooked iron wire used for igniting the piece of phosphorous.

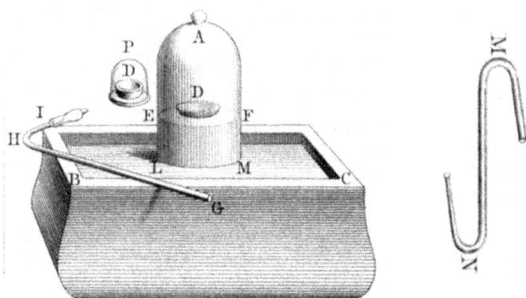

From Antoine-Laurent Lavoisier, *Traité élémentaire de chimie* (Paris, 1789), plate 4, figs. 3 and 16.

This assay developed by Lavoisier and Séguin was not without difficulties. Thus, testing pure vital air made burning so rapid that the top

---

[29] Séguin, 1791, pp. 297-298; LM, Vol.2, pp. 147-148.

of the bell jar warmed up too abruptly, it could not resist a sudden change of temperature and cracked before finishing the assay. Residual humidity in the mercury could facilitate the accident. With all this, it was preferable to use flat-topped jars of green glass. The assay appeared less uneven with atmospheric air, but the inconvenience of having to pass repeatedly the iron wire led them to improve the assay with the aim of also to determine vital air in respirable fluids with greater accuracy.[30] Another major drawback was the zero portability of the assay since the experimental setup forced to perform it in a laboratory. Despite all this, the method had the advantage of making many trials in a short time without needing temperature or pressure adjustments. It was unlikely that during the short duration of the trial there were such important changes in those parameters as could influence significantly the result.[31]

A fortuitous event led them to redesign a new device. They wanted to analyse too large a volume of a respirable fluid (100 cubic inches) to do it in a single trial given the insufficient capacity of the bell jar. After analysing the first sample of 20 cubic inches, they decided to introduce a second sample believing that they should pass again the crooked iron wire through the mercury to ignite the residual phosphorus content as well as the new piece of phosphorous they should place in the capsule. The astonishing fact was that the remaining phosphorus ignited immediately upon contact with the first bubbles of respirable air that got into the jar. In this way, they were able to proceed with a continuous analysis of the whole air sample taking care to bubble it very slowly to avoid an excessive increase in temperature.

Séguin claimed that this surprising phenomenon awakened them, Lavoisier and him, to developing a new eudiometer preferable to the hitherto existing ones. Nevertheless, Séguin began to describe the new device speaking in the first person, thus confirming that the new eudiometer was primarily a work by him:[32]

'I made various trials, and success surpassed my expectations. Here is the device that I use.'

---

[30] Ibid., p. 299; p. 149.

[31] Fourcroy, 1805, p. 282. Lavoisier had foreseen these kinds of corrections and showed them with explanatory examples (Lavoisier, 1789, Vol. 2, pp. 374-383)

[32] 'Je fis doncs divers essais, et le success surpassa mes espérances. Voici l'appareil don't je me sers.' (Séguin, 1791, p. 301; LM, Vol. 2, p.151.)

The new eudiometrical procedure consisted in determining, first, the volume of the air sample to be analysed by a graduated bell jar. Then, the eudiometrical tube was filled with mercury and placed upside down in a pneumatic trough. This was a glass tube of 1 inch in diameter and 7-8 inches long, closed at the upper end and with the bottom end widened. After that, a small piece of phosphorus was passed through the mercury into the tube. Once the piece of phosphorous had ascended to the top, a red-hot small piece of charcoal was approached to the outside of the tube, without touching the glass, to melt the piece of phosphorous. Blowing on the charcoal to redden it was needed. While the piece of phosphorous was melting, the air sample was bubbled in small portions inside the tube until completing the combustion. The next step consisted of warming the residual fluid - to gain accuracy in the determination - and once cold it was transferred to another graduated smaller bell jar. Finally, the amount of vital air in the respirable air sample was determined from the difference of volumes.[33]

Lavoisier mentioned tubes with similar technical features in his *Traité*, describing them as tubes 'called eudiometers by the pottery makers'[34] (Figure 6.2, left). Louis David's portrait of the Lavoisiers shows a device composed of an upside-down cylindrical glass tube immersed in a tray with mercury (Figure 6.2, right). A close-up examination of the figure reveals that the tube flares at the bottom just above the surface of the mercury in the tray. Since David's portrait was finished at the end of 1788 and Lavoisier's *Traité* was published in April 1789, the painted tube could be the same eudiometrical tube drawn in the *Traité*.[35]

---

[33] Séguin, 1791, pp. 301-302; LM, Vol. 2, pp. 151-152; Bouillon-Lagrange; 1798-1799, Vol. 1, p. 66; Fourcroy, 1805, Vol. 4, p. 278.

[34] '[...] les faïenciers qui les tiennent, les nomment eudiomètres' (Lavoisier, 1789, Vol. 2, p. 346).

Bouillon-Lagrange described this eudiometrical tube as one 'closed at its upper part and flared at its lower part' and Fourcroy as 'an elongated glass bell, flared out at the bottom'. Lacking these eudiometrical tubes, glass funnels with flame sealed nozzles were suggested as substitutes for them.

[35] It has been suggested that this cylindrical tube depicted in David's painting corresponded to Séguin's eudiometer (Capuano & Cavalchi, 2010, p. 316). This conjecture seems rather inappropriate because of Séguin developed his instrument during the year 1790 when the portrait was already delivered to the Lavoisiers. The historian Marco Beretta has pointed out that the instrument portrayed by David was

Figure 6.2 Lavoisier's eudiometrical tube.

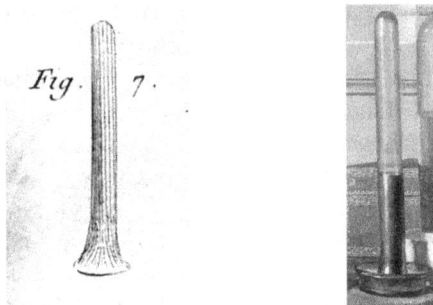

(Left) From Antoine-Laurent Lavoisier, *Traité élémentaire de chimie* (Paris, 1789), plate 5, fig. 7. (Right) Detail of Louis David's portrait of the Lavoisiers. Work released into the public domain. Current location, Metropolitan Museum of Art, New York.

Séguin himself valued his eudiometrical test for its fast execution, high accuracy, low cost and suitability for determining the composition of respirable fluids.[36] However, its main problem was still the risk of breaking the tube when heating it with the red-hot charcoal or during burning phosphorous because of the exposure of the glass to sudden changes in temperature.[37] Séguin announced a second paper, which was never written, on his phosphorous eudiometer using water instead of mercury and also showed his intention to inform with greater detail on the precautions to be taken into account to ensure greater accuracy.

There is no evidence of how Lavoisier and Séguin applied the new eudiometrical test in the respiration experiments. Probably, they first determined the proportion of vital air of the gas mixture before the experiment and then determined the volume change during the experiment. Finally, they would carry out a second eudiometrical analysis of the residual mixture.[38] During the months of April and May 1791, Lavoisier and Séguin read their second paper on respiration where, in addition to the phosphorous eudiometer that Séguin claimed as his own ('my phosphoric eudiometer'), it was manifested that the alkaline sulphide, the nitrous air

---

very likely the device invented by Felice Fontana in 1777 to show how the extinguishing of a red-hot piece of charcoal placed over the surface of mercury inside the tube caused the absorption of a great deal of air (Beretta, 2001c, pp. 37-39).

[36] Séguin, 1791, pp. 302-303; LM, Vol. 2, pp. 152-153.

[37] Fourcroy, 1805, Vol. 4, pp. 278-279.

and the Volta's eudiometers were also used. The choice of a particular test was made on both the nature of the fluid to analyse and to compare results from the different tests.[39] On 15 February 1792. Séguin read his own paper on the salubrity (quality) of the air of different locations of Paris (hospital dormitories, spectacle halls, stalls and even the National Assembly). In this case he used only the phosphorous and the alkaline sulphide eudiometers.[40] If Lavoisier's oxygen theory was the theoretical background against which Séguin's phosphorous eudiometer was developed, an alternative theory to that of Lavoisier favoured the design of another type of phosphorous eudiometer.

Berthollet and the slow combustion of phosphorous eudiometer

After Lavoisier's execution in May 1794, Guyton de Morveau, Fourcroy and Berthollet became the leading French chemists. Of the three, Claude-Louis Berthollet (1748-1822) was who better understand and continued Lavoisier's approach of chemistry as a discipline allied to experimental physics. The paradigmatic evidence of this scholarly perspective was the creation of the Société d'Arcueil.[41] This was a private scientific society set up by Berthollet and Laplace in 1806 that brought together a select group of young scientists to deal with chemical problems in association with experimental physics.

Berthollet's works in the field of applied chemistry such as, for instance, the development of a new bleaching process based on the use of chlorine, made a name for him among industrial chemists. His strict investigative methodology founded on rigorous objectivity and the need of working out hypotheses to control experiences guided him to the construction of a

[38] Holmes, 1985, p. 443. For a discussion about Séguin and Lavoisier's experimental design and the incorporation of eudiometrical tests see Prinz (2005, pp. 48-50)

[39] Lavoisier & Séguin, 1814, p. 322. Actually, Séguin had already adopted Volta's eudiometer jointly with his own in the series of experiments to confirm Lavoisier's large-scale experiment on the synthesis of water. This was before presenting his phosphorous eudiometer in the Academy:

'[...] after making several assays, both with Mr. Volta's eudiometer and with a new eudiometer, which I will have the honour to present shortly to the Academy ...]'

'[...] et après avoir fait plusieurs autres essais, tant avec l'eudiomètre de M. Volta, qu'avec un nouvel eudiomètre que j'aura l'honneur de présenter incessamment à l'Académie [...].' (Fourcroy *et al.*, 1791, Vol. 8, p. 304)

[40] Séguin, 1814, p. 253.

[41] The role of the Société d'Arcueil in the establishment of French science after Lavoisier's death was exhaustively studied by Crosland (1967).

new general theory of chemical affinities. Berthollet proposed an alternative view of chemical change to challenge the traditional conception founded on the system of elective affinities.[42] However, he might also be known much more for a famous controversy that saw him in opposition to the chemist Joseph-Louis Proust regarding the proportions in the combination of chemical substances. Berthollet's belief was contrary to that of the fixed proportions of combination that Proust claimed to have proved. Proust's point of view became more successful and was finally adopted by chemists.

Berthollet's concern with the atmospheric air in its intellectual context

In October 1794, at the beginning of the French revolutionary period, the Comité d'Instruction Publique proposed to the Convention Nationale the creation in Paris of the École Normale to organize courses for instructing people who would become in turn teachers in their own departmental normal schools to train school teachers. Berthollet was in charge of the chemistry course of this École Normale.[43] The tenth lesson of that course (given on 21 April 1795) was devoted to the properties of the atmosphere and included some reflections on different eudiometrical tests. The introductory paragraphs of this lesson give a clue to understanding Berthollet's interest in the study of atmospheric air. His concern was far from any naturalistic approach relating the salubrity of air with its oxygen content. He envisaged the chemical action of the atmospheric air over most of the substances of the surface of the earth as being balanced by the physical conditions that accompanied the atmosphere. This meant admitting that the chemical action of atmospheric air varied with temperature and pressure. For Berthollet, the convergence of the physical qualities with the chemical properties of the atmosphere was bringing closer physics and chemistry, which were responsible for explaining them, to the point that it was difficult to draw a demarcation line between both disciplines.[44] It is in the boundaries of the territories of chemistry and physics, reflecting the alliance between both disciplines, where should be

---

[42] For a reassessment of Berthollet's conception of chemical change in its complex context see Grapí & Izquierdo (1997)

[43] These courses began on 20 January 1795 and ended don 19 May of the same year. For an overview and vicissitudes of Berthollet's chemistry course in its educational context, both as a pedagogical experience and as a part of the scientific creation scenery of his chemical affinities, see Grapí (2014)

[44] Berthollet, 1795, Vol. 5, pp. 67-68.

placed Berthollet's inquiries on the composition of atmospheric air and the methods to determine it.

Berthollet's Egyptian venture in the background of a new eudiometrical approach

Regarding the eudiometrical methods, Berthollet praised Scheele's test based on the wet paste of sulphur with iron filings as his favourite eudiometrical means. He argued that, first, the test indicated the real proportion of oxygen in the air since the diminution in volume could be attributed to oxygen only and, secondly, the absorption of oxygen could be considered as complete. However, there were some setbacks such as the slowness of the assay and the release of ammonium. At the same time, he considered the tests of Achard, Reboul and Séguin as expeditious and accurate enough as long as the air mixture had a high proportion of oxygen. According to Berthollet, the main disadvantages of these phosphorus eudiometers were their difficulty of handling, the risk of breaking due to sudden increases in temperature and the uncertainness of the endpoint of the test. This latter issue was due to the fact that phosphorous stopped burning not by the exhaustion of oxygen but because of the decrease of its proportion in the air sample. Probably for this reason, Berthollet believed that Scheele's test based on the wet paste of sulphur with iron filings was the best available method for that time.[45] On the contrary, he did not show any favour to the nitrous air eudiometers and was a little more enthusiastic with the Volta's eudiometer.

At the end May 1796, Berthollet was commissioned by the Directory to follow the French army in Italy to organize on a large scale the removal of works of art and objects of scientific value to enlarge the collections of the French museums. This Italian campaign was the beginning of a close friendship between Berthollet and Napoleon Bonaparte who was the commander of the military operation. As a result of their relationship, Bonaparte entrusted Berthollet with the organization of the scientific commission that was to accompany him in his expedition to Egypt (1798-1799), a venture undertaken for both foreign and domestic reasons. This expedition combined military and scientific purposes, and to assist the latter ones Bonaparte created in August 1798 the Institute of Egypt of

---

[45] Ibid., pp. 73-74. Later, in his *Essai de statique chimique* of 1803, he found other disadvantages such as the release of hydrogen sulphide and ammonium. He attributed to this release the fact that Scheele had found a significantly high value (27%) for the concentration of oxygen in the air (Berthollet, 1803b, Vol. 1, p. 512).

Cairo, largely founded on the model of the Institute of France. Berthollet, apart from his research on dyestuffs, he carried out important chemical work on eudiometry and, above all, chemical affinities in the chemistry laboratory he had constructed in the rooms of the Institute. He returned to Paris in October of 1799.[46]

While working on eudiometrical procedures in Cairo, he positioned in relation to the alkaline sulphides test.[47] According to Berthollet, the test allowed to determine the proportion of oxygen in an air sample with all the precision that could reasonably be expected at that time. He found the test advantageous for two reasons. First, the assay indicated the real proportion of oxygen because he believed that the observed contraction in volume could be mainly assigned to the absorption of oxygen.[48] The second reason had to do with the fact that this absorption could be regarded as completed since the alkaline sulphide solution was sufficiently concentrated. On the other hand, he also noticed some drawbacks such as that the determination of nitrogen was not entirely accurate since it dissolved partly, although the error was considered very small; the need to introduce corrections when there were differences in pressure and temperature before and after the test; the execution time of the test that could take some days, especially at low temperature, and the detection of the endpoint of the test (i.e. the termination of the contraction in volume) that still required further time.[49]

Göttling's alternative view to Lavoisier's theory of oxygen

During his stay in Egypt, Berthollet was involved in the development of a new version of the phosphorous eudiometer. Actually, his researches on

---

[46] Crosland, 1967, p. 239; Goupil, 1977, p. 48; Grapí & Izquierdo, 1997, pp. 123-125.

[47] It is worth noting that Berthollet could read Lavoisier's paper on the alkaline sulphides as eudiometrical means at an early date, after the printing of the Lavoisier's *Mémoires* in 1793 (Beretta, 2001b, p. 329).

[48] In other eudiometrical tests, such as that of nitrous gas or Volta's test the diminution in volume was apportioned between oxygen and the nitrous or hydrogen gases.

[49] Berthollet, 1800, (a) pp. 76-78; (b) pp. 286-288; 1803, Vol.1, p. 513. Berthollets's paper on eudiometrical observations was published in the *Annales de chimie* and in the *Mémoires sur l'Égypte publiés pendant les campagnes du general Bonaparte*. The publication of the *Annales* reproduces practically the paper published in the *Mémoires sur l'Égypte*. The editors of the *Annales* reported their intention of reproducing the entire paper published in the *Mémoires sur l'Egypte* because it deserved to be communicated to all chemists.

this topic had begun when he was aware of the observations that a few German chemists had presented in 1795 on the action of phosphorus over nitrogen. The primarily responsible of these observations was Johann-Friedrich-August Göttling (1753-1809), who was professor of chemistry at the University of Jena since 1785 and the editor of the journal *Taschenbuch für scheidekünstler und apotheker*. In 1794 Göttling presented an alternative theory to that of Lavoisier from certain observations on the combustion of phosphorus, in the first part of his work *Contribution to the Correction of Antiphlogistic Chemistry Based on Experiments*.[50] Specifically, on the basis of existing experiments he intended to confirm that phosphorus was unable to emit light if it was immersed in pure oxygen whereas it glowed in impure oxygen. On the contrary, phosphorus exhibited good brightness immersed in pure nitrogen even better than in common air. Thus, nitrogen came to be a true light generator while oxygen, which was unable to make glow phosphorus, was enabled to this effect when mixed with nitrogen. Göttling also presented these ideas in 1795 in the article *On the Luminescence of Phosphorus in Nitrogen* (Göttling, 1795) and his results were confirmed by others, especially August-Wilhelm Lampadius and Johann-Friedrich Lempe, who added that the burning of phosphorus in common air left a residue of oxygen.[51]

From their experiments Göttling determined, first, that caloric and light were two different substances. Second, oxygen and nitrogen were compounds of the same base combined with caloric and light, respectively. Therefore, the two components of common air did not differ in their bases but in their surrounding atmospheres, which was exactly the opposite of Lavoisier's theory of oxygen. In this way, it was possible to transform oxygen into nitrogen. Third, simple bodies such as metals, sulphur or phosphorus were anything but compounds of a specific base with light. Finally, fire was a combination of light and caloric.

The combustion of phosphorus could be explained by the mechanism of the double affinities. The base of phosphorus linked with the base of oxygen, which was the same as that of nitrogen, whereas the caloric from oxygen linked with the light from phosphorus, yielding fire. Why phosphorus glowed when immersed in nitrogen but not in oxygen? Göttling argued that at a low temperature, the affinity of the base of phosphorous for its surrounding light was higher than the affinity for the base of oxygen and, simultaneously, the affinity of the base of oxygen for its surrounding caloric

---

[50] Göttling, 1794.

[51] Abbri, 1996, pp. 6-7.

was also higher than the affinity for the base of phosphorus. However, phosphorus glowed immersed in nitrogen, without increasing significantly the temperature, because the base of phosphorus linked to that of nitrogen, which was the same as that of oxygen, producing an acid. As a result, a burst of light coming from phosphorus and nitrogen was emitted. In this phenomenon, no caloric could be released since it lacked both nitrogen and phosphorus. Consequently, the temperature could not increase and there was not at all a true combustion. Phosphorus also glowed when it was in contact with common air thanks to the decomposition of atmospheric nitrogen. But as oxygen also decomposed, caloric was released. Since the affinity of light for the base of oxygen was higher than the affinity of caloric for the same base, Göttling argued that oxygen could be converted into nitrogen, which was observed in some experiments, because of the decomposition of oxygen by light.[52]

Berthollet's approach to the slow combustion of phosphorous as a eudiometrical test.

Berthollet was one of the first chemists interested in repeating and studying Göttling's experiments. He presented his results in a paper on the eudiometrical properties of phosphorus read on 26 January 1796 at the *Institut de France,* four months before leaving to Italy. He noted that when a sample of phosphorus was introduced into the gas nitrogen it glowed in the darkness and a white smoke formed in the daylight.[53] This phenomenon did not last too much and it was less intense than in atmospheric air where it lasted until the total absorption of oxygen. The glow was recovering as fresh oxygen was introduced in the residual air. At low temperature, neither glow nor diminution in volume was observed when the same sample of phosphorus was immersed in pure oxygen.

Berthollet explained these findings, arguing that nitrogen dissolved phosphorus but, on the contrary, at a low temperature phosphorus was not immediately soluble in oxygen. This slow combustion of phosphorus in atmospheric air was due to, first, the dissolution of phosphorus in nitrogen and, second, the simultaneous action of oxygen on this gaseous mixture. Thus, when oxygen was bubbled into nitrogen, with the presence of phosphorus, light glowed until phosphorus had absorbed all the oxygen

---

[52] Fourcroy, 1796, p. 570.

[53] Berthollet had obtained nitrogen by the decomposition of ammonia with chlorine. In a basic medium the following reaction occurs:

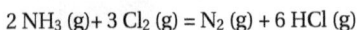

$2\,NH_3\,(g) + 3\,Cl_2\,(g) = N_2\,(g) + 6\,HCl\,(g)$

added. The higher the proportion of nitrogen, the more intense the glow. If the proportion of oxygen was high the glow was initially weak, and it steadily increased until phosphorus melted and the hot combustion happened. Spontaneous ignition only occurred at the melting point of phosphorus.

However, Berthollet's interest in these observations was not only focused on refuting Göttling's ideas, contrary to the new chemistry, but also on exploring the feasibility of a new eudiometrical test based on this slow combustion of phosphorus.[54] Moreover, a phenomenon like the chemiluminescence of phosphorus should be of interest to Berthollet given his understanding of chemistry as a discipline allied to the experimental physics. Berthollet considered the slow combustion of phosphorus as a eudiometrical means equal or superior to those known hitherto as long as the composition of the gaseous mixture was not too different from that of the atmospheric air. Berthollet attributed the glow of phosphorus when immersed in nitrogen both to the existence of a small portion of residual oxygen coming from the production of nitrogen and to the water through which the gas had flowed. Thus, he expressed the need for the existence of oxygen as well as of a certain degree of humidity for a successful slow combustion of phosphorus.[55]

When Berthollet presented this research on the slow combustion of phosphorus, the chemist Bertrand Pelletier (1761-1797) was the person who had worked the most with phosphorus and its compounds in France, publishing five paper on the issue between 1785 and 1792. In February

---

[54] At room temperature, the waxy white phosphorus undergoes a slow or cold combustion giving off faint greenish glow. This glow spreads away from the surface of phosphorus and is associated with the peculiar smell of ozone. The term "phosphorescence" is often wrongly associated with this slow phosphorus oxidation. Phosphorescence, like fluorescence, is a strictly physical phenomenon. "Phosphorescence" in its scientific usage now refers to "photochemiluminescence", while the glow observed in the case of phosphorus is called "chemiluminescence". The surface of phosphorus is extremely sensitive to the action of traces of impurity and the moisture appears to be the primary cause of the phenomenon. Dry oxygen with low moisture content shows a very feeble glow but if oxygen is very dry the oxidation proceeds very slowly, and probably if the system were entirely free from water no glow would take place. The fascination that "phosphorescence" exerted in late seventeenth century experimental philosophers fostered the phenomenon exploitation in public demonstrations. This way, phosphorus became a resource for extending the public appreciation of the new experimental philosophy, as well as being criticised by those who opposed the public status of the new science (Golinski, 1989, p. 11)

[55] Berthollet, 1796, pp. 275-276; Fourcroy, 1796, p. 571.

1792 he presented the last paper where he explained that the manuscript of his first paper, presented at the Academy in 1785, had an attachment concerning certain changes on Sage's method for preparing phosphoric acid that were not published at all. The fact that phosphoric acid was beginning to be used in medicine at that time [56] made Pelletier decide to publish those comments in the paper of 1792, which referred to the insensitive combustion of phosphorus (*combustion insensible du phosphore*) accelerated by moist air.[57] Berthollet and Pelletier had met at various institutions such as the Bureau de Consultation des Arts et Métiers, the École Polytechnique, the editorial board of the *Annales de chimie* and in the chemistry section, first, at the Académie des Sciences and later, at the Première Classe of the Institute de France. In addition, Berthollet reviewed Pelletier's second paper of 1788 on phosphorus and metallic combinations in the *Annales de chimie*. All that can be of help to better understand and contextualize Berthollet's knowledge on the slow combustion of phosphorus.

Berthollet's description of his new eudiometrical test based on this phenomenon has been reconstructed from various sources since none in particular provide a fairly accurate description of the procedure. The first of these sources is the paper *Observations sur les propriétés eudiomètriques du phosphore*.[58] As already mentioned, the analysis of atmospheric air constituted, along with chemical affinities, one of his lines of research at the Institute d'Égypte. There, Berthollet renewed his commitment to the slow combustion of phosphorus as eudiometrical means. His research on this topic was published in the *Observations eudiomètriques*.[59] This paper and the article *Chimie* written by Fourcroy for the *Encyclopédie Méthodique* and the corresponding description by Bouillon-Lagrange are the other sources that have allowed reconstructing the eudiometrical procedure grounded on the slow combustion of phosphorus.[60]

---

[56] Phosphoric acid began to be prescribed as an antiseptic and to remove external lumps. On the other hand, it was used to prepare some salts administered as laxatives.

[57] Pelletier, 1798, Vol. 2, pp. 138-140.

[58] Berthollet, 1796, p. 276.

[59] Ibid., 1800: (a) pp. 76-78; (b) pp. 286-288.

[60] Fourcroy, 1796, p. 571; Bouillon-Lagrange, 1788-1789, Vol. 1, pp. 65-66; 1801, Vol.1, pp. 212-213.

A stick of phosphorus was fixed in a glass stem with a pedestal. This stick should be shorter than a graduated cylindrical glass tube with the bottom end widened of 40 cm in length and 1 cm in diameter (Figure 6.3). After noting the barometer and thermometer readings, the air sample was collected in the graduated tube placed vertically in a vessel with water. After measuring the volume occupied by the air sample, the tube was covered with the thumb, transferred to a larger vessel with water and then uncovered.

Afterwards, the stick of phosphorus fixed in the glass stem was introduced in the graduated tube by passing it through water. This way, the stick of phosphorus once slightly moistened began its slow combustion in contact with the air sample. A cloud was first formed, but it disappeared at the end of the operation. Once the slow combustion was completed, the tube was covered again with the thumb, transferred to the first small vessel with water, uncovered and eventually immersed until to level the surface of water inside the tube with its external surface in the vessel. This latter operation ensured a right measure of the volume of the residual air.

Figure 6.3 Eudiometrical device based on the slow combustion of phosphorous.

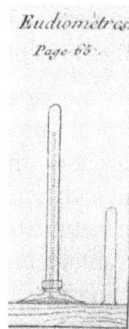

From Edme J.B. Bouillon-Lagrange, *Manuel d'un cours de chimie* (Paris, 1788-1789), Vol.1, plate 6.

Berthollet ensured that, using a narrow tube, the assay could be finished in two hours. Moreover, according to Fourcroy the test could be accomplished in less than an hour. The phosphorus dissolved in nitrogen could increase the residual air volume altering therefore the outcome of the test. The test was very sensitive to heat, at a low temperature, applying the heat from the hand to the glass tube might accelerate the process. However, too high a temperature could ignite the phosphorus and melt it.

It is worth noting that the experimental device and procedure bore some resemblance to Scheele's test based on the wet paste of sulphur with iron filings so admired by Berthollet. The test based on the slow combustion of phosphorus had two important advantages: a true indication of the endpoint of the test, with the vanishing of the glowing cloud, and its duration (a minimum of 2 hours and a maximum of 6 to 8 hours). As a counterpart, two corrections should be introduced. The first one was due to the differences in pressure and temperature before and after the assay. The second correction was attributable to the presence of fumes of unburned phosphorus that increased the final residual volume. Berthollet estimated that this extra volume should represent 1/40 of the total volume to determine the proportion of oxygen with the same accuracy as with the alkaline sulphides method. According to Berthollet, his method combined good accuracy with a reasonable runtime. Using both the alkaline sulphides as the phosphorus tests and after making the corresponding corrections, Berthollet found that the proportion of oxygen in the air of Egypt as well as in that of Paris was no more than 22 %. Berthollet showed his discomfort because these two types of eudiometrical tests, which provided uniform and consistent results and that could not be withdrawn on an experimental basis, had not been sufficiently taken into consideration in front of other tests - such as the nitrous gas test - which provided no comparable results in being dependent on too many factors.[61] In 1803, Berthollet continued appreciating in the same terms the virtues of the method based on the slow combustion of phosphorus.[62]

Bouillon-Lagrange also praised the virtues of the slow combustion of phosphorous as a eudiometrical procedure. He considered this eudiometrical device as more accurate than that of the nitrous air and Volta's and less cumbersome and with a more rapid response than that of the potassium sulphide and the wet mixture of sulphur with iron filings. The test had the additional advantage of presenting a precise endpoint when the air surrounding the phosphorus stick no longer glowed and became transparent.[63] A similar preference for the slow combustion of phosphorus as a eudiometrical test came from the periphery of Eastern Europe in 1800. This was the contribution of the Baltic physicist and chemist Georges-Frédéric Parrot.

---

[61] Berthollet, 1800, (a) pp. 80-85; (b) pp. 291-294.

[62] Ibid., 1803b, Vol.1, p. 514.

[63] Bouillon-Lagrange, 1801, Vol.1, pp. 212-213.

The oxygenometre of Georges-Frédéric Parrot

In the early nineteenth century, far from the main European scientific production centres, in the Baltic provinces of imperial Russia, Georges-Frédéric Parrot (1767-1852) had begun from 1798 an active research on the use of phosphorus as eudiometrical means. Parrot was born in Montbéliard[64] when it was still under the control of the Dukes of Württemberg. He studied physics and mathematics at the Hohe Karlsschuhle in Stuttgart and from 1786 until 1788 he was engaged as a private tutor in Normandy. After returning to Germany, he left for the Baltic provinces[65] in 1795 as a private tutor in the family of a liberal Baltic German nobleman in Riga, the capital of Livonia, which at that time was a Baltic province of imperial Russia.[66] Parrot was impressed by the region's backwardness in comparison with his experiences in Western Europe. During those early years in Riga, Parrot suffered true intellectual and political isolation due to the lack of a university in the region since the Academia Gustaviana - founded by King Gustaf II Adolf of Sweden in 1632 [67] - had ceased to exist 1710 when Tsar Peter the Great conquered the Baltic provinces.

In 1796 Tsar Paul II succeeded his mother Catherine the Great to the throne of Russia and decided to cut off Russia from what he considered the destabilizing and revolutionary influence of Western Europe, particularly France. To prevent his subjects studying abroad to be acquainted with pernicious western ideas, they were prohibited from leaving the country. And to compensate the German Baltic nobility whose sons had hitherto studied in Germany, he planned to establish a new Lutheran university in Mitau. Parrot was appointed professor of this university but in 1801 the Tsar was assassinated and his successor Alexander I, more open-minded to European ideas, changed the location of the new university to Tartu (Dorpat). Parrot was initially appointed to the chair of pure and applied mathematics, and later to the chair of physics. He was the first rector of the

---

[64] A small French principality since 1806 located near the junction of today's French, German and Swiss borders.

[65] This was the case of many other inhabitants of Montbéliard who had moved to Russia benefiting from the marital links between the tsar Paul I and the court of Stuttgart (Langins, 2004, p. 298)

[66] Livonia remained within the Russian Empire until the end of the First World War, when it was divided between the newly independent states of Latvia and Estonia.

[67] After the 1626–1629 Polish–Swedish War Livonia became a very important Swedish dominion until 1721.

re-founded university from 1802 until 1816 when he was elected member of the Academy of Sciences of Petersburgh.[68]

In 1798, Parrot had already made determinations on the air quality of the asylum of Riga using the test of the slow combustion of phosphorous and in 1800 he published his first paper in this regard that served as a calling card of his own eudiometer, *On the Eudiometrical Characteristics of Phosphorus along with a Description of a Proper Phosphorus Eudiometer.*[69] Parrot's economic situation did not favour the acquisition of an exemplar of Fontana-Ingenhousz's eudiometer and, besides, the lack of consensus existed on the results provided by this instrument was not a good incentive to buy it. The device based on the slow combustion of phosphorus resulted in being the most convincing for Parrot, because of its low cost as well as for its accuracy and usefulness. He echoed the bad reputation attributed to this procedure because of Humboldt's criticism that Parrot called on him to rectify, as he did later.[70] Parrot managed to build up to five eudiometers during those two years, but only two (one large and another small) were described in his paper (Appendix VI). He also proposed the name "oxygenometre" as a more appropriate term to describe his instrument since the oxygen content was not the only indication for the quality of atmospheric air.[71]

Parrot had carried out his first eudiometrical experiments while Berthollet in Cairo was engaged with the development of his own eudiometrical test grounded also on the slow combustion of phosphorus. However, Berthollet had been interested in this phenomenon not only regarding his eudiometrical researches but also, and primarily, for rebutting Göttling's ideas contrary to Lavoisier's oxygen chemistry. Similarly, the work of the Italian naturalist Lazzaro Spallanzani centred in disproving Göttling's experiments on the chemiluminescence exhibited by phosphorous, led him to recover Giobert's portable phosphorous eudiometer in a slightly different version.

---

[68] Müürsepp, 2013, pp. 16-20.

[69] Parrot, 1800.

[70] Parrot, 1800, p. 154. See chapter 7; Nitrous gas test vs. Phosphorous test.

[71] The original term "sauerstoffmesser" coined by Parrot has been translated by "oxygenometer" according to Partington (1961-1970, Vol. 3, p. 325). Accum followed this terminological approach by proposing that eudiometers, in general, should be more properly called "oxymeters" (Accum, 1803, Vol. 1, p. 225).

Lazzaro Spallanzani. Between animal respiration and the updating of Giobert's eudiometer

Göttling´s interpretation of the phenomenon of the phosphorus chemiluminescence had a significant reception in Italy, being supported by leading chemists such as Louis-Valentino Brugnatelli. In 1795, when Göttling ideas had been already published, the Italian naturalist Lazzaro Spallanzani, who was busy at work in his researches on animal and plant respiration, became also involved in refuting experimentally Göttling's interpretation of the emission of light from phosphorus during its combustion in different gases. It was in 1796 when he published in Modena his *Chimico Esame (Chemical analysis of the experiments of Mr. Göttling, professor at Jena over the light of Kunkel's phosphorus observed in the common air and in several aerial permanent fluids)*, advocating for the chemistry of oxygen to interpret both the combustion of phosphorus and its consequent emission of light and heat.

Lazzaro Spallanzani (1729–1799) obtained a doctorate in philosophy and embraced the ecclesiastical state after studying law at the University of Bologna. In 1754 he was appointed professor of logic, metaphysics and Greek at the University of Reggio Emilia and in 1768 accepted the chair of natural history in the University of Pavia. He was an indefatigable traveller. In the year 1785, after visiting Turkey, he stayed in the Cyclades islands and in the strait of Bosphorus to study marine life, visiting mines and collecting geological findings. In 1788 he carried out a scientific expedition in southern Italy, where he deepened his interest in volcanology.

Spallanzani had begun his studies on respiration in January 1795 and concluded them in February 1799, the same year of his death. His approach to the study of respiration was due largely to the development of his research on bats carried out in the period 1793-1794 and that led him to continue more systematic investigations following the chain of living beings. The context of these enquiries was not alien to Lavoisier and Séguin's studies on animal respiration and to the debate about the location of respiration in the body of animals. Lavoisier and Séguin located respiration in the lungs whereas for others such as Girtanner, Hassenfratz, Lagrange and the same Spallanzani respiration was not exclusively localized in the lungs but, rather, in the blood flow circulating throughout the body.[72] It is not easy to place neither Spallanzani's first approaches to chemistry nor his acceptance of Lavoisier's chemical theory. Probably, his journeys to Constantinople and Sicily, as well as

---

[72] Ciardi, 2010, pp. 10-11.

Lavoisier's researches on animal respiration might have contributed significantly.[73]

Spallanzani's portable phosphorous eudiometer. Redesigning Giobert's apparatus

Spallanzani focused his study on the combustion of phosphorus at the end of 1795, and he decided to use a somewhat different version of Giobert's eudiometer for his researches since he considered unreliable Göttling's experimental device. This study was not the first Spallanzani's contact with eudiometers. From 1785, during his stay in Constantinople, he had already used the nitrous air eudiometer among others.[74] Spallanzani's version of Giobert's eudiometer was not only useful to him to contrast Göttling's alternative explanations to Lavoisier's chemistry but also proved to be a centrepiece in his studies on animal respiration. Therefore, the implementation of Spallanzani's eudiometer should be placed both in the context of the review of Göttling's theories as in the understanding of the phenomenon of respiration.[75] Figure 6.4 depicts the plate of the *Chimico Esame* showing Spallanzani's eudiometer (*fig. II^a*) as a development of that of Giobert (*fig. I^a*). This instrumental evolution can be better appreciated in Spallanzani's hand drawings of his manuscript *Opuscolo sui fenomeni del fosforo* (*Opuscle on Phosphorus Phenomena*) (Figure 6.5).

Figure 6.4 Spallanzani's phosphorous eudiometer.

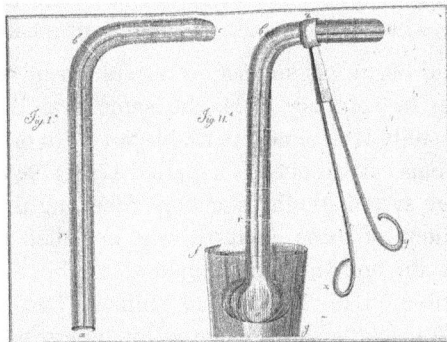

From Lazzaro Spallanzani, *Chimico Esame degli esperimenti del Sig. Göttling* (Modena, 1796).

---

[73] Beretta, 1989, pp. 131-133; Abbri, 1996, pp. 4-5.

[74] Manzini, 1996, p. 60; Capuano *et al,* 2010, p. 324.

[75] Ibid., 2010, p. 303.

Figure 6.5 Sketch drawings of Giobert (Left) and Spallanzani's (Right) phosphorous eudiometers.

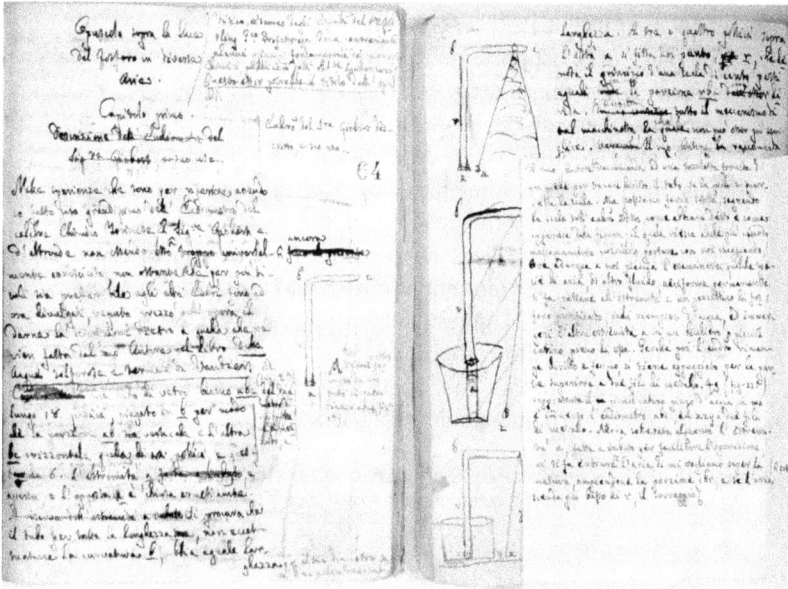

From *Manoscritti di Lazzaro Spallanzani*, Mss. Regg. B148, 64r and 64v, [1790-1796], Biblioteca Panizzi, Reggio Emilia, Italy.

The core of the eudiometer was the glass tube (abc) of a half inch in diameter and 18 inches long bent at a right angle at (b), so that the vertical part (ab) was 12 inches long and the horizontal part (bc) was 6 inches long. The lower end (a) was funnel-shaped, and the opposite end (c) was sealed airtight. At 3-4 inch above the lower end (a), the signal (r) indicated the zero point of a scale (rbc) engraved on the glass and divided into 100 equal parts. It was needed a uniform inner width over the tube length. The tube was hold by a wire tongs (xyz) to keep it firmly standing, leaving the end (a) slightly raised to ease the operations and with a view to making the device portable and easier to use. (Figure 6.6 shows a modern replica of Spallanzani's eudiometer).

The test was executed placing a piece of phosphorus at the end (c) of the tube, which was then filled with water keeping the piece of phosphorus at the bottom. Afterwards, the open end (a) of the tube was submerged in the vessel (fg) with water and the air sample to be examined was passed through the water into the tube so filling the part (rbc). In case of exceeding the lower mark (r), the excess of air was removed sucking it through a curved thin tube introduced into the eudiometer until the water level was at that mark. After that, the end (c) was heated approaching the flame of a burning paper underneath.

The piece of phosphorus started burning and smoking until the oxygen inside the eudiometer was exhausted. First, the water descended beneath the mark (r) due to the expansion of the air. Then, after cooling, water rose inside the tube replacing the consumed oxygen, the base of which was absorbed by phosphorus. Finally, the reading of the level of water in the graduated scale provided the percentage of oxygen in the air sample. This way the amount of nitrogen in the air sample could be also determined. The residual air was mainly nitrogen, but it was often mixed with carbonic acid gas that should be removed and quantified. To this end, Spallanzani proposed to follow the same procedure pointed out by Giobert.[76]

Figure 6.6 Modern replica of Spallanzani's eudiometer exhibited at the Centro Studi Lazzaro Spallanzani, Scandiano.

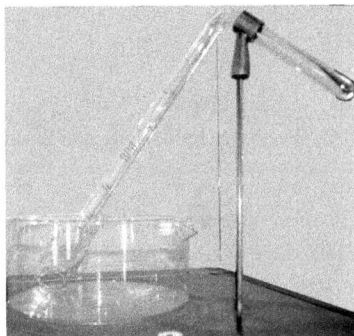

Courtesy of Centro Studi Lazzaro Spallanzani, photography by the author.

Spallanzani made notice Giobert that his analysis of the air of Pavia had given proportions of oxygen between 19% - 21%, lower than those obtained by Giobert with the air of Vaudier. Giobert argued that he had been able to get better results - between 20% and 22% oxygen - using glass tubes of uniform section graduated with mercury.[77] This is a feature showing how instrument reliability meant to use heat resistant glassware but also fine handcrafted. The series of experiments performed by Spallanzani helped him to address the important choice of carrying out

[76] Spallanzani, 1796 (a), pp. 6-9; (b), pp. 352- 354.

[77] Davoli, 1996, p. 53.

the burning of phosphorus over water or mercury. If mercury was used instead of water remarkable differences appeared: the combustion of phosphorus was less intense, and the test required one or more days to be completed. If mercury and the air sample were previously dried, the test could be completed in three or four days, with a weak emission of light and without producing smoke.[78]

As in the case of Giobert, Spallanzani provided his eudiometer with the additional function of measuring the proportion of carbonic gas in the air sample. Neither Gioberti nor Spallanzani was interested in eudiometry as a measure of the goodness of the air. However, it was well known that the carbonic gas released from fermentations, marshy waters or underground exploitation was unhealthy. In this sense, therefore, to enable the phosphorus eudiometer to determine the amount of carbonic gas supposed to somehow give back the instrument to its original idiosyncrasy of measuring the salubrity of the air. To determine the content of carbonic gas in exhaled air, Spallanzani had to modify the experimental device of his eudiometer replacing water by mercury due to the solubility of carbonic gas in water.

Spallanzani also used his phosphorus eudiometer to explore Vauquelin's proposal of using a certain type of slugs and snails as eudiometrical means, believing that they could separate oxygen from nitrogen in the air during respiration.[79] To analyse the expired air by snails, a sample of this kind of air was introduced in the eudiometrical tube full of mercury until measuring 100 parts. Then, the sample was transferred to a bowl with limewater to absorb the carbonic gas. After that, the remaining air sample was transferred back to the eudiometer to observe the new level of the mercury, which indicated the percentage of carbonic gas in the air exhaled by snails. When this latter sample was analysed with the phosphorus eudiometer, only slight air absorption was detected, which denoted that snails had not exhausted all the available oxygen in the air. Eventually, he dismissed this proposal when observing that these beings did not fully absorb oxygen from the air. In this case, as in others in which the eudiometrical test of phosphorus was not sensitive enough to detect low proportions of oxygen, Spallanzani used alternative tests such as the one with the nitrous air.[80]

---

[78] Spallanzani, 1796a, pp. 26-27.

[79] Vauquelin, 1792, pp. 290-291.

[80] Spallanzani, 1803 (a); Vol.1, pp. 18-19, 35, 41-43, 143-147; 1803 (b), pp. 121-123, 146-147, 154-155; 2010, pp. 79, 85-86, 88-89; pp. 183, 190-191, 193. Between October 1798 and early February 1799, Spallanzani was involved in the publication of three

Final observations on the combustion and chemiluminescence of phosphorus

Spallanzani carried out intensive and extensive research on the chemiluminescence and combustion of phosphorus in different gaseous media such as common air, nitrogen, hydrogen, carbon dioxide and oxygen. Regarding the eudiometrical assay, his observations on these phenomena in atmospheric air are especially appropriate. These observations not only reminded Giobert's warnings on the eudiometrical test but also provided some conclusive results about the slow combustion of phosphorus that allowed a better understanding of its suitability as a basis for a eudiometrical test.

Spallanzani noted that it was not necessary to heat the piece of phosphorous with a flame as long as the air temperature was not too low. Thus, in the darkness, at a low temperature (about 20 ºC) a bright halo around phosphorus and an increase of temperature could be observed. In daylight, the formation of a whitish cloud or smoke was attributed to the decomposition of phosphorus and the subsequent formation of phosphorous acid. At lower temperatures (about 15 ºC), these phenomena could hardly be observed until becoming non-existent (below 12 ºC). However, if the temperature rose by holding the end of the horizontal part of the tube - containing the piece of phosphorus - with the fingers, a slight glow and smoke recovered momentarily but faded away. At higher temperatures (between 27-38 ºC) the burning of phosphorus was very intense as well as the emission of light and smoke. In these circumstances the assay could be completed in less than six hours while at lower temperatures its duration extended. Furthermore, humidity favoured the combustion of phosphorus, which could give chemiluminescence without releasing fumes (i.e. without igniting) and vice versa. This emission of light depended on the presence of oxygen as well as on a sufficiently high

---

papers on animal respiration worked out from his researches of the last four years of his life (he died on 12 February 1799). These papers were published four years later in two Italian and French editions (Spallanzani, 1803). For the vicissitudes of this publication see Stefani (2010)

temperature. Finally, Spallanzani was able to discard the emblematic tenet of Göttling's theory that light was able to transform oxygen into nitrogen.[81]

The limits of error of the eudiometers based on the combustion of phosphorous were not an issue of minor importance. Guyton de Morveau, for instance, had already observed that the burning of phosphorus indicated an amount of oxygen in common air lesser than that given by the nitrous gas test.[82] Göttling also confirmed this suspicion in 1797 using a modified version of Reboul's eudiometer with an elongated graduated tube. Göttling's researches on this issue had a discernible influence in the future chemical carrier and eudiometrical pursuits of the naturalist Alexander von Humboldt after their encounter in Jena in the spring of 1797.

Summary

Spallanzani brought Giobert's minimalist design of his phosphorous portable eudiometer to the limit, replacing the wooden board of Giobert's eudiometer by simple wire tongs, which improved even more its portability. The resemblance of Spallanzani's eudiometer with that of Giobert was extensible to the measurement of the proportion of carbonic gas in an air sample. This functionality can be interpreted as revisiting early beliefs of eudiometers as instruments to ascertain the salubrity of the air.

Spallanzani's updating of Giobert's eudiometer in 1796 should be placed both in the context of a theoretical dispute, the refusal of Göttling's theories, and the comprehension of animal and plant respiration. Spallanzani developed his eudiometer to disprove experimentally Göttling's interpretation of the light emission from phosphorus during its slow combustion in different gases (i.e. phosphorous chemiluminescence). Göttling had suggested an alternative theory to that of Lavoisier, and he particularly sought to confirm that phosphorus was unable to emit light when immersed in pure oxygen whereas it radiated some light if oxygen was not totally pure. In contrast, phosphorus exhibited a good light emission when immersed in pure nitrogen, even better than in atmospheric air.

---

[81] Spallanzani, 1796a, pp. 11-12, 25-27, 143-144; 1796b, 356; Abbri, 1996, pp. 10-15. At about 19 °C the test lasted 16 hours and at about 11 °C 30 hours. Spallanzani noted that at about 6 °C there was no absorption of oxygen by phosphorus, against the opinion of those chemists who claimed that phosphorus ignited at any temperature.

[82] Guyton, 1789, p. 709.

Animal respiration was also the context in which Séguin developed, during the year 1790, his bench phosphorous eudiometer. Although undoubtedly he must be credited for designing the instrument, Lavoisier's preliminary experiments on the combustion of phosphorous and on animal respiration played also a significant part in this project. The triggering point for the development of that phosphorous eudiometer was Lavoisier's belief that fixed air could no longer be the only product of animal respiration, but also water was exhaled. Therefore, measuring the rate of respiration based on oxygen intake needed to determine the amount of oxygen in both inhaled and exhaled air, and with this end in mind Séguin worked out his phosphorous eudiometer. The unexpected discovery of the permanent ignition of a piece of phosphorous while bubbling uninterruptedly an air sample into a glass jar, encouraged Lavoisier and Séguin to be confident with phosphorous as their preferable eudiometrical option.

A controversial issue in early phosphorous eudiometers had been the choice of water or mercury as the medium upon which the burning of phosphorous was carried out. Apart from the exceedingly different costs of both liquids there also existed theoretical matters at stake. According to an early interpretation of combustion in terms of the theory of phlogiston, during the burning of materials, such as a piece of phosphorous, phlogiston was released from phosphorous saturating the atmospheric air whereas certain hypothetical fixed air flowed from the common air. Thus, if combustion was carried out upon water, this absorbed the fixed air, thereby causing a contraction in volume of the residual aerial sample. However, from June 1773 onwards, when Lavoisier began to be persuaded that no fixed air dropped from the common air during combustion, experiments involving combustion were preferably performed over mercury in the pneumatic trough. This event, as already occurred in the case of Achard's phosphorous eudiometer, is closely akin to the interplay between a theoretical framework and an experimental design. Séguin was working as Lavoisier's assistant in his laboratory at the Arsenal. Therefore it is perfectly understandable that he used mercury, and not water, in the making of his eudiometer. Séguin and Spallanzani's phosphorous eudiometers were mainly involved in researches on the physiology of animal respiration, although they were also used to ascertain the quality of the air in different kinds of crowded spaces as well as in open locations.

The determination of the composition and quality of the air was also the principal utility of the eudiometers based on the slow combustion of phosphorous. Berthollet was the main driver of this kind of eudiometer developed between 1796 and 1800. He repeated Gottling's experiments on the chemiluminescence of phosphorous not only to disprove Göttling's ideas contrary to Lavoisier's chemistry but also to investigate the viability of the slow combustion of phosphorus as the basis for a new kind of eudiometer. Berthollet's approach to eudiometry and particularly to the phenomenon of the chemiluminescence should be placed in his view of chemistry and experimental physics as allied disciplines with common areas of research that included, for instance, the study of atmospheric air and its related phenomena. During those same years, in Riga, in the Baltic countries, Parrot developed his own eudiometer also grounded in the slow combustion of phosphorous.

Over the last decade of the eighteenth century, all the different kinds of eudiometers were in action to a greater or lesser extent. Because of the constraints of each one, it became a common practice to use them in tandem to compensate for their limitations. This way, Volta used the nitrous air test to check out the purity of the inflammable air for his eudiometer. Saussure combined the alkaline sulphide test with Giober'ts phosphorous eudiometer in substitution of the nitrous air eudiometer. Achard used both the nitrous air and the Volta's eudiometers in his project of assessing the salubrity of air in different places and conditions. Guyton de Morveau carried out a series of eudiometrical experiments using in conjunction the nitrous air with the Volta's eudiometers. Lavoisier recommended using the alkaline sulphides test combined with the eudiometer based on the fast combustion of phosphorous to determine the purity of air accurately, and he and Séguin combined these two latter tests with the nitrous air and the Volta's eudiometers in their researches on animal respiration. Berthollet used his eudiometer based on the slow combustion of phosphorous along with the alkaline sulphides test for determining the oxygen content in common air.

Bethollet's introspection with regard to eudiometry provided an assessment of the different eudiometers in practical use in the late eighteenth century. He commended the eudiometrical test based on the wet paste of sulphur with iron filings despite its slowness. He also found the excessive period of time for the execution of the alkaline sulphides test as a serious setback, although the test was advantageous and reasonably accurate. Berthollet also considered the eudiometers based on the fast combustion phosphorous accurate and expeditious with highly oxygenated air samples. However, the test was uncertain with other kinds of air samples since the decrease of the oxygen proportion, when approaching the assay endpoint, could stop the phosphorous burning without exhausting the oxygen. On the contrary, the nitrous air eudiometers were not at all pleasing to Berthollet who was a little more in favour of the Volta's eudiometer. He praised his own eudiometer based on the slow combustion of phosphorous for combining a precise endpoint with a reasonable duration. As a counterpart, the final reading of the test required a correction factor for an accurate determination of the proportion of oxygen in the air sample.

The road of the phosphorous eudiometers, like in the development of any other instrument, was not free of risks and material constraints. A permanent concern of most phosphorous eudiometers makers was the breakability of the glass eudiometrical tubes because of temperature increases due to sudden warming if the combustion of phosphorous was not properly controlled. To ameliorate the problem flat-topped tubes of green glass could be used. The risk of breaking was also present when heating the piece of phosphorous by approaching a flame or a red-hot piece of charcoal to the outside of the tube. To control the heating of the eudiometrical tube was also of concern for eudiometers based on the slow combustion of phosphorous since the heat from the hands could hasten the process and even burn out the stick of phosphorous. The proper performance of the phosphorous eudiometers was a matter of having within reach heat resistant glassware, fine handcrafters and transmitted tacit knowledge.

7.

The metamorphoses
of the nitrous gas test

Alexander von Humboldt: naturalist, explorer and chemist

Alexander von Humboldt's contribution to chemistry has been often underrated, despite his extensive research in the field of eudiometry and agricultural chemistry. Humboldt's chemical endeavours reached a peak during his short but intensive stay in Paris in 1798 when he designed a variant of the nitrous gas test and was involved in eudiometrical disputes. His innovative approach to the nitrous gas eudiometer left its imprint beyond the borders of the continent. These types of eudiometer underwent diverse transformations throughout the first decade of the nineteenth century, especially in Great Britain.

Alexander von Humboldt (1769-1859) arrived in Paris in April 1798, and left the French capital in October of the same year. During these five months he made the acquaintance of many leading scientific figures, such as Jussieu, Cuvier, Laplace, Delambre and Fourcroy. He arrived too late in Paris to have any contact with Berthollet, who had already joined the French expedition to Egypt under Bonaparte.[1] His closest contact was undoubtedly Vaquelin, with whom he had already established a good relationship that was strengthened during his stay in Paris.[2] Nicolas-Louis Vaquelin (1763-1829) was a reputed pharmacist and analytical chemist who started work as Fourcroy's assistant and gradually became his collaborator. He left Fourcroy's laboratory before 1792 when he became the manager of a pharmacy in Paris, and in 1793 was for a time a military pharmacist at Melun. On his return to Paris he took up a post as an assistant professor at the École Centrale des Travaux Publiques (later renamed as the École Polytechnique), and in 1794 he was appointed as

---

[1] Crosland, 1967, p. 107.

[2] Hein, 1987, p. 164.

inspector of mines and professor of assaying of the École des Mines in Paris.[3]

Humboldt had previously been in touch with Vauquelin to inform him of his galvanic researches and his experiments on plant germination, which he resumed in Paris. They eventually met in person when Humboldt visited Paris in 1798 and were able to use the laboratory facilities of the Agence des Mines to carry out exhaustive experimental work. These experiments gave rise indirectly to a subsequent collaboration between Humboldt and Gay-Lussac. Humboldt's foray into the world of chemistry should be placed in the general context of his eudiometrical researches as well as in his personal interest in improving farmland. Furthermore, both of these pursuits were linked to his intellectual and professional development.

A view of the chemical facet of Humboldt's scientific career

Between 1787 and 1788, Humboldt studied at the University of Frankfurt-der-Oder, where he received training in cameralistic science (*cameralwissenschaft*) and economics in order to acquire a post in the civil administration. The following year he returned to Berlin and during that time he extended his studies on mathematics and science. In general, Humboldt was interested in questions concerning climate, vegetation and plant physiology. He spent the academic year 1789-1790 at the University of Göttingen where he acquired an education of a more intellectual nature. He learned the basic principles of physics and chemistry and became interested in botany. In March 1790, at the end of the second semester in Göttingen, he embarked on a tour around Europe that took him to England and Paris. In England he met Henry Cavendish, who introduced him in Lavoisier's new chemistry, and during his first visit to Paris he was attracted by the revolutionary ideals that shaped him into a liberal who embraced the human rights proclaimed at that time for the rest of his life. He spent his time in Paris hard at work in laboratories and libraries and came into contact with luminaires of French chemistry such as Fourcroy, Berthollet and Chaptal.

---

[3] The École des Mines was founded in 1783, but five years later it was in a state of decline. In 1794, the school was re-established by the Comité de Salut Publique as an establishment of the Agence des Mines, created in the same year, in a building near to the École Polytechnique. The courses in the school began in November 1798; in the meantime the Agence des Mines had become the Conseil des Mines.

In August 1790, he continued his studies on trade and commerce in Hamburg, and in April 1791, he returned to Berlin where he decided to specialize in mining, one of the most prestigious fields of civil administration, motivated by his inclination towards mineralogy and operations saltworks. In May 1791, he applied to the Prussian minister and director of the Department of Mines and Foundries (Bergwerks und Hünttendepartement) for a post, the result being that he was allowed to complete his studies at the School of Mines in Freiberg. Although these mining engineering studies took up most of his time, he never lost sight of his interest in plant physiology and chemistry,[4] which eventually led to the publication of one of his early works, *Aphorisms of the chemical physiology of plants*.[5] Much of his knowledge and many of his instrumental skills were the fruit of this stay in Freiberg.[6] One of Humboldt's most enduring friendships during his time in Freiberg was with the Spanish chemist and metallurgist, Andrés Manuel del Río, who was a follower of Lavoisier and whose influence on Humboldt was such that he adhered to the new chemistry when most German chemists were still advocating the chemistry of phlogiston. Among the friends and acquaintances he made in the field of the German chemistry, it is worth mentioning three prominent influences: Carl Ludwig Willdenow, Martin Heinrich Klaproth and Sigismund Friedrich Hermbstaedt.[7]

After leaving Freiberg in February 1792, Humbolt was appointed advisor to the Prussian Department of Mines, and between September 1792 and January 1793 he made trips to Bavaria, Austria, Poland and Silesia to report on the method for improving water evaporation from brine in the saltworks. In his official position as advisor, he also inspected the mines of Bayreuth and Ansbach, where he promoted the revival of the local mining economy and concerned himself with the miners' precarious social situation. In a letter to Christoph Girtaner, dated February 1793, Humboldt told him that his reading of Girtaner's article on the subject of chemical physiology had provided him with a better understanding of the new antiphlogistic chemistry. In this same letter, Humboldt explained that

---

[4] McCrory, 2010, pp. 37-38.

[5] *Florae fribergensis specimen, plantes cryptogamicas praesertim subterraneas exhibens. Accedunt aphorismi ex doctrina physiologiae chemicae plantarum*, Berlin, 1793. In 1794 it was translated into German as *Aphorismen aus der chemischen physiologie der pflanzen*, Leipzig

[6] Klein, 2012, p. 34.

[7] Hein, 1987, pp. 29, 154-155.

he was devoting his leisure time to the chemistry of plant physiology, experimenting on the germination and growth of plants in environments assumed to be unfavourable to vegetable life.[8]

During his time in Bayreuth, from December 1795 until the autumn of 1796, he conducted experiments on galvanism and expressed his concern about the accidents in mines caused by explosions due to gaseous emanations or by asphyxiation due to harmful gases, and ultimately about the working conditions that jeopardized the life and health of miners. These conditions led him to analyse the various gases to be found in mines and to investigate the causes of their locations, for which he invented different lamps for miners and a kind of respirator. The lamps were designed to burn in inflammable firedamps without igniting the gas, while the respirator was an apparatus for use in rescue operations in case of emergency. Humbolt also laid plans for a national network of eudiometrical stations to determine the composition of the atmosphere in different locations.[9] His interest in the study of the air in mines and its salubrity resulted in the construction and use of different eudiometers, thereby turning the mine into a chemistry laboratory. Thus, in September 1796, he made a replica of Guyton's eudiometer, and in early October 1796, he analysed mine air by using a simplified version of Reboul's phosphorus eudiometer that he made himself and for which he had a high regard, emphasizing its excellent portability and robustness for use in the mine galleries, which was not the case for the nitrous air eudiometer.[10] Humboldt's contributions to chemistry have often been underestimated or ignored, but it is worth stressing that, for him, with his training as a mining engineer, the chemistry of the atmosphere, the air circulation above and below ground, and the effects of the release of subterranean gases on living beings were of paramount importance in his research.[11]

---

[8] Bruhns, 1873, Vol.1, pp. 132-133.

[9] Ibid., pp. 151-153, 157-178.

[10] Klein, 2012, pp. 54, 57; 2015, pp. 180-182, 363-364.

[11] McCrory, 2010, pp. 52-53. In this respect, he published two important works. One on the chemical composition of the air and its circulation *Versuche über die chemische zerlegung des luftkreises und über einige andere gegenstände de naturlehre* (Humboldt, 1799c). The other was on subterranean gases and the resources for reducing their dangers (*Ueber die unterirdischen gasarten und die*

Humboldt's intensive work on chemistry in Paris

From 1795 onwards, Humboldt made a series of trips through northern Italy and Switzerland. He worked at the Department of Mines until the death of his mother in 1796. After this event, and from March 1797 to June 1799, he was devoted to preparing his renowned expedition to America. This process entailed a tour of different cities: to Jena, Vienna, Salzburg and Paris, before arriving in Spain. During his stay in Jena (between March and May 1797), he completed a series of experiments on galvanism and examined the effect of certain chemicals on plants, and while he was there he contacted Göttling, who had been a professor of chemistry at the University of Jena since 1785. Humboldt reported that Göttling's studies, especially those devoted to gases and the chemistry of phosphorus compounds, had been helpful for his future investigations with Vauquelin.[12] He spent three months in Vienna (from July 1797 to October 1797) before moving to Salzburg (from October 1797 to April 1798), where he experimented with nitrous gas and the absorption of atmospheric oxygen by different kind of soils and its influence on the cultivation of soils.

The iron sulphate variant of the nitrous gas test

As mentioned above, once in Paris in April 1798, Humboldt continued this research in the laboratories of the Agence de Mines working jointly with Vauquelin and also receiving advice from Fourcroy. The results were presented at the Institut de France and published in the *Annales de chimie*. The first paper he read on May 25th and 30th was on the experiments with nitrous gas and its combinations with oxygen,[13] in which he addressed the controversial question about the amount of nitrous gas required to saturate one part of oxygen. This was an important factor for ascertaining the composition of atmospheric air. The dispersion of data with respect to this information (with ratios ranging between 1.7 and 4.5) was for Humboldt an unexpected outcome, given the solid foundations of pneumatic chemistry at that time. He proposed a resumption of experiments that had hitherto been lacking in accuracy, with the aim of resolving issues such as the nature of the nitrous gas and its degree of nitrogenation, the formation of nitric acid from the reaction

---

mittel, ihren nachtheil zu vermindern. Ein beytrag zur physik der praktischen bergbaukunde, Braunschweig, 1799)

[12] Hein, 1987, p. 161.

[13] Humboldt, 1798d.

of the nitrous gas with atmospheric air over mercury, and the saturation ratio of nitrous gas with respect to oxygen. Humboldt had already carried out most of these experiments in February and April 1798, during his visit to Salzburg, but the most conclusive were those that he performed and repeated in Paris.

Although the core of this research was to determine the exact amount of nitrous gas required to saturate a portion of oxygen, its development allowed a better understanding of the nature of the nitrous gas as well as the peculiarities involved in the procedural manipulations of the test. Humboldt found that the average saturation ratio of nitrous gas with respect to oxygen was 2.55, which differed greatly from the value of 1.725 used by Lavoisier. This part of the research was conducted according to two different methods. The first one followed Lavoisier's procedure by reacting nitrous gas with oxygen, while the other consisted in reacting nitrous gas with atmospheric air and then removing the exceeding nitrous gas with iron sulphate.[14] This second method enabled him to determine the amount of oxygen in the air - from the amount of residual nitrogen - and therefore the proportion of the combination of nitrous gas with oxygen. Humboldt also used this procedure with iron sulphate during his stay in Paris in order to determine the amount of oxygen in air samples collected at different heights during an ascent in an air balloon.[15]

Both methods were prone to errors that needed to be controlled or avoided. One was the presence of nitrogen mixed with the nitrous gas,

---

[14] The reaction between the iron (II) sulphate and the nitrous gas is a chemical trial known as the brown ring test. It is based on the reaction:

NO (g) + [Fe $(H_2O)_6$]$^{+2}$ SO_4 $^{-2}$ (aq) = [Fe $(H_2O)_5(NO)$]$^{+2}$ SO_4 $^{-2}$ (aq) + H_2O (l)

The brown ring formed is an earthy solution that becomes greenish and can eventually form a precipitate (probably a Fe^{+3} compound). Priestley had already mentioned the quick absorption of nitrous gas by a solution of iron sulphate in the preface to the third volume of his *Experiments and Observacions of Different Kinds of Air* (Priestley, 1777, p. xxxiii). He later described the particulars of that observation in the first volume of his *Experiments and Observations Relating to various Branches of Natural Philosophy*, (Priestley, 1779, pp. 48-50)

[15] Humboldt, 1798c. The air samples collected at certain heights were more impure than the samples at ground level. Humboldt attributed this difference to the winds and air currents, and especially to the decomposition of water, which altered the purity of the air.

which could be determined by means of a solution of iron sulphate.[16] The other was due to the portion of nitrous gas that dissolved in water without combining with oxygen. For that reason, the contraction in volume could not be attributed to a complete combination of the nitrous gas with oxygen. Humboldt confirmed that the nitrous gas test should be performed over water rather than mercury. He observed that when the gases were mixed over mercury the mixture reddened, which could be attributed to the nitric acid that, once formed, remained as a vapour before becoming a liquid acid solution. On the other hand, if the test was conducted over water, the nitric acid was absorbed, and a diminution in volume occurred. The size of the tubes had a significant influence on the contraction in volume. Thus, in narrow tubes with a diameter of 25 mm, most of the nitric acid remained gaseous at a distance from the surface of the water rather than being dissolved in it. Conversely, the same gaseous mixture in a vessel with a diameter of 10-15 cm underwent a greater contraction in volume.   However, this diminution in volume also depended on the order in which the gases were mixed. For instance, if common air occupied the upper part of the mixture, the diminution in volume was smaller than when nitrous gas was at the top of the mixture.[17]

Nitrous gas test vs. phosphorous test

On July 19th, 1798, Humboldt read a second paper in which he presented the experiments carried out on June 19th and July 3rd. In fact, this paper echoed the experiments performed with Göttling during his three months stay in Jena. These experiments were focused firstly on those gaseous

---

[16] Humboldt suggested shaking the nitrous gas with a hot saturated solution of iron sulphate for 4-5 minutes. Vauquelin and Humboldt devoted a special paper to the interpretation of this reaction (Humboldt & Vauquelin, 1798; Humboldt, 1799c, pp. 55-62). However, Humbolt had found that the gas chlorine obtained from potassium chlorate was more efficient in this regard:

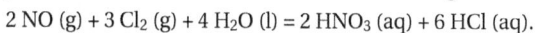

$2 NO (g) + 3 Cl_2 (g) + 4 H_2O (l) = 2 HNO_3 (aq) + 6 HCl (aq)$.

The presence of nitrogen in the nitrous gas was due to the nitric acid used in the reaction with copper. It was thought that a portion of the acid was deoxidised, thus releasing nitrogen. To control this source of error, the first suggestion was always to use the same copper wire to generate the nitrous gas and, secondly, to use a solution of diluted nitric acid with distilled water to prevent an excessively quick reaction. Bouillon-Lagrange recommended both copper and silver as the best metals for obtaining the nitrous gas, and also the use of nitric acid of 25 degrees Baumé (1788-1789, Vol. 1, pp. 125-126).

[17] Humboldt, 1798d, pp. 151-171, 177-180.

mixtures that could dissolve solid substances such as sulphur, phosphorus and iron. The reason for this was that Humboldt had observed the existence of iron in the mines gas deposits and wished to contribute to the study of hypothetical ternary combinations consisting of two gases and one solid. Another issue of interest that had occupied him in recent years concerned the limits of error of the different eudiometers; particularly those based on the combustion of phosphorus, since it was known that the gaseous residue still contained uncombined oxygen after burning. Guyton de Morveau had already observed that the burning of phosphorus indicated an amount of oxygen in common air that was less than that yielded by the nitrous gas test. In Jena, Göttling and Humboldt had confirmed this supposition by using a modified version of Reboul's eudiometer with an elongated graduated tube.[18]

The first part of that summer paper was a continuation up of the preceding paper on the development of the iron sulphate variant of the nitrous gas test. The modified procedure consisted first in determining the proportion of nitrogen using solutions of iron sulphate and chlorine.

> At that time he was using a sample of nitrous gas with a content of 13% nitrogen. 100 parts of common air were then mixed with 100 parts of nitrous gas in a eudiometric tube, thereby obtaining a contraction in volume of 97 parts. After that, the residual gas mixture was washed with a saturated solution of iron sulphate to remove any excess of nitrous gas. A new residue of 85 parts was obtained, from which should be subtracted the pre-existing nitrogen (85 – 13). Thus, the air sample contained 72% of nitrogen.

To calculate the percentage of oxygen, Humboldt took into account the fact that the saturation ratio of nitrous gas with respect to oxygen had previously been determined as 2.55. Thus, the air sample should contain nearly 27 % of oxygen. He confirmed this result with the alkaline sulphide test.[19] Humboldt then proceeded to check the same air as in Berthollet's test based on the slow combustion of phosphorus. He found oxygen absorptions between 12% and 22% but without reaching 27%. However, on analyzing the residue with the nitrous gas test, he found an unexpectedly low oxygen content, particularly in the first two out of the

---

[18] Guyton, 1789, p. 709; Humboldt, 1798a, pp. 150-151.

[19] Humboldt, 1798a, pp. 145-147. A contraction in volume of 3.55 parts involved the absorption of 1 part of oxygen and 2.55 parts of nitrous gas. Then 97 / 3.55 = 27.3 parts of oxygen had to be involved in a contraction in volume of 97 parts.

five tests.[20] Humboldt attributed this finding to the fact that the oxygen had only partially combined with phosphorus, and thus postulated the formation of a double oxide of phosphorus and nitrogen that made uncombined oxygen undetectable by the nitrous gas test. He repeated this type of experiment in Paris and obtained analogous results.[21]

In Humboldt's opinion, this triple combination of nitrogen, phosphorus and oxygen should not be considered as exceptional, especially bearing in mind that the mixture of atmospheric nitrogen and oxygen approached the status of a true chemical combination. Many of the most important phenomena depended on this peculiar between the two constituent parts of the atmosphere.[22] This was a view that Humboldt had already expressed in his first paper, in which he considered that a mixture of 27% oxygen with 73% nitrogen was different from atmospheric air. He believed that although the oxygen obtained from the potassium chlorate did not differ from that of atmospheric air, it did not act in exactly the same way. By regarding gaseous mixtures coming from nature as presenting more homogeneous and uniform affinities than those from an artificial mixture, formed by gases released from different substances at very different temperatures, he was placing the "natural" in opposition to the "artificial". As a physical mixture of nitrogen and oxygen, artificial air still retained the ability to absorb a significant amount of nitrous gas. Nevertheless, oxygen and nitrogen in atmospheric air approached the status of chemical combination. Atmospheric oxygen was more strongly attracted by

---

[20] The test was carried out at 16ºC – 20ºC over nine days. The stick of phosphorus was washed several times a day to ensure its reactivity with atmospheric oxygen. The final figures were calculated after the corresponding corrections, because of variations in pressure and temperature (Table 2).

Table 2 Experimental results of the determination
of oxygen of a sample of common air.

Test #	Air absorption (%)	Expected oxygen content in the residue (%)	Oxygen content in the residue using the nitrous gas test (%)
1	12	27-12 = 15	4
2	17	27-17 =10	3
3	18	27-18 = 9	5
4	20	27-20 = 7	6
5	22	27-22 = 5	1

[21] Ibid., pp. 147-155.

[22] Ibid., p. 158.

nitrogen and linked in quantities far less with nitrous gas. Humboldt therefore concluded that the question of eudiometry could not be posed by experimenting with samples of pure oxygen or a mixture of oxygen and nitrogen, but with atmospheric air directly.[23] Indeed, he did not cease to collect air samples from the summits of the highest mountains and in balloon ascents.

The dispute with Berthollet on the slow combustion of phosphorous

A summary of Humboldt's paper, which he read on July 19th, 1798, and was published in the *Bulletin des sciences par la Société Philomatique de Paris*,[24] triggered Berthollet's rejection of the uncertainties attributed to the slow combustion of phosphorus test.[25] Berthollet reacted by announcing that he was about to start a study with the intention of refuting the nitrous gas method. Unfortunately, Berthollet was obliged to leave Egypt without having completed this study and leaving behind his laboratory notes, and on his return to Paris he had to begin the research all over again. He confined himself to stating that he had not obtained the same results with the hypothetical triple combination and expected that further research would settle the question between his opinion and that of Humboldt. For Berthollet, Scheele's test based on the wet mixture of sulphur with iron filings, as well as the slow combustion of phosphorus test, provided the most uniform and constant results.[26] On May 21st, 1800, Berthollet read a paper at the Institut de France in which he described Humboldt's opinions about the action of iron sulphate on the nitrous gas

---

[23] Ibid., 1798d, pp. 162-164.

[24] During the French revolutionary period many institutions of the ancien régime were supressed and, consequently, they stopped publishing their journals. Thus, the *Mémoires of the Académie des Sciences* for the year 1789 were published when the institution was closed down in August 1793. One of the institutions that could circumvent the revolutionary turmoil was the Société Philomatique, which became a refuge where academicians who were also members of the Society could continue to meet to discuss scientific matters. This resulted in a revitalization of the institution that in 1797 decided to expand its journal - *Bulletin des sciences par la Société Philomatique de Paris* – previously confined to members only, by putting it on public annual subscription (Crosland, 1994, pp. 123-125). It is not surprising, then, that Berthollet, despite being in Egypt, had the *Bulletin de la Société Philomatique* as a means to keep abreast of the latest scientific research in France (Berthollet,1801, p. 3).

[25] Humboldt, 1798b.

[26] Berthollet, 1800a, pp. 74-75, 82, 84.

as paradoxical, and showed in particular that the iron sulphate was unable to absorb all the nitrous gas.[27] He would later wonder why so much effort had been expended on perfecting the use of the nitrous gas test, the many errors in which had been recognized, as well as its inability to provide conclusive indications about the proportion of oxygen. What this amounts to is that Berthollet was actually advocating for Volta's eudiometer.[28]

Nevertheless, the uncertainties raised by Humboldt on the use of phosphorus as a eudiometrical means had already had their impact. For example, Giobert's comparative analyses of Vaudier's Turin air performed with his phosphorus eudiometer were criticized as a result of Humboldt's reflections. The reviewer of Giobert's book recommended the repetition of those analyses by using alternative eudiometrical tests.[29] Furthermore, Fourcroy believed that Séguin's eudiometrical procedure did not warrant the trust that had been placed in it in the light of Humboldt's research, which proved that the gaseous residue still contained between 2-3% of oxygen[30] combined with a gaseous compound of nitrogen and phosphorus.[31] On the other hand, Humboldt's criticism of the test based on the slow combustion of phosphorous went some way to strengthening Parrot's resolution to develop his own phosphorous eudiometer

Eudiometry for agricultural purposes

On October 12th, 1798, shortly before leaving for Spain, Humboldt read a paper on the absorption of oxygen by different kind of soils.[32] This paper enables Humboldt's interest in the nitrous gas test to be situated in the fields of eudiometrical as well as agricultural research.[33] Humboldt

---

[27] Ibid., 1801.

[28] Ibid., 1803b, Vol. 1, p. 510.

[29] Giobert, 1789, pp. 198-199, note 1.

[30] In fact, Humboldt had found even higher values (between 6% and 7%) of oxygen in the residual gas. Moreover, he had determined absorptions of oxygen from 17% to 25%, depending on the celerity of the combustion and the shape of the recipient, which could help to ensure that oxygen did not come into contact with phosphorus (Humboldt, 1799b, p. 134)

[31] Fourcroy, 1800, Vol 1, pp. 191-192.

[32] Fourcroy and Vauquelin reviewed this paper at the Institut de France on February 14th, 1799. *Process-verbaux des séances de l'Académie tenues depuis la foundation de l'Institut jusqu'au mois d'août*, 1835 T. 1, An IV - VII (1795-1799), (1910), Hendaye, Imprimerie de l'observatoire d'Abbadia, pp. 525-527.

[33] Humboldt, 1799b.

considered soil cultivation as one of the most important challenges of agriculture and plant chemical physiology. Despite discoveries on the germination, nutrition, secretion and respiration of plants, Humboldt foresaw that the major problems in agriculture were still to come. Among these problems were the nature of animal manures or the influence of simple earths, such as lime, on vegetation. It was also necessary to increase soil fertility, which could not be achieved by simple reliance on the good behaviour of the local climate. Humboldt believed that only experimental research could perfect plant physiology and successfully address the challenges facing agriculture.

It was known that humus – the organic component of soil – could decompose atmospheric air by absorbing oxygen. Humboldt had found in the Austrian Alps a kind of clay and other types of earth, such as alumina, that also possessed this property. This phenomenon led to the problem of determining how the soil acted on the lower layers of the atmosphere after it had been ploughed and exposed to air.[34] Humboldt provided some initial information on this line of research in a letter to Ingenhousz, who had shown interest in problems concerning agriculture, such as plant nutrition and the action of sunlight on plant oxygenation.[35] Humboldt suggested a series of experiments that he described later in a paper published in the *Annales de chimie*, in which he concluded that freshly ploughed fields needed a certain period of contact with the air before planting began.[36] The ploughing of fields increased soil fertility and favoured its oxygenation; in short, air behaved like a true fertilizer.[37]

Humboldt did not explicitly mention the use of the nitrous gas test for determining the absorption of oxygen by soils exposed to common air. However, Nicolas-Théodore de Saussure's comments on Humboldt's letter to Ingenhousz, Humboldt's reply to Saussure[38] and his paper in the *Annales de chimie* served to endorse it. Saussure was doubtful about

---

[34] Ibid., pp. 125-128; 1799c, pp. 117-149.

[35] Ibid., 1798e.

[36] Humboldt experimented with different kinds of earth (alumina, baryte, lime, silica and magnesia). Alumina and magnesia were the best oxygen absorbents and Humboldt considered them as new eudiometrical means that were simpler and more active than the phosphorus and the alkaline sulphides (Ibid., 1799b, p. 140).

[37] Ibid., p. 142.

[38] Nicolas-Théodore de Saussure (1767-1845) was the second child of Horace-Bénédict de Saussure. Most his published papers dealt with the chemistry and physiology of plants, the nature of soils, and the conditions of vegetable life.

Humboldt's experimental results, since in his opinion Humboldt had assigned an inaccurate limit of error to the phosphorus eudiometer, which had led him to reject it in favour of the nitrous gas eudiometer. Indeed, Saussure believed that the errors arising from the handling of this apparatus would persuade any inexperienced experimenter to use the phosphorus eudiometer. Saussure was in effect referring to the iron sulphate variant of the nitrous gas test. He alleged that the addition of iron sulphate was counterproductive for the test because, in addition to the manipulative complications that it entailed, it hardly absorbed the nitrous gas without correcting the sources of error attributable to the Fontana-Ingenhousz eudiometer, especially those due to the immediacy in mixing gases.[39] Humboldt defended his case by saying that he had never advocated the use of the nitrous gas test with iron sulphate, but in any event a knowledge of how the iron sulphate acted on the nitrous gas would have contributed to improving the test. Although he did not settle the question raised by Saussure about the suitability of the two eudiometers, Humboldt pointed out that ten trials performed on the same air sample with the Fontana-Ingenhousz eudiometer had given quantities of oxygen that did not differ by more than 3 thousandths.[40] Indeed, Humboldt used both the nitrous gas test alone and combined with the iron sulphate.

Humboldt left Paris on October 20th, 1798, with the botanist Aimé-Jacques-Alexandre Bonpland as his companion. They went to Madrid to obtain permission to travel through the Spanish colonies in South America for the purpose of acquiring scientific information. He embarked a complete collection of the latest instruments of measurement, such as chronometers, telescopes, sextants, compasses, magnetometer, barometers, hygrometers, thermometers and electrometers, among which was a Reboul phosphorous eudiometer and a Fontana-Ingenhousz nitrous gas eudiometer with a eudiometrical tube capable of detecting a 1/1000 of a cubic centimetre change in gas volume.[41] Volta's eudiometer was troublesome for those travelling through damp countries on account of the small electric discharge needed for the ignition of the oxygen and hydrogen mixture.[42] Humboldt used the nitrous gas test to analyze the air not only from different locations but also coming from experiments on animal respiration. In May 1801, in the locality of Santa Cruz de Mompós

---

[39] Saussure, 1798.

[40] Humboldt, 1799a.

[41] Dettelbach, 1999, p. 477.

[42] Humboldt & Bonpland, 1814-1825, Vol. 1, pp. 111-113; 1814-1829, Vol. 1, pp. 35-36.

(Colombia), Humboldt set up an experimental device using large gourds as pneumatic troughs in which he placed three young crocodiles bound to bamboo frames in the form of a cross and covered by glass bell jars. He released the crocodiles forty-three minutes after the beginning of the experiment and determined the amount of the remaining oxygen by means of the nitrous gas test.[43] Humboldt's innovation of introducing iron sulphate as a mediating reagent in the nitrous gas test paved the way for Humphry Davy (1778-1829) to devise a new eudiometer.

Back on the other side of the channel: Davy, Hope, Pepys and Henry

Humphrey Davy (1788-1829)

In December 1794, Davy's family moved to Penzance where Humphry was apprenticed to a local apothecary-surgeon. There he pursued an impressive program of self-education that consisted of ten topics, one of which under the heading of "My profession" included: botany, pharmacology, nosology, anatomy, surgery and chemistry.[44] In late 1797, Davy began to study chemistry with Lavoisier's textbook *Elements of Chemistry* and also apparently with William Nicholson's *Dictionary of Chemistry*. In addition to his autodidacticism, Davies Giddy, a Member of Parliament with scientific interests, played an important part in Davy's career. He had met Davy in the Penzance apothecary where he learned about his interest in chemical experiments.

At Oxford, Giddy had attended the chemistry lectures of Thomas Beddoes, who was extremely interested in the medical applications of chemistry, and particularly on the usefulness of the new gases (factitious airs) obtained by Priestley in the treatment of respiratory diseases like tuberculosis. In April 1798, with Giddy's help, Beddoes received Davy's account of his experiments on heat and light. Beddoes was so delighted with this work that when he needed an assistant at his Pneumatic Institution in Clifton he arranged through Giddy for Davy to take up that post. In October 1798, Davy set off from Penzance for Clifton. The Pneumatic Institution had originated in Beddoes' house in Clifton, a suburb of Bristol, but by early 1799 the Institution's well-equipped laboratory was housed in its own premises. It was there that Davy started

---

[43] Humboldt, 1811, pp. 253-259.

[44] Davy left the grammar school in December, 1793, just before his sixteenth birthday.

investigating the gas nitrous oxide[45] in response to the claims made by the American physician Samuel Mitchill, who feared that this gas might be a principle of contagion that would prove fatal to anyone who breathed it. Davy's careful experimental work on the isolation and the effects on animals of the various oxides of nitrogen was published in 1800 in the monograph *Researches Chemical and Philosophical*, chiefly concerning nitrous oxide.[46] Davy's work at the Pneumatic Institution has been regarded in some quarters as a late manifestation of the pneumatic medicine program.[47] In March 1801, after little more than two years, he gave up his post of superintendent at the Pneumatic Institution and moved to the Royal Institution in London. His initial appointment was as an assistant lecturer in chemistry, director of the Chemical Laboratory and editor of the Journal of the Royal Institution.[48]

The first *Research* of that work concerned the production of nitrous oxide and the analysis of nitrous gas and nitrous acid. Davy recognized that although little of his own research was involved in this part, by repeating the experiments of other chemists he had sometimes been able to draw different conclusions. As regards the absorption of nitrous gas by a solution of sulphate of iron, he mentioned that Priestley had identified this phenomenon a long time ago and that it had been applied by Humboldt to determine the nitrogen generally mixed with nitrous gas. He devoted two complete articles to the absorption of nitrous gas by solutions of green sulphate and muriate (chloride) of iron. The solutions of both green salts of iron were good nitrous gas absorbents, but not the solutions of the corresponding red salts.[49] Since a sample of solid sulphate of iron was most probably the red or mixed sulphate, Davy chose to obtain the

---

[45] Dinitrogen oxide

[46] The *Researches* was published in June, 1800, but by February Davy had already communicated a short account of the way in which he had prepared his experiments on nitrous oxide (Davy, 1800a). Davy was able to refute Mitchill's claims that the nitrous oxide was poisonous.

[47] Golinski, 2016, p. 19.

[48] Davy, 1836, Vol. 1, pp. 1-133. The first two chapters of this biography of Davy are devoted to the early years of his life before moving to London in 1801.

[49] The discovery of the difference between the sulphates and muriates (chlorides) of iron was owed to Proust [«Extrait d'un Mémoire intitulé: Recherches sur le Bleu de Prusse», *Journal de physique, de chimie, d'histoire naturelle et des arts* (1794), Vol. 45, pp. 334-341; *Annales de chimie* (1797), vol. 23, pp. 85-101; «Abstract of a Memoir Entitled: Enquiries Concerning the Nature of Prussian Blue», *A Journal of Natural Philiosophy, Chemistry and the Arts* (1797), Vol. 1, pp. 453-458]

solution of green sulphate of iron by dissolving iron filings in diluted sulphuric acid. Similarly, when iron filings were dissolved in pure muriatic acid, and the solution preserved from the contact of air, it was of a pale green colour.[50]

In July 1800, after the publication of his *Researches*, Davy had already developed his own eudiometrical test based on the nitrous gas impregnating a solution of green sulphate of iron. This eudiometer, fruit of his abovementioned researches, was described in a paper published in August 1801, the title of which, *An Account of a New Eudiometer*, emulated Cavendish's paper of 1783.[51] Davy praised the slow combustion of phosphorous and the alkaline sulphide tests as the only two of all the eudiometrical methods that had managed to survive. However, he found their operation time to be extremely slow, and it was often difficult to ascertain their endpoints. On the other hand, he found that a saturated aqueous solution of green chloride or sulphate of iron impregnated with nitrous gas was more suitable as the eudiometrical agent.

> The eudiometrical reagent was prepared by dissolving in water as much of the green sulphate or muriate of iron as the water would take up, or by reacting iron filings in diluted sulphuric or muriatic acid, leaving an excess of the iron in order to ensure complete saturation of the acid. Once a wide-mouthed bottle was filled with this solution and inverted in a cupful of the same solution, the nitrous gas from nitric acid and mercury was passed into the inverted bottle, shaking it frequently. The colour of the solution changed to black, and the production of gas and agitation were continued until absorption could be carried no further. The eudiometrical instrument (Figure 7.1) was very straightforward and portable. It consisted of a glass tube approximately 18 inches long and three-quarters of an inch diameter closed at one end and divided into

---

[50] Davy, 1800b, pp. 152-186; 1836-1840, Vol. 3, pp. 92-111. The pale green iron (II) sulphate exists commonly as a heptahydrate salt. Upon exposure to air, it oxidizes to form a brown-yellow coating of basic iron (III) sulphate, which is an adduct of iron (III) oxide and iron (III) sulphate. The greenish iron (II) chloride is commonly encountered as a tetrahydrate salt and it also oxidizes upon exposure to air to give the dark green or purple-red iron (III) chloride, which exists in the anhydrous and hexahydrate forms. Only the iron (II) solutions are able to absorb the nitric oxide according to the following equilibrium (see also note 14):

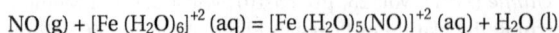

$$NO \ (g) + [Fe \ (H_2O)_6]^{+2} \ (aq) = [Fe \ (H_2O)_5(NO)]^{+2} \ (aq) + H_2O \ (l)$$

[51] Davy, 1801.

cubic inches and tenths of inches. The eudiometrical device was complemented with a small measure of about two cubic inches and similarly graduated. For employing the solution of iron sulphate impregnated with nitrous gas, glass tubes about five inches long and half an inch wide, divided decimally, were also necessary.[52] After being filled with the air to be examined, the tube was introduced into a recipient containing the prepared eudiometrical reagent, gently moving it from a perpendicular to a horizontal position. The air rapidly diminished and the volume of the residual air (nitrogen) could be easily read due to the dark colour of the solution.

Figure 7.1 Davy's eudiometer.

From William Henry, *Elements of Experimental Chemistry* (London, 1810), plate II, fig. 24.

The eudiometrical reagent absorbed atmospheric oxygen without acting upon nitrogen. The test was completed in a few minutes, and the oxygen combined with the nitrous gas was dissolved in the aqueous solution.[53] The impregnated solution with green chloride was more rapid in its operation than the solution with green sulphate.[54] However, the period

---

[52] Davy did not provide any detailed description of his eudiometrical device or the preparation of the eudiometrical reagent. This information has been obtained from the 1803 third edition of William Henry's *An Epitome of Chemistry* (Henry, 1803, pp. 18, 76). Likewise, Figure 7.1 belongs to Henry's *Elements of Experimental Chemistry* of 1810.

[53] In a letter to Giddy, dated October 1800, Davy stated that the eudiometer that he had lately employed gave in a few minutes the proportion of oxygen without correction (Paris, 1831, p. 73).

[54] As mentioned in note 18 of chapter 1, the combination of oxygen and nitric oxide involved a set of chemical equilibriums (Usselman *et al*, 2008). In Davy's test, nitric oxide impregnated a saturated solution of iron (II) sulphate and was released from the solution on coming into contact with common air. Nitric oxide reacted with atmospheric oxygen and then the equilibrium:

during which the diminution in volume was stable could be accurately observed, since shortly after this period the volume of the residual gas began to increase.

A number of comparative trials conducted on the constitution of the atmosphere in Bristol in July, August and September, 1800, using the phosphorous, the alkaline sulphides and the impregnated solution of nitrous gas tests, demonstrated the accuracy of the latter. In no instance was it found that 100 parts in volume of air contained more than 21 parts of oxygen.[55] Davy was firmly convinced that the atmosphere contained very nearly the same proportions of oxygen and nitrogen in all locations. Thus, if the different degrees of salubrity of air did not depend on the differences in the proportions of its principal constituent parts, researches were to be addressed to the different substances dissolved or suspended in the air, which were noxious to the human body.[56]

Although Humboldt's iron sulphate variant of the nitrous gas test was present in Davy's mind, his eudiometrical procedure was much closer to Martí's calcium sulphide test. Humboldt used the iron sulphate solution to first determine the nitrogen mixed with the nitrous gas, and then to remove the excess of nitrous gas from the residual mixture after executing the test. In essence, the nitrous gas was the eudiometrical reagent. In Martí's eudiometrical test, a solution of calcium sulphide impregnated with nitrogen was used to absorb only the atmospheric oxygen. Likewise, in Davy's eudiometrical procedure, a solution of iron sulphate impregnated with nitrous gas was employed also to absorb the atmospheric oxygen without absorbing the atmospheric nitrogen. Basically, the solution of iron sulphate impregnated with nitrous gas was the eudiometrical reagent. In addition, both procedures gave an indirect determination of the oxygen content of an air sample, and employed a simple graduated glass tube as the sole eudiometrical instrument. As previously stated, Davy had praised the solution of potassium sulphide as a eudiometrical reagent, because it was able to absorb all the oxygen of atmospheric air at common temperatures

NO (g) + [Fe $(H_2O)_6$]$^{+2}$ (aq) = [Fe $(H_2O)_5(NO)$]$^{+2}$ (aq) + H_2O (l)

shifted to the left until all the oxygen in the sample of atmospheric air was combined with the nitric oxide.

[55] In the lecture *On the Chemical Composition of the Atmosphere*, belonging to the course "The chemistry of nature" delivered in January, 1807, Davy assessed Martí's, Berthollet's, Hope's and his own eudiometrical method as error free (Davy, 1836-1840, Vol. 8, p. 249).

[56] Davy, 1801, p. 58.

without altering the volume of the residual nitrogen. By 1804, this kind of eudiometrical test had already been reinvented by Thomas Charles Hope, William Pepys and William Henry.

Thomas Charles Hope (1766–1844)

In 1799, Thomas Beddoes received a visit from Thomas Charles Hope, professor of chemistry at Edinburgh University, who had witnessed the work of the young Davy at the Pneumatic Institution. Soon afterwards a lecturer in chemistry was required for the Royal Institution which was then under the management of Count Rumford. Rumford consulted Hope, who strongly recommended Davy for the post, to which he was appointed in 1801. Hope combined the practice of medicine with the teaching of chemistry. In October 1787 he was appointed lecturer in chemistry in the University of Glasgow, where he was to spend the next eight years. In 1789, he was appointed assistant Professor of Medicine at the same institution and continued to teach both chemistry and medicine. Although in 1791 he became Professor of Medicine and had to resign his lectureship in chemistry, he continued his private chemical research. Hope had attended Joseph Black's course in the 1781-1783 sessions, and in 1795 he was appointed his conjoint Professor, and from October 1797 he was the sole teacher of chemistry. After Black's death in 1799, he succeeded him as full Professor of Chemistry in the Chemistry Department of Edinburgh University, where he lectured for nearly 50 years.

Through his lectures Hope became the first university teacher in Britain to abandon unambiguously the phlogiston theory. In November 1793, he read a paper to the Royal Society of Edinburgh about a 'hitherto unknown kind of earth', being one of those who declared it to be a compound of a new element, strontium. In January 1804, he read the paper *On the Contraction of Water by Heat, at Low Temperatures*,[57] concerning the singular fact that water expanded as it cooled towards its freezing point. In this respect, he devised an apparatus to measure the temperature at which water attains its maximum density.[58] An account of a new eudiometrical

---

[57] Transactions of the Royal Society of Edinburgh, 1805, Vol. 5, pp. 379-405.

[58] Traill, 1849; Doyle, 1982; Anderson, 2015. Hope's opposition to the introduction of a Chair of Practical Chemistry in the Edinburgh University was a blot on his professional career. Meanwhile, Germany was developing teaching laboratories and British students went abroad for this kind of education. Hope's intransigence contributed to the decay of Edinburgh as a world leader in chemistry education (Anderson, 2015, p. 159).

device with the habitual use of calcium sulphide as the eudiometrical means was published in 1803. This apparatus was designed and used by Thomas Charles Hope in his lectures and in experiments. Hope's biographers hardly mention his contribution to eudiometry.[59]

His eudiometer consisted of two bottles (A) and (B) connected together (Figure 7.2). (A) was a small bottle of nearly two inches in external diameter and three in length, with a neck and a stopper at (D) and another neck at (C). This bottle contained the eudiometric means. (B) was a larger bottle of nearly of the same diameter and 8 ½ inches long.[60] The neck of (B) was fitted accurately by grinding into the neck of (A) at (C). To execute the test, the bottle (B) was first filled with the gaseous sample, and then the bottle (A) was also filled with a eudiometrical means such as a solution of calcium sulphide, which Hope commonly employed. Afterwards, the mouth of (A) was covered with a flat piece of glass, plunged under the surface of water and there inserted into the neck of (B).

This dual-chamber recipient (AB) was then removed from the water and inclined until a sufficient quantity of the sulphide solution had flowed into the upper part (B). While the recipient was being agitated, the liquid sulphide absorbed the oxygen and formed a partial vacuum. To prevent a complete or rapid absorption, the bottle (A) was plunged beneath the surface of common water in a plate, and the stopper (D) was slowly opened. The atmospheric pressure forced a quantity of water in to replace the absorbed gas. In this way, the sulphide solution was diluted, but not to such a degree as to interrupt the process. The absorption was deemed to be completed when, after agitation and opening the stopper (D), that the liquid solution was observed to rise no higher. The operation was concluded by allowing the instrument to recover its original temperature. The degree of absorption could be determined, if the bottle (B) were graduated, by plunging the apparatus into water to the level of the inner liquid solution and removing the stopper. Otherwise, the residual gas may have been

---

[59] Traill (1849, p. 427) is the exception.

[60] According to the author of this account, the size referred to here was very well adapted to the purposes of public exhibition, although the instrument ought to have been made considerably smaller for routine eudiometric experiments.

transferred into a tube that was expressly graduated for measuring gases.[61]

Figure 7.2 Hope's eudiometer.

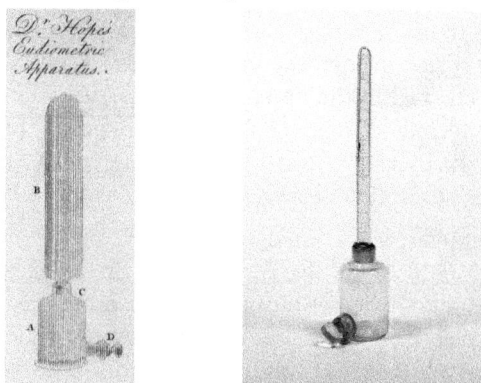

(Left) From *A Journal of Natural Philiosophy, Chemistry and the Arts*, 1803, Vol. 6, plate 12. (Right) Copy (ca. 1825) at the Museum of the History of Science, University of Oxford. Inv. 4056 © Museum of the History of Science, University of Oxford.

The reason for devising this new eudiometer on the basis of the alkaline or calcium sulphides cannot be disentangled from that description. Although by 1803 Scheele's, Guyton's and Martí's contributions on the use of sulphides as eudiometrical agents were already known, there was no reference to them in the report.[62] On the other hand, Lavoisier's paper on the use of a solution potassium sulphide as a eudiometrical means did not begin to circulate until 1805. However, it cannot be ruled out that Hope obtained first-hand knowledge of it. In 1783, after being admitted as a Fellow of the Royal Society of Edinburgh, he decided to spend his summer vacation in Paris, where he was well received by Lavoisier and Berthollet. Hope regarded these meetings as an important event in his life and was

---

[61] Hope, 1803. The description of Hope's eudiometer was presented in two accounts in the sixth volume of the *Nicholson's Journal* of 1803. The first appeared in the September issue (pp. 61-62) and the second in the November issue (pp. 210-212). The reporter (W.N.) of both accounts mentioned that because of the first description was in some respects inaccurate, he gave an entirely new description, and a drawing of the instrument in a second description.

[62] The German abridged translation of Martí's paper described Hope's eudiometer as being similar to that by Martí, completed by the former one in some details only (Martí, 1805, p. 389)

greatly impressed by Lavoisier's cordiality and his abilities.[63] Lavoisier's paper dealing with the use of potassium sulphide as a eudiometrical agent was written after 1789, and Hope might have become acquainted with it during the summer of 1783. The salient difference from the preceding eudiometrical trials using alkaline or calcium sulphides was the equipment employed in the experimental device. While simple tubes or bottles had been used in the other tests, Hope's test required a tailored two-chamber recipient that could be connected to each other. Furthermore, the device incorporated a mechanism to prevent an unwanted quick absorption of oxygen.

Hope's eudiometer was chosen by the Scottish physician Alexander Henderson (1780–1863) for his experiments in order to address the controversial point of whether nitrogen was absorbed during respiration. After several trials made with the eudiometers of Séguin, Guyton and Davy, he concluded that they were all prone to several sources of error. According to Henderson, the problem with the alkaline and calcium sulphides test was not its accuracy but the time elapsed before completing the absorption of oxygen. Nevertheless, he found that Hope's eudiometer overcame these obstacles to a great extent combining the virtue of neatness and simplicity with those of accuracy and expedition. After conducting his trials with this eudiometer, Henderson decided to use the calcium sulphide instead of the potassium sulphide as the eudiometrical agent in order to make the trial more expeditious.[64] He was aware of the objection that the solutions of calcium sulphide absorbed not only oxygen but also small proportions of atmospheric nitrogen, but discounted it as lacking a solid foundation, while recognizing that such nitrogen absorption could be observed in fresh solutions of calcium sulphide.[65]

The problem with Hope's eudiometer was that if the neck of the larger bottle (B) and the stopper (D) were not both very accurately ground, air made its way into the instrument to counter the partial vacuum occasioned by the absorption of oxygen. This absorption generated a diminution of the pressure within the instrument and, consequently, towards the end of each agitation the absorption took place very slowly. In addition, the eudiometric solution became increasingly more dilute by the admission of water through (D). It was William Henry who pointed out

---

[63] Traill, 1849, pp. 420-421; Doyle, 1982, pp. 2-3.

[64] Most of these tests were completed in about twenty minutes, providing sufficient agitation was employed.

[65] Henderson, 1804.

these shortcomings, which he and William Pepys intended to rectify with new designs for eudiometers.

William Haseldine Pepys (1775 –1856)

Pepys was apprenticed to his father as a cutler and became a liveryman of the Worshipful Company of Cutlers of London in 1796. He was one of the founders of the Askesian Society (a debating club for scientific thinkers in London), and in April 1799 he became part of the group that founded the British Mineralogical Society, whose members were middle-class chemists, physicians, and owners of businesses such as instrument makers. He was from 1800 a Proprietor of the Royal Institution where, motivated by his friend Humphry Davy, he became a successful researcher in chemistry, a developer of scientific instruments and a reputable analyst. He was skilled in the constructing powerful electric batteries, blowpipes for use in mineral analysis and soda-water apparatuses. Pepys was one of the first privileged witnesses in 1807 to Davy's crucial experiment on the isolation of the alkali metals potassium and sodium by electrolysis of their fused hydroxides. In 1808 and 1809 he was involved in the subscription to the great battery installed at the Royal Institution.[66] Pepys own application of electrolysis, linked to his business, was evident in his attempts to melt platinum by the discharge from his very large batteries.[67]

In 1807, Pepys and his close friend William Allen, who was active in pharmaceutical business, embarked on projects for determining the composition of carbon dioxide and for investigating the chemical phenomena of animal and plant respiration. Guyton's contribution to this first subject in 1799 had raised uncertainty about the proportion of carbon, and it was against this background that Pepys and Allen undertook their research.[68] As for their second project, they were aware that some important matters concerning the process of respiration were still in dispute. For instance, the question of an accurate method for separating and ascertaining the exact proportion of the different gasses in any given mixture, or the quantity of residual gas in the lungs after a forced expiration.[69] The key apparatuses for both research projects were mercury

---

[66] Kurzer, 2003, pp. 163-167.

[67] Weindling, 1982; Knight, 2009, pp. 98-99.

[68] Allen & Pepys, 1807; Kurzer, 2003, pp. 156-158.

[69] They finally reported a carbon proportion close to 28.6% in the carbon dioxide. In addition, after a series of fifteen experiments they found that the expired air consisted, by volume, of 8.5% of carbon dioxide, 12.5% of oxygen and 79% of nitrogen.

gasometers; a platinum combustion tube passing through a furnace to burn completely a known quantity of carbon; a respirator in which they could breathe from three to four thousand cubic inches of gas, and a polyvalent eudiometer that Pepys had already described in the same year 1807.[70] Having had occasion to repeat many of the eudiometric experiments of others, and to perform some new ones, Pepys came to the conclusion that a more sizeable instrument capable of providing correct results with the greatest accuracy was still a desideratum in eudiometry. It was in this context that on June 4th, 1807, he read a paper at the Royal Society in which he announced that he had succeeded in contriving an instrument possessing the above properties to a prominent degree. Before describing Pepys' apparatus, it is important to state that the new instrument was not only a eudiometer for measuring the proportion of oxygen in a gaseous sample, but also a gas mixture analyser. For this reason, the following description refers to a general liquid reagent rather than to any particular eudiometrical reagent or any gaseous sample.

The apparatus (Figure 7.3), regarded by Pepys as being extremely portable, consisted of a glass measure (*Fig. 1*) graduated into one hundred parts; a small rubber (otherwise called gum elastic or caoutchouc) bulb (*Fig. 4*) capable of holding approximately twice the quantity of the measure and furnished with a perforated glass stopper (S), which was fastened into the neck of the bulb by means of tightly-wound waxed thread. This glass stopper had its exterior end ground to fit the mouth of the measure. The last component was a thin glass tube (T, *Fig. 3*) graduated into one thousand parts of the measure and with a small steel stopcock cemented to its lower end. This stopcock could be fastened into the neck of a very small rubber bulb by means of a waxed thread (*Fig. 2*). The other extremity of the thin tube was conical-shaped so as to provide a very small orifice. This tube could be slid easily up and down inside a moveable glass cistern (C, *Fig. 3*) to be filled with water or mercury without leakage. This was accomplished by means of a cork fitted into the mouth of the container and perforated through its axis to receive the thin tube. The test was performed by filling the measure with the gas sample over mercury in a pneumatic trough, and the rubber bulb with the solution of the liquid reagent. The orifice of the stopper (S) was then connected to the mouth of the measure still in the mercury. Being thus united, the bulb and measure were held firmly at the joint. Upon

---

[70] Allen & Pepys, 1808; Kurzer, 2003, pp. 159-163.

pressing the bulb, a portion of the liquid was squeezed into the measure. On releasing the pressure, the bulb regained its original form and received back the liquid. This process was repeated until any absorption was no longer observed.

In the case of an atmospheric air sample, a large gas residue was left in the measure. To determine the degree of absorption, the hundred parts in the measure were first recorded. To obtain the fractional parts, the measure was removed to the cistern (C) full of mercury in which the graduated thin tube also full of mercury was placed. This tube was slid above the surface of the mercury in the measure, and the stopcock was opened to allow the mercury to descend until it had drawn the mercury in the measure up to a regular division. After that, the hundred parts on the measure and the thousand parts on the graduated tube could be recorded, which gave the united quantities the sum of the residual gas. Before recording the thousand parts, it was necessary to bring the mercury to the same level in both the measure and the cistern. This could be effected by pouring out or adding a portion of the mercury from the cistern. When using the apparatus as a standard eudiometer, the bulb was filled with a solution of sulphate of iron impregnated with nitrous gas, in accordance with Davy's eudiometrical test.

Figure 7.3 Pepys' eudiometer.

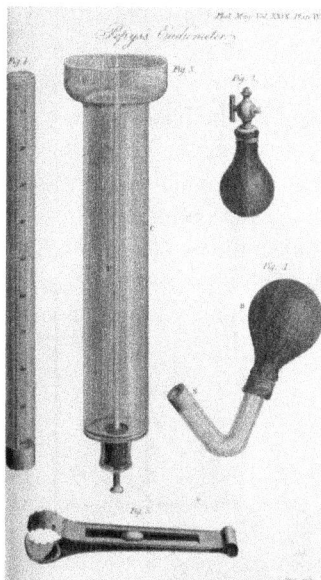

From *The Philosophical Magazine,* 1808, Vol. 29, plate 4.

Pepys was able to find an oxygen content of 21.5% for atmospheric air. However, the same apparatus could also be used to determine the purity of other gases. Thus, for ascertaining the purity of nitrous gas, the bulb was filled with the solution of sulphate or muriate of iron, and for carbonic acid gas with limewater. The instrument could be likewise applied to the analysis of gas mixtures, such as carbonic acid gas with sulphurated hydrogen (hydrogen sulphide), by using a solution of the nitrate of silver or mercury, or nitrous gas with carbonic acid gas by means of a hot solution of the green sulphate of iron.[71]

The paper read on June 18th, 1807, on the composition of carbon dioxide revealed more details about the experimental procedure with the new eudiometer. The purity of oxygen, generally determined within about 10 minutes, was constantly ascertained by the eudiometer before every experiment by using the solution of green sulphate of iron saturated with nitrous gas. Whenever the diminution in volume reached its maximum and the gas began to increase in volume, a simple solution of the green sulphate of iron was substituted by that saturated with nitrous gas. The sulphate of iron absorbed any nitrous gas that may have escaped from the saturated solution, and the residue enabled the quantity of oxygen contained in the gas to be ascertained.[72]

Pepys held his apparatus in high esteem and explained its advantages. The gaseous sample to be examined was completely under control; it could be agitated without fear of intrusion of air from the outside, thereby shortening   the process. The key innovation was undoubtedly the introduction of a new material in a chemical device. This was the rubber, which was highly resistant to chemical agents and, above all, that could tolerate the use of hot solutions, a factor that was important for the examination of some gas mixtures.[73] Furthermore, the flexibility of the rubber bulb enabled the mixing of the liquid reagent with the gas sample to be finely regulated, and thereby also controlling the endpoint of the test. There is no doubt that the new material opened up procedural capabilities previously unattained.[74] Pepys' incorporation of rubber into

---

[71] Pepys, 1807, pp. 250-255.

[72] Allen & Pepys, 1807, p. 270.

[73] Pepys, 1807, p. 258.

[74] Actually, Pepys was not the first to use rubber in a eudiometrical device. In the first of his own eudiometers, Ingenhousz replaced one of Fontana's customary glass chambers for a soft rubber bag. This innovation forced the nitrous and common air to mix together by squeezing the rubber bag containing nitrous air. The purpose was to

his apparatus may have inspired William Henry in his improved version of Hope's eudiometer.

William Henry (1774-1836)

William Henry was the son of a Manchester apothecary. He began to study medicine at the University of Edinburgh in 1795, which at that time was enjoying its highest repute as a school of medicine, and where Joseph Black still occupied the Chair of chemistry. He left the university after a year to join his father in the apothecary business. While assisting his father, he took on an active role in the intellectual life of Manchester and became a member of the Manchester Literary and Philosophical Society. A series of lectures he gave in 1798 and 1799 evolved into a text entitled *An Epitome of Chemistry*. After publishing several other papers on chemistry, he returned to Edinburgh in 1805 to continue his studies, obtaining his diploma of Doctor in Medicine in 1807 for writing a thesis he wrote on uric acid. In 1808 he was elected a Fellow of the Royal Society, and in the following year was awarded the Copley medal. Henry was most influential as a recorder of experimental results and an exponent of the current state of chemical science. Successive editions of his long-lived *An Epitome of Chemistry* kept its content up to date, but with the appearance of the sixth edition in 1810 the title was changed to *Elements of Experimental Chemistry*.[75] The progress of chemistry after the last edition of *An Epitome of Chemistry* in 1808 exemplified not only the number and novelty of the discoveries that had been made but also the importance of the generalizations to which they had led, thus making the former title redundant. Henry remained an active researcher, his most important work being carried out on the solubility of gases. He also engaged in the construction and sale of chemical apparatus, for which he wrote texts that served as an illustrated catalogue of the many types of laboratory apparatus available for research and commercial use. In this regard, Henry's textbook could act as a promotional prospectus for a particular kind of apparatus, i.e. a type of eudiometer, in preference to others.

---

ensure that the same quantity of nitrous air always mixed with the common air sample. Ingenhousz reported this ostensible improvement to Fontana's eudiometer in a letter written to John Pringle in 1775 (Ingenhousz, 1776, pp. 257-262).

[75] The five editions of *An Epitome of Chemistry* were published in the years 1800, 1801, 1803, 1806 and 1808. *Elements of Experimental Chemistry* was also a very successful book that ran to eleven English editions, the last one appearing in 1829.

As far as the use of iron sulphate together with nitrous gas in eudiometrical tests was concerned, Henry began referring to it in the second edition of his *Epitome* in 1801. First, in accordance with Humboldt's variant of the nitrous gas test; that is, to ascertain the amount of nitrogen contained in a given quantity of nitrous gas as the step prior to deducing the proportion of oxygen from its effects on atmospheric air.[76] In later editions, Henry stated that the easiest method of applying the solution of iron sulphate impregnated with nitrous gas was by means of Hope's eudiometer, although he had previously described this eudiometer but with the recommendation of using a solution of calcium sulphide. In the fourth edition of 1806, Davy's eudiometrical test was briefly referred to as 'a happy invention of Mr. Davy, which leaves nothing to be desired in eudiometry'. On the other hand, he made no mention of Pepys' eudiometer.[77] The first edition of the *Elements of Experimental Chemistry* in 1810 constituted an update of the *Epitome*, in particular for its content in eudiometry. The reference to Davy's eudiometer remained unaltered, but Pepy's paper on his own eudiometer was quoted, and the new instrument introduced as an ingenious device that enabled absorptions of up to one-thousandth of a part of the gas employed to be measured.[78] Henry's genuine contribution to eudiometry consisted of the identification of a few of the aforementioned shortcomings in Hope's eudiometer and the proposal of substantial procedural as well as material modifications.

He first replaced the larger bottle (B) of Hope's eudiometer (Figure 7.2) with a graduated tube (a) accurately fitted into the neck of the bottle (b) (Figure 7.4, *Fig. 20*).[79] The tube held precisely a cubic inch and was divided into 100 equal parts. When no further diminution in volume took place, the tube was withdrawn (with the neck of the bottle being underwater) and held inverted in water for a few minutes. After that, the diminution in volume was already apparent and could be measured. Henry used as the eudiometrical means a calcium sulphide solution prepared by boiling a mixture of quicklime and sulphur with water, which was filtered and agitated for some

---

[76] Henry, 1801, p. 64; 1803, p. 75.

[77] Ibid., 1808, pp. 53, 68, 155. This is the first American edition of Henry's *Epitome* from the fourth English edition of 1806.

[78] Ibid., 1810, Vol. 1, pp. 192-193.

[79] The reference in the text (*Fig. 28*, plate II) is a misprint; it should be (*Fig. 20*, plate II). This bottle (b) was shown in Figure 7.2 of Hope's eudiometer by (A).

time in a bottle half-filled with common air.[80] In order to overcome the difficulties observed in Hope's eudiometer, he substituted a rubber bulb for Hope's glass bottle (*Fig. 21*, right, b). The tube (a) was accurately ground into a short section of very strong tube with a wider bore (*Fig. 21*, centre, c), the outer surface of which was made rough by grinding in order to retain more effectually the neck of the gum bottle when tied by a string. According to Henry, the only difficulty was in returning all the residual gas to the tube, an exercise that turned out to a matter of practice.

Figure 7.4 Henry's eudiometer.

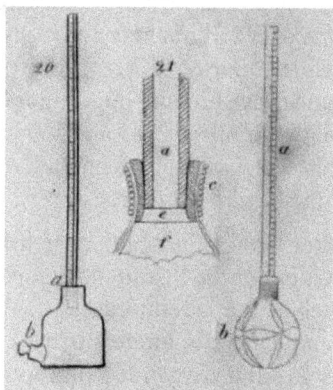

From William Henry, *Elements of Experimental Chemistry* (London, 1810), Vol. 1, plate II, figs. 20, 21.

Although Henry had included the novelty of Pepy's eudiometer, he still insisted that the simplest procedure for applying the solution of iron sulphate saturated with nitrous gas was by means of Hope's eudiometer. In fact, in his eudiometer Hope had left open the possibility of using an alternative eudiometrical means to the calcium sulphide.[81] Henry's work on gases reached its peak with the formulation of the law establishing that the amount of any gas dissolved in water was proportional to its partial pressure in the gas phase. This finding lent support for John Dalton's theory of mixed gases, which eventually led him to the chemical atomic theory.[82]

---

[80] In this way, the solution became saturated with atmospheric nitrogen.

[81] Ibid., pp. 192, 421.

[82] Rocke, 1984, pp. 24-25.

Summary

Alexander von Humboldt's multifaceted personality as a naturalist, traveller and explorer has often overshadowed and even hidden his contributions to chemistry. His devotion to mining in the different posts he occupied during the period 1791-1797 provided him with instrumental skills and awakened his concern for the precarious social situation and the working conditions of the miners. Throughout this period he devoted part of his leisure time to chemistry and plant physiology. His concern about accidents in mines due to fire and black damps transformed the mine into a chemistry laboratory, as a result of which he studied the chemistry of the atmosphere, invented apparatus (a lamp and a respirator) to improve mine safety, remade his own eudiometers and even planned a network of eudiometrical stations. His tour of different locations in preparation for his expedition to America gave him the opportunity to contact Göttling in Jena and to experiment with nitrous gas and soil cultivation in Salzburg. With all that behind him, he was ready to embark on exhaustive chemical research during his short-term stay in Paris.

From May to October 1798, Humboldt read three papers containing eudiometrical connotations at the Institut de France, all of which were the result of his researches at the laboratories of the Agence des Mines in collaboration with Vauquelin. His interest in confirming the saturation ratio between nitrous gas and oxygen led him to establish a procedural variant of the nitrous gas test by introducing a solution of iron sulphate to determine, first, the proportion of nitrogen mixed with the nitrous gas used as eudiometrical reagent and, second, to remove the exceeding nitrous gas after completing the test. Humboldt's innovation also proved useful for demonstrating that the test should be performed over water instead of mercury, as well as the influence of the experimental outcomes of the size of the eudiometrical recipient and the order in which gases were added to it.

Humboldt's mastery of the different eudiometrical methods was patent in his criticisms of Berthollet's test concerning the slow combustion of phosphorous. Humboldt's experiments, recalling his work with Götling in Jena, led him to dismiss Berthollet's test, since a portion of the oxygen in the air sample was masked by the hypothetical formation of a double oxide of phosphorous and nitrogen, making it undetectable. Humboldt reached this conclusion as a result of the use of the iron sulphate variant of the nitrous gas test, and Berthollet counterattacked by casting doubt on the effectiveness of the iron sulphate as a nitrous gas absorbent. Later, in 1805, Gay-Lussac ascribed Humboldt's criticisms to the poorly grounded work of his youth. Humboldt's social concern with agriculture was focused on the factors affecting soil fertility, particularly the earthy composition of soils. His research into this matter was a continuation of his earlier studies in Salzburg and also contributed to the dispute surrounding the phosphorous eudiometer. Humboldt regarded the introduction of the iron sulphate variant as a significant improvement for the nitrous gas test, although he also performed the test without it.

The association of iron sulphate and chloride with the nitrous gas test underwent its greatest development in Great Britain from 1800 onwards. Humphrey Davy's investigations into the oxides of nitrogen (mainly the nitrous oxide) at the Pneumatic Institution also took into account the absorption of nitrous gas by those iron salts from the perspective of their eudiometrical application. The fact was that in July 1800, a new eudiometrical test based on a solution of iron sulphate impregnated with nitrous gas was already available to Davy. He held the slow combustion of phosphorous and the alkaline sulphide tests in high regard, against which he compared his new eudiometrical test to confirm its accuracy. Although an iron sulphate solution was used in the experimental procedure for Davy's eudiometrical test and for Humboldt's variant of the nitrous gas test, from a procedural point of view the former was much more comparable to Martí's test than to the latter. The main resemblance was that both provided an indirect determination of the oxygen content of an atmospheric air sample using a reagent (solutions of iron sulphate impregnated with nitrous gas or calcium sulphide impregnated with nitrogen) that absorbed oxygen without absorbing atmospheric nitrogen.

In 1803, Thomas Charles Hope presented a new eudiometrical device to be commonly used with a solution of calcium sulphide. It consisted of two connected vessels with an ingenious contraption to prevent the oxygen from being absorbed too rapidly. This apparatus was used in experiments to check whether nitrogen was absorbed during human respiration. William Henry considered Hope's eudiometer as the simplest way of using the solution of iron sulphate impregnated with nitrous gas (i.e. Davy's eudiometrical reagent). Nevertheless, Hope's eudiometer suffered from two operational problems: the leakage of air and the gradual dilution of the eudiometrical reagent. To solve both drawbacks, in 1810 Henry proposed the replacement of Hope's glass bottle with a rubber bulb and a means of making the connection of this bulb with the eudiometrical tube airtight. The incorporation of a component made from an elastic material into a eudiometer had in fact been an idea of William Haseldine Pepys and William Allen for their research projects on plant and human respiration. In 1807, they had jointly designed a polyvalent gas mixture analyser that could function as a eudiometer with the use of a solution of iron sulphate impregnated with nitrous gas. Obviously, the inclusion of a flexible material in a chemical apparatus implied the development of new experimental procedures. William Henry was thus responsible for proposing both material and procedural modifications in Hope's eudiometer by means of design transfer from Davy's and Pepys' eudiometers.

8.

Reinventing and consolidating eudiometers at the beginning of the nineteenth century

Dalton's chemical atomism was inspired by his physical fascination with gases and developed through his chemical investigation.[1] In regard to the latter, Dalton's very first chemical experiments on nitrogen oxides enabled him to identify the first verifiable case of integral multiple proportions of combination, as well as playing a significant role in the process of establishing the basis for the reinvention of the nitrous gas eudiometer. In 1804, Gay-Lussac joined Humboldt in his research on the composition of water and atmospheric air that presaged the success of Volta's eudiometer as an analytical instrument. Five years later, Gay-Lussac, in line with Dalton's suggestions, was eventually able to deliver a reshaped version of the nitrous gas eudiometer.

John Dalton. Laying the foundations for the reconversion of the nitrous gas test

The first public description of these experiments appeared in Dalton's paper *Experimental Enquiry into the Proportions of the Several Gases or Elastic Fluids Constituting the Atmosphere*, read on November 12th, 1802, at the Literary and Philosophical Society of Manchester, and which remained unpublished until 1805. Dalton's account of his experiments is commonly regarded as the confirmation of his understanding of multiple proportions in the nitrogen oxides.

Unfortunately, Dalton's notebook, together with many of his papers held in Manchester, were destroyed as the result of an air raid in 1940 during the Second World War. The only surviving records of the notebook are to be found in the work by Roscoe and Harden, *A New View of the Origin of*

---

[1] Rocke, 2005, p.150.

Dalton's Atomic Theory. In a discussion of Dalton's experimental results on nitrogen oxides, these authors state that the interesting question was not *how* Dalton managed to obtain them, but *when* he obtained them.[2] Undoubtedly, the chronology of these experiments constitutes a crucial part of the origin and development of Dalton's widely studied chemical atomic theory.[3] Even so, a knowledge of how these experimental results were obtained would enable both them and the origin of the atomic theory to be placed in their instrumental and procedural context.

Dalton gave an account of the first clear instance of multiple proportions of combination in the first section of his 1805 paper. This section, entitled *Of the Weight of the Oxygenous and Azotic Atmospheres*, was devoted to assessing different eudiometrical procedures in relation to the composition of common air. Although he had had the opportunity to revise the draft of this paper, which he read in 1802, before its publication, the eudiometrical context of his early experiments on the oxides of nitrogen nevertheless remained unclear.[4] Notwithstanding, there is no doubt that in the published and debatable version of his 1805 paper, Dalton wished to frame these experiments in a eudiometrical context. An examination of Dalton's notebook descriptions as recorded by Roscoe and Harden may help to shed light on the eudiometrical context of this episode.

A beginner in chemistry and eudiometry

In order to gain an understanding of Dalton's knowledge in the field of eudiometry, it would first be worthwhile to become acquainted with the extent of his training in chemistry before 1805. During his stay in Kendall (1781-1793), his post as an assistant teacher at a boarding school provided him with access to the vast library of his tutor and friend, the natural philosopher John Gough, as well as to the impressive library of the school, where he became familiar with Boyle's and Boerhaave's works. In 1793, he moved to Manchester to teach mathematics and natural philosophy at the New College but soon found himself obliged to teach chemistry as well. While in Manchester he entered a more challenging scientific world than the one he had known in Kendall, and in 1794 went on to become an elected member of the Manchester Literary and Philosophical Society, of which he became Secretary in 1800 and President in 1817. Dalton's involvement in the

---

[2] Roscoe & Harden, 1896, p. 33.

[3] Rocke, 1984, pp. 27-33.

[4] Ibid., 2005, pp. 136-137.

activities of this Society and his close friendship with William Henry considerably extended his scientific knowledge and experience.

It was in Manchester, in early 1796, where Dalton received his first formal education in chemistry thanks to a series of thirty chemical lectures given by Thomas Garnet, who was to become a professor at the Royal Institution in London. After these lectures he felt confident enough in his expertise in chemistry to agree to give some six lectures on chemistry the following summer in Kendall. In 1800, he resigned his teaching position and opened his own Mathematical Academy, where he offered tuition in mathematics, experimental philosophy and chemistry. In March 1803, he informed his brother that in his leisure time he had been very busily engaged in his chemical and philosophical enquiries.[5] It would not be presumptuous to say that prior to 1805 Dalton had had access to the foremost chemistry books and scientific journals, first, in Kendall, in Gough's private library, and then in Manchester in the extensive Chetham's Library as well as in the Society's library, not to mention his own private library.[6]

There exists no published trace of Dalton's involvement in eudiometrical tests before the publication of his 1805 paper on the proportions of the several atmospheric gases. Nevertheless, it seems that Dalton was well acquainted with the current eudiometrical methods after attending the chemical lectures of 1796. Thus, in the sequel to a paper on the constitution of the atmosphere, published in 1837 and devoted entirely to Volta's eudiometer, the nitrous gas test, and especially the calcium sulphide test,[7] Dalton affirmed:[8]

'[...] I shall now proceed to state the means by which the proportions of oxygen and azote in mixtures of these two gases may best be determined. Having been engaged in this investigation occasionally **for more than forty years**, I may be entitled to give my opinion on this important subject in practical chemistry.'

---

[5] Henry, 1854, p. 47; Thackray, 1972, pp. 48-51, 64-66.

[6] Oliver & Carrier, 2006, pp. 9-14, 28-30.

[7] Dalton called it the quadrisulphuret of lime. He praised Martí's test as the most successful attempt at removing the oxygen from the atmospheric air and recommended the French version of Martí's paper as still being of interest (Dalton, 1837, p. 351)

[8] Ibid., p. 348. The words in bold are by the author.

The eudiometrical context of Dalton's law of multiple proportions

Dalton began the first section of this 1805 paper by listing the five eudiometrical tests widely used at that time: nitrous gas, alkaline or calcium sulphides, hydrogen ignition, green sulphate or muriate of iron impregnated with nitrous gas, and phosphorous fast combustion. He then made it clear that he regarded the finding that atmospheric air contained 21% oxygen as an accepted fact, explaining past discrepancies as a misunderstanding of the nature of the different tests and of the circumstances influencing them. He was convinced that if each of those tests was conducted skillfully, the results from all of them would be the same.

It was nothing unusual for Dalton to focus his attention on the nitrous gas test because of his work on meteorology and mixed gases. While he acknowledged the discredit attaching to the nitrous gas test, he valued it for being not only the most elegant and expeditious of all the existing eudiometrical tests but also as accurate as any other when properly conducted. It appears that Dalton had not been fully aware that the reliability of the test depended on skilful, trained experimenters to conduct it. His intimate friend William Henry had discarded it because the sources of error inherent in the employment of the test had caused him to mistrust the results obtained by it. Henry's reference to Humboldt's researches on the nitrous gas published in the *Annales de chimie* may have influenced Dalton's experimental design and textual presentation of his enquiries into the combination of nitrous gas with atmospheric oxygen.[9] Dalton began his conclusions by criticizing the nitrous gas test in four comments addressed basically to some material and procedural aspects of the test already addressed by Humboldt in his paper:[10]

'I shall, on this occasion, animadvert upon it [the nitrous gas test]'

In the first comment, Dalton pointed out the need of using nitrous gas that was virtually free of azotic gas (nitrogen), with less than 2-3% at most, and nitrous oxide (N_2O in Daltonian terms). The remaining comments were devoted to summarizing his experiments in the form of two eudiometrical trials.

The first trial consisted of adding 100 measures of common air to 36 of nitrous gas (NO in Daltonian terms) in a tube $3\,^1/_{10}$ inches wide and 5 inches long. After waiting for a few minutes, the whole mixture

---

[9] Henry, 1803, p. 74.

[10] Dalton, 1805a, p. 249. See chapter 7; Humboldt's intensive chemistry work in Paris: The iron sulphate variant of the nitrous gas test.

was reduced to 79 or 80 measures, without exhibiting signs of either oxygen or nitrous gas.

In the second trial, 100 measures of common air were added to 72 of nitrous gas in a wide vessel over water so as to form a thin stratum of air.[11] After an immediate momentary shaking, as before, a residue of 79 or 80 measures of pure azotic gas was found.[12]

Finally, if fewer than 72 measures of nitrous gas had been used, there would have been a residue containing oxygen, but if more, then some residual nitrous gas would have been found.[13]

At this point, all the foregoing facts led Dalton to state what has been regarded as a key step in the development of his atomistic reasoning; the discovery of multiple combining proportions:[14]

---

[11] Dalton gave no details of the size of this wide vessel.

[12] Although Dalton had stated that the common air contained 21% of oxygen, the fact that he reported that the trials yielded 79 or 80 measures of residual azotic gas was indicative of his doubt as to whether oxygen comprised 20% or 21% of common air (Usselman *et al.*, 2008, p. 110, note 15)

[13] This is a means of confirming that 72 parts of nitrous gas saturated precisely the oxygen contained in the common air.

[14] It is plausible that Dalton already had the desired results in mind when he stated his first case of integral combining proportions. In 1811, he reported how impressed he had been on observing the proportion of oxygen to nitrogen as 1, 2, and 3 [*sic* for 4] in nitrous oxide, nitrous gas and nitric acid, respectively, in accordance with Davy's analysis, which approximately indicated multiple proportions for oxygen and nitrogen in these oxides (Rocke, 1984, p. 29). Davy gave the following compositions in weight of three oxides of nitrogen (Table 3):

Table 3 Composition of three oxides of nitrogen.

	Nitrogen	Oxygen	Ratio oxygen to nitrogen
Nitrous oxide	63.30	36.70	0.58
Nitrous gas	44.05	55.95	1.27
Nitric acid	29.50	70.50	2.39

(Davy 1800b, pp. 38, 138, 324-325, summarized p. 565; 1836-1840, Vol. 3, pp. 22, 84, 192, summarized p. 335)

Setting the ratio at 0.58 as a reference, the other ratios become 1.27 = **2.2** x 0.58 and 2.39 = **4.1** x 0.58. On rounding off, the ratios of oxygen to nitrogen would be **1** for nitrous oxide, **2** for nitrous gas and **4** for nitric acid (Rocke 1984, p. 45, note 28).

'The elements of oxygen may combine with a certain portion of nitrous gas, or with twice that portion, but with no intermediate quantity'

In order to account for the diversity of the results obtained with the nitrous gas test, Dalton suggested that nitric acid (NO_2 in Daltonian terms) had been formed in the first trial, and in the second nitrous acid (N_2O_3 in Daltonian terms).[15] However, since both acids could be formed at the same time, one part of the oxygen went to one of nitrous gas, while another part of oxygen went to two others of nitrous gas.[16] Therefore, the quantity of nitrous gas absorbed had to be variable across a range of 36 to 72 parts for 100 parts of common air. Regarding the size of the tube used, he concluded that the wider the tube the quicker the test could be completed, and the more exposed the mixture to water the greater was the quantity of nitrous acid and the lesser of nitric acid yielded.

Sometime between October and November 1803, Dalton carried out a series of experiments on the oxides of nitrogen that he reported in the two trials in his paper of 1805. According to Dalton's notebook, nitrous gas and common air should be suddenly mixed in the second trial.[17] In earlier trials, Dalton had calculated the corresponding nitrous gas–oxygen ratio. Actually, this ratio was nothing but the proportion between nitrous gas and oxygen at the point of saturation.[18]

Davy's analyses, published in 1800, were carried out regardless of the nitrous gas test. He had learned about it by December 1803 at the latest, and it was to his credit to assume that the test performed on the basis of a procedure similar to Priestley's would provide him with valuable data and an argument in favor of his multiple combining proportions.

Although the law of multiple proportions was conceived by Dalton, and was supported by Thomas Thomson and experimentally established by William Hyde Wollaston, the credit for the best experimental work - regarding the predictions of atomic theory – should go to Jacques-Etienne Bérard (Usselman, 2000, pp. 254-266).

[15] At that time, nitrous acid was not known to be a distinct and less oxygenated acid, but rather regarded as a mixture of nitric acid and nitrous gas (Davy, 1880b, p. 31; 1836-1840, Vol. 3, p. 22)

[16] The first trial yielded the most oxygenated nitric acid [$O + NO = NO_2$], while the second yielded the least oxygenated nitrous acid [$O + 2\ NO = N_2O_3$]. The latter reaction took place rapidly in the thin gas stratum of the wide vessel, but slow enough in the narrow tube to enable nitric acid to be the sole product (Usselman *et al.*, 2008, p. 107). See chapter 1, note 18, for the chemical equations of the nitrous gas text over water.

[17] Roscoe & Harden, 1896, p. 35.

[18] Thus, assuming that the residual gas was composed of pure nitrogen gas, the

As regards eudiometrical purposes, Dalton recommended attempting to form either nitric acid (first trial) or nitrous acid (second trial) entirely alone rather than a mixture of both. Nevertheless, he decided on the first experiment because it appeared to be the most easily and accurately performed. To this end, he recommended the use of a narrow tube, but wide enough to allow nitrous gas to be absorbed by water without the need for any shaking.[19]

> The test was executed by providing a little more nitrous gas to the oxygen gas than was sufficient to form nitric acid. As soon as the diminution in volume appeared to be over, the gaseous residue was transferred to another tube. 7/19 of the loss was due to oxygen.[20] This was necessary to prevent the nitric acid, formed and combined with water, from absorbing the remainder of the nitrous gas to form nitrous acid.

On October 21st, 1803, nearly a year after the reading of the 1802 paper, Dalton read a paper at the Literary and Philosophical Society of Manchester *On the Absorption of Gases by Water and Other Liquids*, which also remained unpublished until 1805 in the Memoirs of the Society. By that date, therefore, Dalton had already arrived at the conclusion that the

---

amount of oxygen absorbed in the first trial was 21 [100-79] or 20 [100-80], and therefore the ratios would be 1.7 [36/21] or 1.8 [36/20]. In the second trial, the amount of oxygen absorbed was also 21 [100-79] or 20 [100-80], and the ratios would be 3.4 [72/21] or 3.6 [72/20].

In early experiments carried out between March and August, 1803, Dalton reported for both trials the ratios 1.7 - 2.7 and 1.7 – 3.4.Under the date April 1st, 1803, Dalton's notebook provided a list of experimental results detailing in almost every case whether the gas mixture had been made rapidly or slowly. The results indicated that more nitrous gas was absorbed when the mixture was rapidly made. Between May – June 1804 and June – September 1805, Dalton continued his experiments on the combinations of oxygen with nitrous gas. (Roscoe & Harden, 1896, pp. 34, 61, 68). It has been suggested that Dalton obtained the value 1.7 from Lavoisier's paper of 1783 on the combination between nitrous and respirable air. (Rocke, 1984, p. 30). See chapter 3; The nitrous air test in the hands of Lavoisier.

[19] Dalton, 1805a, pp. 247-251. Dalton's option of not shaking the air mixture was not that chosen by early eudiometrists such as Ingenhousz and Cavendish. In principle, the more energetic the shaking the greater diminution in volume.

[20] The 136 [100 +36] parts of the mixture gave a residue of 79 parts in the first test (assuming 21% of oxygen in common air). Therefore, the loss was 57 [136-79] and 7/19 [21/57] of the diminution was oxygen.

rapid mixture of oxygen and nitrous gas over a broad surface of water occasioned a greater diminution in volume than otherwise.[21]

From 1806 onwards, more details emerged about the development of Dalton's nitrous test thanks to the new editions of Henry's work *An Epitome of Chemistry*. A personal communication to Henry provides the conclusions of Dalton's study regarding the influence of the size of the tube and the manner in which the gases were mixed on determining the proportion of oxygen in an air sample.

> If pure nitrous gas was admitted to pure oxygen gas in a narrow tube so that the oxygen gas was uppermost, the two gases united very nearly in the proportion 1.7 [First trial]. If, on the other hand, the nitrous was the upper gas, a much smaller quantity of it disappeared [1 oxygen/1.24 nitrous gas]. If nitrous gas was admitted to pure oxygen gas in a wide vessel over water, the whole effect took place immediately and one measure of oxygen united with 3.4 of nitrous gas [Second trial]. To render this rule more intelligible, Dalton gave as an example the case of 100 measures of common air that were delivered to 100 measures of a mixture of nitrous gas with an equal proportion of azotic or hydrogen gas, which after standing for a few minutes in the eudiometer were found to give 144 measures. When this loss of 56 was divided by 2.7, it gave a measure of almost 21 for the oxygen gas present in 100 measures of common air.[22]

As regards the experimental equipment, two graduated tubes with funnel-shaped extremities (Figure 8.1, **1** and **2**) were employed, each from 3 to 4 tenths of an inch in diameter and 8 or 9 inches long.[23] When

---

[21] Dalton, 1805b, pp. 274-275, note. In this way, the nitrous acid was formed, whereas when water was not present the nitric acid was obtained, which required just half the quantity of nitrous gas. The replications of Dalton's experiments reported in his 1805 paper proved that in the narrow tube conditions (first test) the greatest diminution in volume occurred at a ratio of 1.7, as Dalton reported. Nevertheless, the contraction in volume reported at a ratio of 3.4 (second test) was significantly different from the replicated value (Usselman *et al.*, 2008, p. 108-109).

[22] Henry, 1808, pp. 153-154. In general, in a mixture of **x** parts of oxygen and **y** parts of nitrogen, the oxygen should be saturated with **1,7 x** parts of nitrous gas (assuming that the proportion at the point of saturation between nitrous gas and oxygen was 1.7) and the reduction in volume would be **2.7 x** [x+1.7x]. Therefore, **10/27** [x/2.7x] of the loss of volume was due to oxygen and **17/27** [1.7x/2.7x] to nitrous gas.

[23] These tubes were of the same width but o shorter length than the one described by Dalton in 1805.

analysing atmospheric air samples, it was scarcely necessary to dilute the nitrous gas with any other gas prior to its use. The recommendation was to wait for a certain period of time - 10 minutes, for instance - before noting the diminution in volume, without the need to transfer the residue to another vessel. If the gas sample under examination contained much more oxygen than in atmospheric air, then it was appropriate to dilute the nitrous gas with an equal volume of hydrogen, in which case the narrower the tube, the more accurate would be the result.

Figure 8.1 Apparatus that belonged to Dalton.[24]

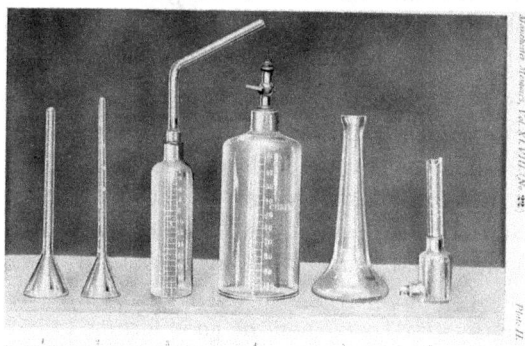

1, 2. Glass funnels with long graduated stems closed at the ends used by Dalton as eudiometrical tubes. 3. Graduated bell jar with bent tube attached for collecting and measuring gases. 4. Graduated bell jar with brass cap and stopcock for measuring gases. 5. Conical glass vessel containing mercury. 6. A fragment of Hope's eudiometer. From *Memoirs and Proceedings of the Manchester Literary & Philosophical Society*, 1904, Vol. 48 (No. 22), plate 2.

By 1806, Henry had changed his mind about the employment of nitrous gas for determining the purity of air. He came to prefer Dalton's method to all the others because of its facility, quickness and accuracy, at least for gaseous mixtures of a very similar standard to the atmosphere. Notwithstanding this constraint, the method was valued because it could be applied to determining the proportion of oxygen in some gaseous mixtures to which other eudiometrical tests were not applicable, such as mixtures of hydrocarbons and oxygen gases. The application of nitrous gas to eudiometrical purposes would still admit of further accuracy when used by Gay-Lussac.

---

[24] In 1904, the Council of the Manchester Literary and Philosophical Society resolved that photographs of the apparatus that belonged to Dalton should be taken for reproduction in the papers of the Manchester Literary & Philosophical Society.

After spending five years away from Europe, Humboldt arrived back in Bordeaux on August 3rd, 1804, and then went to Paris just a few days before the first balloon ascent made by Biot and Gay-Lussac (August 24th, 1804). The main objective of this ascent was to check if the magnetic force at the earth's surface diminished with altitude. They also calculated their altitude from the barometer reading and tested the electricity of the atmosphere. Gay-Lussac made a second balloon ascent alone on September 16th, 1804, and on this occasion he took two samples of atmospheric air at over 6,000 meters that he analyzed jointly with Thenard at the laboratory of the École Polytechnique. The analyses carried out using Volta's eudiometer and the potassium sulphide test showed that the proportion of oxygen at that altitude was identical to that of common air in Paris.[25]

Joseph-Louis Gay-Lussac (1778-1850) was appointed to the post of demonstrator (répétiteur) at the École Polytechnique in September, 1804, a situation that would enable Humboldt to benefit through his training in chemistry at this institution, where Gay-Lussac had attended the chemistry lectures given by Fourcroy and Vauquelin in the first year (1798), by Chaptal in the second year, and by Guyton de Morveau and Berthollet in the final year. All were luminaries of French chemistry in the late eighteenth century. Gay-Lussac had the good fortune to be recruited for the Arcueil group by Berthollet, who eventually became the supervisor of his scientific career. His volumetric approach to matter, i.e. his concern with gases, volatile liquids and volumes rather than condensed matter and weights, was largely due to the influence of Berthollet and Laplace, the patrons of the Arcueil group. The study undertaken with Humboldt in 1805 was one of a number of research projects that exemplifies this volumetric approach to matter.[26]

Collaborative work with Humboldt on the composition of water and atmospheric air

Humboldt met Gay-Lussac a few weeks after the first balloon ascent at Berthollet's country house in Arcueil. Although Humboldt spent much of his time in Paris preparing the publication of an account of his travels and attending the meetings at the Institut de France and the Société d'Arcueil, he found the time to join Gay-Lussac in the laboratory at the École Polytechnique.

---

[25] Gay-Lussac, 1804, pp. 459-461; Crosland, 1978, pp. 28-31.

[26] Ibid., pp. 92-95.

Gay-Lussac and Humboldt became involved in a comparative study of various eudiometrical methods that lasted from November 17th to December 22nd. The corresponding paper of this research was read at the Institut de France on January 21st and 28th, 1805. Their initial stance on the different eudiometrical means implied that all of the eudiometrical procedures could still be improved in terms of accuracy. In the case of the nitrous gas, while it appeared to be the most uncertain eudiometrical means, it was nevertheless capable of indicating very accurately the amount of oxygen contained in the air when combined with the action of the iron sulphate. Gay-Lussac and Humbolt believed that all the eudiometrical means would be able to give the same results if they were all equally known. However, since it was extremely difficult to make all the necessary corrections, preference was given to those methods that had the least need of them, even though they were not always the easiest to operate. According to the authors, they were obliged to interrupt their research and thus were unable to assess all the eudiometrical procedures, but only with those of Martí and Volta. Regrettably, no comparative study with the phosphorous eudiometers accompanied the research.

As mentioned in chapter 5, Gay-Lussac and Humboldt considered Martí's eudiometer to be more inaccurate than Volta's because it determined the proportion of vital air in the atmospheric air as being between 21% and 23%. The fact was that, after perfecting the experimental procedure, Martí succeeded in reducing the margin of uncertainty to between 21% and 22%. They also mistook Martí's eudiometrical means, calcium sulphide, for an alkaline sulphide (i.e. potassium sulphide).[27] Berthollet had attributed to Martí the erroneous idea that alkaline sulphides were responsible for the absorption of nitrogen from the air. Accordingly, Gay-Lussac and Humboldt carried out their experiments on the absorption of atmospheric nitrogen by solutions of alkaline sulphides by employing potassium sulphide instead of calcium sulphide. This confusion not only spread that mistaken notion but also contributed to a biased assessment of Martí's eudiometer.[28] It is obvious that the authors adopted a proactive attitude towards Volta's eudiometer, to which they devoted half of the paper:[29]

---

[27] This is an unexpected mistake, since the abridged French translation of Martí's paper, published in the *Journal de physique*, refers to the correct term of "sulfure calcaire liquid" (liquid sulphuret of lime) (Martí-Franquès, 1801a, p. 177).

[28] Humbolt & Gay-Lussac, 1805, pp. 131-134.

[29] Humboldt & Gay-Lussac, 1805, p. 13.

'Some people accused this instrument of being inaccurate, of indicating too low a quantity of oxygen in the air; but it seemed to us that with the introduction of some corrections, and taking into account the law of their variations, we could make it very exact and very practical [...].'

They therefore addressed four crucial aspects of Volta's eudiometrical method: the complete absorption of one of the two gases after ignition, the uniform nature of the product of their combination, the exact proportion of the two gases to form water, and the limits of error inherent to Volta's eudiometer. Very few details were provided about the operational procedures, except for the production of both gases. Oxygen was obtained by the thermal decomposition of potassium chlorate (*muriate sur-oxigené de potasse*) using an experimental device intended to prevent the absorption of nitrogen. Hydrogen was released by reacting zinc with muriatic acid or diluted sulphuric acid.[30]

After igniting different mixtures of oxygen and hydrogen, they observed that the absorption of both gases could only be completed in certain proportions and that proportions existed such that it was impossible to ignite them (Table 4).

Table 4 Summary of the diminutions in volume observed in the ignition of mixtures of oxygen and hydrogen.

Hydrogen (parts in volume)	Oxygen (parts in volume)	Diminution (in volume)
100	200	146
100	300	146
100	600	146
100	900	146
100	950	68
100	1,000	55
100	1,200	24
100	1,400	14
100	1,600	0

'On avait accusé cet instrument d'être infidèle, d'indiquer dans l'air de trop petites quantités d'oxygène; mais il nous avait paru qu'en supposant qu'il exigeait de corrections, on pouvait en les appréciant, ainsi que la loi de leurs variations, le rendre très exact et très commode [...]'

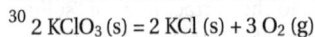

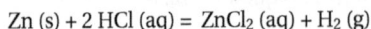

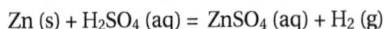

[30] $2 KClO_3 (s) = 2 KCl (s) + 3 O_2 (g)$

$Zn (s) + 2 HCl (aq) = ZnCl_2 (aq) + H_2 (g)$

$Zn (s) + H_2SO_4 (aq) = ZnSO_4 (aq) + H_2 (g)$

After carrying out a number of careful quantitative experiments with an excess of first one gas and then the other, they found that the combining proportions of both gases were constant, despite variations in the original proportions before reacting. Consequently, they were able to state that the product of the combustion of hydrogen (i.e. water) was of a uniform nature, as well as to calculate that 100 parts by volume of oxygen combined with 199.89 parts of hydrogen, practically 200 parts.[31]

When it came to the fourth aspect, Gay-Lussac and Humbolt were faced with questions regarding the limits of error of Volta's eudiometer and the lowest quantities of oxygen or hydrogen that the instrument was able to detect. Volta's test was so instantaneous that the results were practically independent of the outside temperature and atmospheric pressure variation. In this respect, it had a very distinct advantage over phosphorus and alkaline sulphides tests in that it yielded highly comparable results. Accuracy for the quantity of oxygen could be of nearly three-thousandths of the quantity of the analysed air.

Gay-Lussac and Humboldt foresaw the future analytical power of Volta's instrument for gasometry. They pointed out that regardless of its capacity of determining the entire quantity of oxygen contained in the air, the instrument was the only one with which the proportion of hydrogen in a gas mixture could be assessed, and in this sense it still deserved much more attention and greater commitment to the study of its functioning. More than enough reasons to warrant praise of Volta for his contribution to chemistry:[32]

---

[31] Humbolt & Gay-Lussac, 1805, pp. 145-149. In May, 1790, Fourcroy, Vauquelin and Séguin had already determined that 100 parts by volume of vital air (oxygen) combined with 205.36 parts of hydrogen at 10ºR and with 206.97 at 14ºR (Fourcroy *et. al*, 1791, Vol. 9, pp. 37-41). Circa 1785, Volta doubted whether inflammable (hydrogen) and dephlogisticated air (oxygen) mixed in a proportion of 100 to 48-50, respectively, completely combined after ignition. See chapter 2; Interpreting what occurred inside the eudiometer.

In this respect, historians have insisted on recognizing this 1805 paper as an antecedent of Gay-Lussac's law of the combining volumes of gases. Nevertheless, it is worth noting that the paper was largely ignored until four years later, when Gay-Lussac's paper on that law was published (Crosland, 1978, p. 59).

[32] Humbolt & Gay-Lussac, 1805, pp. 149-151.

'Ainsi l'illustre physicien Volta qui a enrichi la physique des plus belles découvertes, aurait encore la gloire d'avoir donné à la chimie l'instrument le plus exact et le plus précieux pour ses analyses.'

'So the illustrious physicist Volta who has enriched physics with the finest discoveries, would still have the glory of having given to chemistry the most accurate and most valuable instrument for its analysis.'

It would have been regrettable if, with such a fine instrument, Gay-Lussac and Humboldt had not brought their investigations into the nature and composition of atmospheric air to a conclusion. They analysed nineteen air samples collected over the river Seine on different days and under different meteorological conditions and came to the conclusion that the composition of atmospheric air was generally invariable, that it contained 21/100 of oxygen and that it did not contain a measurable amount of hydrogen. When addressing the apparent differences in the composition of atmospheric air in local circumstances (fermentations, marshes, volcanoes, hospital dormitories or theatres) they observed no significant variation in the composition of these local airs. Thus, the anxiety experienced by some people in closed and crowded rooms, as well as the specific lake and marsh diseases, could not be explained by variations in the air composition.[33]

Humboldt's eudiometrical researches were the object of some severe criticism from Berthollet and Saussure, and this paper provided an opportunity for some reconsideration. The introductory section seemed an appropriate place for Humboldt's rectification of his early work on eudiometry. Gay-Lussac reported that Humboldt:[34]

'In the year VI he [Humboldt] presented to the Institute two papers on the analysis of air, containing a large number of experiments that he now regards – as he himself states - not only as highly inaccurate but as rightly opposed by Mr. Davy, and by a chemist who honours us both with particular kindness, Mr. Berthollet. Zealous for the advancement of science, Mr. Humboldt wished to replace the work of his youth by another based on more solid foundations [...]'

---

[33] Ibid., pp. 156-157.

[34] Ibid., p.13.

'En l'an VI il [Humboldt] avait présenté à l'Institut deux mémoires sur l'analyse de l'air qui renferment un grand nombre d'expériences qu'il regarde aujourd'hui - c'est lui même qui le déclare - non seulement comme très inexactes, mais encore comme justement combattues par M. Davy et par un chimiste qui nous honore tous deux d'une bonté particulière, par M. Berthollet. Zélé pour le progrès de la science M. Humboldt a voulu remplacer ce travail de sa première jeunesse par un autre fondé sur des bases plus solides [...]'

On March 11th, 1805 Gay-Lussac and Humboldt set out from Paris for Rome to meet Humboldt's brother, who was serving as the Prussian envoy. After this journey they travelled to Berlin in September. Humboldt did not return to Paris until 1808, while Gay-Lussac returned home in early 1806.

The study on the oxides of nitrogen

The combinations of nitrous gas with oxygen constituted one of the issues in chemistry about which little agreement existed at the beginning of the nineteenth century. On March 13th, 1809, Gay-Lussac read the *Mémoire sur la vapeur nitreuse et sur le gaz nitreux considére comme moyen eudiométrique* at the Institut de France, where he reported on his research work that aimed not only at establishing the theory of the formation of nitrous and nitric acids using nitrous gas and oxygen, but also the transformation of the nitrous gas eudiometer into an instrument of accuracy.

In his landmark paper of 1808 on the law of combining volumes of gases, he had ascertained that the nitrous gas was composed of equal parts in volume of oxygen and nitrogen.[35] In other words, 100 parts of oxygen and 100 of nitrogen produced 200 parts of nitrous gas without any diminution in volume. He also recalled that nitric acid was composed of 100 parts of nitrogen and 200 of oxygen. Nitric acid could therefore be regarded as composed of 100 parts of oxygen and 200 of nitrous gas, because the latter contained as much oxygen as nitrogen without any diminution in volume.

---

[35] From the data provided by Davy on the composition in weight of three oxides of nitrogen (see note 14), Gay-Lussac calculated the corresponding ratios in volume (Table 5):

Table 5 Ratios in volume oxygen to nitrogen.

	Nitrogen	Oxygen
Nitrous oxide	100	49.5
Nitrous gas	100	108.9
Nitric acid	100	204.7

According to Gay-Lussac, the first and last ratios differed little from 100 to 50 and from 100 to 200, and only the second ratio differed a little more from 100 to 100. Although this difference was not very great, he wanted to ensure that it was wholly nil. From the data for the density of nitrous gas determined by Bérard at the laboratories of the Société d'Arcueil, Gay-Lussac could establish that nitrous gas was composed of equal parts by volume of nitrogen and oxygen (Gay-Lussac, 1809a, pp. 215-216).

He also found that 100 parts of nitrogen required 50 parts of oxygen to form nitrous oxide.[36]

To obtain the nitric or the nitrous acids by combining nitrous gas with oxygen was not simply a matter of first introducing one gas and then the other, but of which gas predominated in the mixture. When oxygen and nitrous gas were mixed in the appropriate ratios, the absorption of the vapour formed thereby was prompt and complete. Thus, by using a narrow tube, nitric acid containing 100 parts of oxygen and 200 of nitrous gas was obtained. However, when both gases were mixed in a slightly larger tube, absorption did not vary significantly providing that no shaking took place, because water would dissolve the nitrous gas. In this case, the acid obtained was nitrous acid gas containing 100 parts of oxygen and 300 of nitrous gas (Table 6). On the other hand, if either of the two gases predominated to excess, the nitrous gas was prevented from coming into contact with the water and dissolving easily. Thus, with an excessive amount of oxygen, nitric acid was produced, while on the other hand an excessive amount of nitrous gas produced nitrous acid.[37]

Table 6 Ratios in volume of four oxides of nitrogen recognized by Gay-Lussac in 1809 (*) Modern formulas.

	Nitrogen	Oxygen	
Nitrous oxide (N_2O)*	100	50	
Nitrous gas (NO)*	100	100	
Nitric acid (N_2O_5)*	100	200	
	Nitrous gas	Oxygen	Volume reduction
Nitric acid (N_2O_5)*	200	100	
		In excess	300
Nitrous acid gas (NO_2)*	300	100	
	In excess		400

[36] Gay-Lussac, 1809a, pp. 216, 218, 284.

[37] Gay-Lussac, 1809b, pp. 236-239. The study of the oxides of nitrogen was a complex field of research for early nineteenth century chemistry. Gay-Lussac successfully re-examined his conclusions on this issue in 1816 after Dalton and Davy's criticisms (*Annales de chimie et de physique*, 1816, Vol.1, pp. 394-410).

Gay-Lussac's results on the composition of the oxides of nitrogen did not agree with those that Dalton had published in 1805.[38] According to Dalton, 21 parts of oxygen could unite with 36 of nitrous gas or with twice 36, i.e. 72 parts. In other words, 100 parts of oxygen united with 171.4 or 342.8 parts of nitrous gas. In Gay-Lussac's opinion, these results were inaccurate because the first ratio of nitrous gas was too small and the second was too large, in addition to which the two gases did not combine in simple ratios. It should be remembered that Gay-Lussac's law of combining volumes established that gases combined in very simple ratios and that the volume reduction they underwent on combination also had a simple ratio to their volume, or at least to the volume of one of them.[39]

Despite the discrepancy of Gay-Lussac's results with Dalton's on the composition of the oxides of nitrogen, he did not refrain from stating his conclusions on the influence of the size of the tube in which the gases combined - a key factor in the design of his eudiometrical device. Gay-Lussac's volumetric approach to matter was not the only influence of his mentor Berthollet, whose experience with procedures in large scale chemical productions was probably decisive for his view that chemical phenomena were to a large extent conditioned by their surrounding circumstances. From this perspective, the fact that Gay-Lussac gave so much importance to the size of the reaction tube and its effect on the outcome of the combination of gases may be better understood.

---

[38] Actually, Gay-Lussac did not quote Dalton's paper published in the *Memoirs of the Literary and Philosophical Society of Manchester* of 1805 but in the *Philosophical Magazine* of 1806 (Vol. 23, pp. 349-356)

[39] Ibid., 1809b, p. 244. The laws of multiple proportions and combining volumes were experimental generalizations concerning the composition of chemical compounds. The law of multiple proportions came to establish that when two chemical species (element, compound, etc.) combined in more than one proportion, then the quantities (usually, but not necessarily, measured by mass) of one species combined with a fixed quantity of the other species are expressed in ratios of whole numbers. Furthermore, the law of combining volumes can be reviewed in terms of volumes of gases, either elements or compounds, corrected to standard conditions, taking part in, or being formed, or both taking part in and being formed, by chemical reactions, bear small whole number ratios to each other. Dalton's law essentially relates to gravimetric composition, whereas Gay-Lussac's law relates exclusively to the volumes of gaseous elements and compounds (Bradley, 1992, pp. 32, 52-53).

Reshaping the nitrous gas eudiometer

Since the aim of eudiometrical analysis was to remove all the oxygen in an air sample, an excess of nitrous gas was needed in order to obtain a volume reduction four times larger than the volume of oxygen in the sample. Thus, possible errors corresponded to only a quarter of the oxygen, and since it was not possible to go wrong by four degrees, the oxygen content in a gas mixture could be estimated by much less than one-hundredth. The only precautions to be taken were to avoid shaking the mixture and to ensure that nitrous gas was always predominant without too much excess, since the more it was absorbed, the less it would be mixed. Even in this case, however, the error would never reach a hundredth part of oxygen. In addition to these precautions, two sources of error also had to be taken into account. First of all, if the gases were mixed in a very narrow tube, nitrous acid would scarcely be absorbed by water because of the lack of contact, which would necessitate shaking, in which case nitrous gas would also be absorbed. It was for this reason that by mixing 100 parts of common air with 100 parts of nitrous gas very variable absorptions were obtained. Secondly, the question of whether to introduce the nitrous gas into the tube before or after the air sample was also important, because if it was introduced first, both nitrous and nitric acids might be formed. To avoid these two shortcomings, Gay-Lussac conducted a test that employed an apparatus very similar to that used by Humboldt for assessing carbonic acid in a gas mixture or for analyzing common air by means of nitrous gas and chlorine.[40] This test was performed in the following manner (Figure 8.2):[41]

> The sample of the air to be analysed was collected in the measure (N), equivalent to 100 parts of the tube (K) graduated in 300 parts. The air sample was then introduced into this tube (K) with the copper funnel (M) coupled to the ferrule (HI) of the tube. The number of parts of the air sample contained in the tube was noted. Afterwards, the air sample was transferred to a wide glass vessel (A) with a flat bottom, containing about 250 parts and closed by a copper component (BFGC). This component consisted of a slightly funnelled part (BC), a funnel (FG) and a sleeve (DE) abraded with emery so that

---

[40] Gay-Lussac, 1809b, pp. 246-247.

[41] Ibid., pp. 249-251.

the ferrule (HI) of the tube (K) fitted exactly. The nitrous gas[42] was measured in the same way and rapidly mixed with the air sample by coupling the tube to the sleeve (DE) without agitating. A red vapour appeared immediately and then disappeared very quickly. After half a minute, or one minute at most, absorption could be regarded as complete. The device was then turned upside down and the residual gas ascended in the tube. After that, the tube (K) was removed from the vessel (A) to restore the pressure equilibrium and the residue was assessed. The total absorption divided by 4 gave the quantity of oxygen.

Gay-Lussac reported having performed many varied analyses, always finding a perfect agreement among them. In 1818, William Henry still regarded Gay-Lussac's application of nitrous gas to eudiometrical purposes as an accurate procedure, provided certain precautions suggested by his theoretical views of the constitution of nitrogen oxides were taken into account.[43]

Figure 8.2 Gay-Lussac's nitrous gas eudiometer.

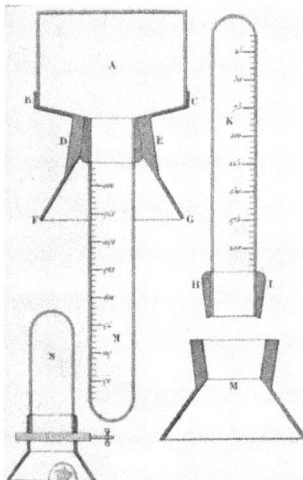

From *Mémoires de physique et de chimie de la Société d'Arcueil,* 1809, Vol. 2, plate 2.

---

[42] To ensure the quality of the nitrous gas, Gay-Lussac analyzed air samples in which Humboldt had made various animals breathe and which he had analyzed with Volta's eudiometer.

[43] Henry, 1818, Vol. 1, pp. 393-394.

Summary

In Manchester, William Henry befriended John Dalton who was a newcomer to chemistry. Henry's reference in the 1803 edition of his *Epitome of Chemistry* to Humboldt's researches on the nitrous gas test probably drew Dalton's attention to that test with the idea of presenting an initial case of multiple proportions of combination. Although no published trace of Dalton's involvement in eudiometry seems to exist before the publication of his 1805 paper, it appears that he was well acquainted with the current eudiometrical methods at that time.

Dalton came to favour the nitrous gas test because he found it elegant, expeditious and straightforward, in the sense that the Priestleyan version of the test was characterized by simplicity of materials, apparatus and experimental procedures. He showed no interest at all in the nitrous gas test as a eudiometrical method for verifying the oxygen content in common air, because he clearly believed that this content was 20-21%. His apparent ambivalence to the nitrous gas test oscillated between disapproval and praise, with an occasional expression of animadversion in the form of two eudiometrical trials. The sources of Dalton's criticisms of the nitrous gas test were very precise: the influence of the size of the eudiometrical vessel and the shaking of the gas mixture in its volume reduction. He opted for the use of a narrow tube that allowed nitrous gas to be absorbed by water without shaking in an attempt to obtain nitric instead of nitrous acid. Nevertheless, he was aware that a greater reduction in volume was obtained if the test was performed over a broad surface of water.

His conclusions on these two factors were in principle concerned with the justification of his statement on the multiple combining proportions, rather than with the improvement of the nitrous gas test, and he was in fact obliged to conduct the test in vessels of different sizes vessels and with variable procedures until he obtained the results he desired. His interest in achieving the value 1.7 for the nitrous gas–oxygen ratio in the tests carried out with the narrow tube is especially significant, and cannot be disassociated from the fact that Lavoisier had established the same numerical value for the proportion of both gases at the point of saturation.

From 1806 onwards, Dalton contributed material as well as procedural improvements to the nitrous gas test employed as a eudiometrical method. Thus, in addition to recommending the use of narrow tubes, he emphasized the advantage of adding the nitrous gas once the oxygen gas was already in the tube and not the other way round. Arguably, the nitrous gas test that in the hands of Dalton had evolved from a eudiometrical method to an iconic case of multiple combining proportions was returned to eudiometrists in a simpler and more trustworthy version of the eudiometrical test than those performed with the latest nitrous air eudiometers. In a certain sense, it was as if Priestley's conception of the nitrous air test had won out in the end.

Dalton's investigations on the nitrous gas test went on to induce further development in 1809 at the hands of Louis-Joseph Gay-Lussac, who had already begun his research work on eudiometry some years earlier. Humboldt had arrived at Paris in 1804 after his American expedition, and once there began a joint investigation with Gay-Lussac on the composition of water and atmospheric air at the laboratory facilities of the École Polytechnique, with the intention of conducting a comparative study of different eudiometrical methods. For reasons unknown, they were obliged to abandon this study, their investigations being limited to Martí's and Volt'as eudiometers. With the former device they persisted in mistaking the genuine eudiometrical reagent (calcium sulphide) for an alkaline sulphide, which contributed to their unfair assessment of Martí's eudiometer.

Undoubtedly, the fact that they came down in favour of Volta's eudiometer definitely strengthened its analytical power. An important contribution of the study was the procedure to remedy the nullity of Volta's eudiometer with mixtures of low oxygen content. The solution was formed by bringing such gas mixtures to another composition by adding known quantities of oxygen and/or hydrogen to make the gas mixture ignitable. The study underscored two outstanding advantages of this eudiometer: the swiftness of its execution, remarkably superior to that of the other eudiometers, and its high degree of accuracy. Nevertheless, the decisive experimental result of their research was that it established the volumetric composition of water in two parts of hydrogen to one of oxygen. The other feature of this joint work was the recurring topic of the composition of atmospheric air. Humboldt and Gay-Lussac confirmed that common air generally contained 21/100 parts in volume of oxygen and no measurable trace of hydrogen. They regarded the nitrous gas test as the most uncertain of the eudiometrical methods, despite recognizing that it could be improved if combined with iron sulphate. However, five years later this perspective underwent a radical change.

In 1809, Gay-Lussac undertook the first attempt at establishing a theory of the formation of the oxides of nitrogen, which subsequently led to a reinvention of the nitrous gas eudiometer. He did not agree with Dalton's experimental results on the proportions of a combination of nitrous gas with oxygen, mainly because these proportions did not match his law of combining volumes. However, this discrepancy proved to be no obstacle to

Gay-Lussac's acceptance of Dalton's conclusion on the influence of the size of eudiometrical recipients on the experimental outcomes. His own researches on the oxides of nitrogen, together with Dalton's recommendations on the size of the recipients, guided him in the reshaping of the nitrous gas eudiometer that culminated in the definitive version of this type of instrument.

9.

From eudiometry to gasometry

Volta's eudiometer as a gas mixture analyser

Some early attempts

At the very beginning of the nineteenth century, Volta's eudiometer was already being used to analyse gas mixtures other than common air samples. In 1801, the French chemists Charles-Bernard Desormes and Nicolas Clément established the identity of carbon oxide (carbon monoxide) by differentiating it from other inflammable gases. William Cruickshank had also achieved the same result shortly before. The fact that this gas was able to detonate with oxygen prompted Desormes and Clément decide to analyze it with Volta's eudiometer using a mixture of carbon oxide and oxygen in equal proportions. It was possible to determine the composition of this new gas because the composition of the carbonic acid gas (carbon dioxide) produced was previously known.[1]

When dealing in 1803 with the issue of the composition of carburetted hydrogen compounds (hydrocarbons) and oxy-carburetted hydrogen (a supposed ternary combination of carbon, hydrogen and oxygen in variable proportions), Berthollet addressed the general question of the analysis of inflammable gas mixtures using Volta's eudiometer. These analyses were based on the fact that hydrogen and carbon combined with oxygen to form water and carbonic acid gas, both being compounds of constant composition. In this way, the mass quantity of carbon and

---

[1] Desormes & Clément, 1801, pp. 36-45. Their analyses showed that the weight proportion of carbon in carbon oxide varied from 46% to 52%. Desormes and Clément obtained carbon monoxide by means of different methods: passing carbon dioxide over red-hot charcoal, strongly heating charcoal with zinc oxide or barium carbonate, or distilling charcoal. The detonation of the gas occurred according to:

$$2 CO \text{ (g)} + O_2 \text{ (g)} = 2 CO_2 \text{ (g)}$$

They calculated a weight proportion of 71.65 of oxygen to 28.35 of carbon for carbon dioxide. The actual proportion is 72.7 of oxygen to 27.3 of carbon.

hydrogen in the inflammable gas could be determined once its specific weight was known. The method turned out to be more accurate for carburetted hydrogen than for oxy-carburetted gases.[2] By 1809 Berthollet had resumed his early work on this matter with the assistance of Bérard, who was an expert in eudiometrical analysis. Bérard carried out these experiments in a Volta eudiometer with iron armour over mercury. Berthollet found that no pure carburetted hydrogen gas existed and that those considered to be such always contained some oxygen; that is, they were all oxy-carburetted gases.[3]

Dalton had already discovered the method of analysis by firing in Volta's eudiometer by September 1803.[4] He employed it in his experiments on the construction of his chemical atomic theory. In particular, his analysis of marsh gas (methane) and olefiant gas[5] (ethene) using that instrument, carried out by August 1804, allowed him to establish a second case of multiple proportions.[6] In 1808, Henry summarized the experimental procedure followed in the analyses of these inflammable gases on the basis of their rapid combustion with oxygen:[7]

> A mixture of the inflammable gas with oxygen in known proportions was admitted over mercury into a Volta eudiometer, ignited by an electric spark and the diminution in volume ascertained. The residual gas mixture was washed with caustic potash or limewater, and a second diminution in volume indicating the quantity of carbonic acid formed by the combustion was noted. The quantity of nitrogen, accompanying the oxygen employed and in the residue left by potash or limewater, was determined by an appropriate eudiometrical test.

---

[2] Berthollet, 1803b, Vol. 2, pp. 64-65. For a full account of these researches see (Ibid., 1803a)

[3] Ibid., 1809. An iron armor was used to prevent the eudiometrical tube from breaking when detonating highly explosive gas mixtures.

[4] Dalton, 1819, p. 476.

[5] The term "olefiant" described the remarkable property of the gas to form an oil (ethylene dichloride) when combined with chlorine. This olefiant gas was popularly known as "Dutch liquor" or "*liqueur des hollandaise*"

[6] Dalton, 1810, pp. 440, 445, 448; Roscoe & Harden, 1896, pp. 29, 62-64, 68; Farrar, 1968, p. 179; Rocke, 1984, pp. 27, 34. Dalton established the formulae CH for ethene and CH_2 for methane.

[7] Henry, 1808b, p. 284. In 1819, Dalton was still employing this method in his analysis of sulphuric ether (Dalton, 1819, pp. 476-479)

The quantity of oxygen absorbed by the detonation was deduced from all this information.

Since it was assumed that oxygen gas underwent no change of volume by conversion into carbonic acid gas, then in the formation of each volume of this gas an equal volume of oxygen was employed.[8] Thus, after deducting the volume of the carbonic acid gas formed from the volume of oxygen consumed, the remaining number indicated the amount of oxygen employed in the saturation of hydrogen. In these calculations, it was also assumed that by combustion the carbon acquired all the oxygen necessary for its acidification and that no part of it existed previously in the state of carbon oxide.[9]

Nevertheless, reasons still existed to be dissatisfied with the aforesaid method and to distrust the results obtained thereby. The products of the combustion of one same gas varied considerably in different experiments, and in some cases it was evident that not all of the carbon was oxygenized as a result of the formation of charcoal during detonation. The quantities of inflammable gases that could be submitted to experimentation in this way were extremely small, and the ignition of highly combustible gasses such as the olefiant gas involved the considerable danger that the eudiometrical tube might burst.[10]

Nicolas-Théodore de Saussure's paper on his researches, read at the Société de Physique et d'Histoire Naturelle in Geneva on August 31st, 1809, concerned the subject of whether previously heated, red-hot charcoal furnished hydrogen gas on burning, and enabled him to reach some conclusions on the most usual eudiometrical processes. Saussure employed the process ideated by Humboldt and Gay-Lussac for determining very small quantities of hydrogen in another gas, with the aim of ascertaining the small proportion of hydrogen in the residue obtained after burning common charcoal, wood or oil or any vegetable substance in pure oxygen

---

[8] Henry, 1805, p.66, note.

[9] Henry used Dalton's experimental results from the detonation of carburetted hydrogen (methane) to exemplify these calculations. Dalton had determined that 100 volumes of carburetted hydrogen consumed 200 of oxygen and gave 100 of carbonic acid gas. Assuming that in the formation of each volume of carbonic acid gas, an equal volume of oxygen was employed, it followed that the carbonic acid held in combination 100 volumes of the oxygen consumed, and that the remaining 100 volumes of oxygen saturated the hydrogen (Roscoe & Harden, 1896, p. 63; Henry, 1808b, p. 285).

[10] Ibid., p. 286.

obtained from potassium chlorate. However, he also warned that Volta's eudiometer was liable to error when hydrogen gas was burnt slowly in a mixture of oxygen and nitrogen, since the nitrogen was in part combined either with the oxygen alone or with oxygen and hydrogen.[11]

To ascertain the proportion of oxygen before and after the combustion of charcoal, Saussure used Martí's test, employing a concentrated solution of potassium sulphide impregnated with nitrogen, and then compared this test with that of Volta's eudiometer. Saussure followed Martí's process (except for using potassium instead of calcium sulphide) quite closely, but allowed the mixture of alkaline sulphide and gas under analysis to stand at rest for five days, thereby obtaining more consistent results than those achieved in a few minutes by agitation according to Martí's original method. The eudiometrical process conducted with potassium sulphide was more accurate than Volta's process for determining the proportion of oxygen when mixed with nitrogen only, but when the mixture was made with oxy-carburetted hydrogen gases it was preferable to employ Volta's eudiometer. The reason for this was that the solution of potassium sulphide sensibly absorbed all oxy-carburetted hydrogen gases, making the proportion of oxygen appear larger than it really was.[12]

In April 1810, Saussure read another paper at the same Genevan Society that turned out to be conclusive on the question of the composition of olefiant gas as well as on the methodology for analyzing compound inflammable gases with Volta's eudiometer. Saussure took care to see that the products of the combustion of olefiant gas were exclusively water and carbonic acid gas, so that the whole part of the oxygen, which had not been employed in forming carbonic acid gas, had been employed to form water. Taking into account that two parts by volume of hydrogen consumed one of oxygen to produce water, and that in the formation of each volume of carbonic acid gas an equal volume of oxygen was employed, Saussure found that the volume of hydrogen gas contained in the olefiant gas was twice the difference between the volume of the oxygen consumed and the carbonic acid gas produced.[13] He disagreed with

---

[11] Saussure, 1809, pp. 270-271, 279-280; 1810, pp. 170, 174.

[12] Ibid., 1809, pp. 265-267; 1810, pp. 166-167.

[13] $V_{(oxygen\ consumed)} = V_{(oxygen\ consumed\ to\ form\ water)} + V_{(oxygen\ consumed\ to\ form\ carbonic\ acid\ gas)}$

$V_{(oxygen\ consumed\ to\ form\ water)} = \frac{1}{2} V_{(hydrogen\ in\ olefiant\ gas)}$

$V_{(oxygen\ consumed\ carbonic\ acid\ gas)} = V_{(carbonic\ acid\ gas\ produced)}$

$V_{(oxygen\ consumed)} = \frac{1}{2} V_{(hydrogen\ in\ olefiant\ gas)} + V_{(carbonic\ acid\ gas\ produced)}$

$V_{(hydrogen\ in\ olefiant\ gas)} = 2\ [V_{(oxygen\ consumed)} - V_{(carbonic\ acid\ gas\ produced)}]$

Henry's analysis of the olefiant gas and decided to carry out his own analysis and calculations.[14] Accordingly, he proceeded to the analysis of the gas by its combustion over mercury using Volta's eudiometer.[15] In this way, Saussure was able to conclude that olefiant gas, when properly prepared, contained no sensitive quantity of oxygen, so it could only be composed of hydrogen and carbon and should therefore be termed carburetted hydrogen. Omitting fractions, the gas contained by weight 86% of carbon and 14% of hydrogen.[16]

Gas lighting manufacturing in Great Britain

The gaseous compounds of hydrogen and carbon were objects of scientific research, not only for their intrinsic interest but also because of their application for a major economic purpose. At that time, the light obtained by the combustion of gas derived from coal had begun to be used in large-scale manufacturing. The inflammability of these compound gasses in a closed vessel, and their fitness for the purpose of affordable light, were directly proportionate to the quantity of oxygen required for their saturation. Therefore, olefiant gas burnt with the greatest brilliancy and carburetted hydrogen, though inferior produced a dense and compact flame. On the other hand, carbonic oxide and hydrogen were entirely unfit to be employed as a means of artificial illumination.[17]

---

[14] Henry, 1808b, pp. 292-294.

[15] A mixture of 100 parts of olefiant gas with 500 parts of oxygen, deprived of carbonic acid by potash, which contained 23.5 of nitrogen and 476.5 of pure oxygen, was detonated. The mixture was reduced to 409.5 parts that were deprived of carbonic acid by potash and analyzed with Martí's test (using potassium sulphide). The residue contained 201 parts of carbonic acid gas, 184.5 of oxygen and 24 of nitrogen. After separating the carbonic acid from this residue, a small portion of hydrogen was added to examine whether the original olefiant gas was burnt. On a second detonation of the mixture, not more than one hundredth of carbonic acid gas at most was formed. The condensation of the gasses by the combustion was equal within a hundredth to what should have resulted from the action of the hydrogen gas added. The first detonation had therefore affected the combustion of the olefiant gas. (Ibid., 1811, pp. 61, 64; 1812, p. 72.)

[16] Ibid., 1811, pp. 66-67; 1812, pp. 73-74. The actual composition of ethene by weight is 85.7% carbon and 14,3% hydrogen

[17] Henry, 1808b, p. 285-286.

In Great Britain, the scientific context of research into the composition of carburetted hydrogen compounds cannot be disengaged from the economic and social context of providing artificial illumination by the combustion of gas from coal distillation.[18] Therefore, the need for an analytical procedure that was not susceptible to the drawbacks mentioned above should be understood within these contexts. In this respect, after many attempts Henry was able to contrive the following apparatus described in the paper he read on June 23rd, 1808, at the Royal Society.[19]

> The principal components of the apparatus (Figure 9.1) were two cylindrical glass receptacles: the larger one (bb) for containing oxygen and acting as a eudiometrical tube, and the smaller one (oo) for containing the inflammable gas submitted to experiment, both connected by a bent glass tube (ss) measuring $^1/_{10}$ of an inch in diameter. An iron burner (t) with a $^1/_{30}$ of an inch aperture was cemented to the upper end of the tube, while a socket was fixed to the lower end on which the stopcock (r) could occasionally be screwed. The receptacle (oo) was placed in a larger glass jar (nn) closed at the top by the brass cap (p) and the stopcock (q). The oxygen gas receptacle was also closed by the brass cap (e) and the stopcock (f). The chain (h) connected this stopcock to the prime conductor of an electrical machine.[20] The lower aperture of the stopcock (f) was tapped internally for the purpose of receiving a small screw at the end of the copper wire (g). This wire consisted of two parts, both of which screwed into a movable socket connecting them. With this contrivance the wire could be lengthened or shortened at will. To prepare the apparatus for use, the receptacle (oo) was partly filled with the inflammable gas and secured by wedges of cork (vv) in the jar (nn), the level of the water in which was regulated by opening the stopcocks (x)

---

[18] The development of gas lighting in Great Britain is associated with the name of William Murdoch (1754-1839). In 1792, he commenced a series of experiments on the quantity and quality of the gases contained in different substances. He observed that the gas obtained by distillation from coal, peat, wood and other inflammable substances burnt with great brilliancy, and it occurred to him that if the gas was confined and conducted through tubes it might be employed as an affordable substitute for lamps and candles. In 1798, he constructed an apparatus at the Soho Foundry that was applied successively over many nights to the lighting of the building. In 1802, this factory provided the opportunity to mount a public display of the new artificial illumination (Henry, 1805, pp. 73-74; Clow & Clow, 1992, pp. 427-429).

[19] Henry, 1808b, p. 286-291.

[20] A Leyden jar.

or (z). The bent tube (ss) with its stopcock (r) was screwed onto the top of the receptacle and partially immersed in the water of a pneumatic trough (aa) so that the aperture of the burner could rise a few inches above the surface of the water. [...] Once the stopcock (f) had been connected to the electrical machine chain (h), a rapid succession of sparks was triggered between the copper ball at the end of the wire (g) and the aperture of the burner (t). The stopcocks (q) and (r) were then opened, and the stream of gas was ignited to undergo a slow combustion [...]

Figure 9.1 Henry's eudiometrical apparatus.

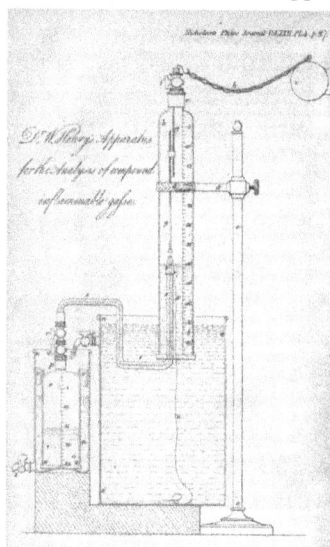

From *A Journal of Natural Philosophy, Chemistry and the Arts*, 1809, Vol. 22, plate 4, p. 87.

Although an obvious objection to this method was that the real proportion of the products resulting from their combustion might be disguised due to the absorption of a part of the carbonic acid gas by the water over which the experiment was performed, Henry found that this was a source of error too trivial to be deserving of consideration. If the operator was sufficiently dexterous, it was actually a matter of practice to ensure that the interval of time between the completion of the combustion and the measurement of the residue would be too short to allow the absorption of any significant amount.

Another concern was that the process of combustion was often complicated by the imperfect combustion of the inflammable gas, since a part of it escaped through the orifice of the burner, either wholly unaltered or only partially burned. As this portion could not be detected by Martí's test, it gave a false appearance of an actual addition of nitrogen to the oxygen gas remaining in the receptacle. Nevertheless, this amount could be diminished by the adoption of certain precautions. In general, the more combustible the inflammable gas submitted to experiment, the more complete was its decomposition. The apparatus was therefore better adapted to the analysis of olefiant and carburetted hydrogen gases, or mixtures of both, than to that of carbon oxide and gases from different kinds of coal. The apparatus could also be adapted for use with mercury instead of water. Henry preferred this method of slow combustion when only a few experiments were required on gases of great combustibility, thanks to the greater safety of the apparatus, and also to its greater accuracy in terms of the quantities that could be consumed. On the other hand, when a great number of experiments were necessary, he favoured the standard rapid method of detonation for the time it saved.[21]

Gay-Lussac. A renewed version of Volta's eudiometer and the search for the nature of prussic acid.

Berthollet's son (Amédée-Barthélemy) concluded his paper on the analysis of ammonia, which he read at the Institut de France on March 24th, 1808, and in which he indicated some precautions that should be taken with the use of Volta's eudiometer. He was especially concerned with the fact that whenever operating over water, the residue was increased by the air released from the water due to the vacuum formed after detonation. This was a shortcoming that could be partially obviated by operating in open mode; that is, ensuring that the inside of the eudiometer could freely communicate with the water during detonation.[22] However, operating in open mode involved the risk of losing gas. Gay-Lussac would later devise a eudiometer to overcome these difficulties.[23]

> The apparatus (Figure 9.2) consisted of a thick glass tube (op) closed at its upper part by a brass ferrule (ab) bearing an inner ball (c) opposite to another ball (d), between which the electric sparks should be triggered. The ball (d) was suspended from a spiral wire that was frictionally engaged in the glass tube (Figure 9.3, left). This arrangement made it possible to bring together or separate the two balls at will. The ferrule (fg) at the lower end of the eudiometer was

---

[21] Ibid., 1819, pp. 421-422.

[22] Berthollet, 1809, pp. 287-294.

[23] Gay-Lussac, 1817.

intended to lend solidity to the instrument. This ferrule was fixed by a screw (q) to a circular plate (ik), movable around the screw that served as its axis. This plate had at its centre a conical opening closed by a valve held by the rod (mn) (Figure 9.3, right). A small pin (n) delimited the extent of the ascent of the valve. At the moment of the explosion, the valve, pressed from top to bottom, was thus closed. But as soon as a vacuum began to form in the eudiometer, the water raised the valve and filled it, thereby preventing the vacuum from being formed. In order to strengthen the plate (ik), it was fitted into a small notch (k) in the prolongation of the ferrule (gh). A metallic clamp (M) held the instrument in place while operating. This clamp terminated with an open ferrule that the screw (V) pressed against the glass tube.[24]

Figure 9.2 Gay-Lussac's eudiometer (1).

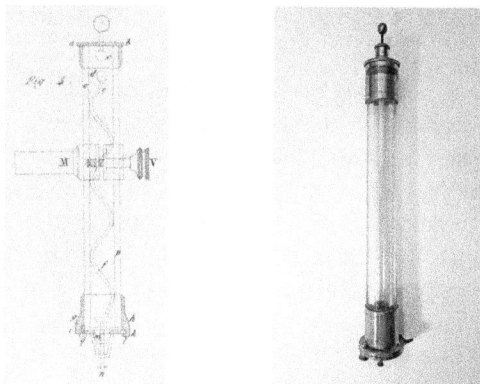

(Left) From *Annales de chimie et de physique*, 1817, Vol. 4. (Right) A copy without the spiral wire belonging to the Lycée Bertran de Born, Périgeux. Photography courtesy of Francis Gires.

---

[24] Later, Gay-Lussac realised that this anti-vacuum valve somewhat disturbed the transfer of gas and decided to remove it from the eudiometer and install it in the water trough. The innovation consisted of a cork stopper with a conical profile in its lower half that was fixed firmly in the trough with mastic. This stopper was pierced longitudinally and a glass or metal tube was introduced inside the hole to prevent the stopper from collapsing under compression. The hole was closed at the top by a small tin disk with a copper wire appendage for replacing the disk in the hole after each detonation. This very mobile valve was held in place under its own weight so that the hole closed exactly during the explosion. The valve was lifted immediately afterwards, allowing the water to pass through and fill the vacuum produced in the eudiometer. Before starting the trial, the eudiometer was placed on the stopper, cut to close exactly and pressed onto its base, which had to be perfectly flat (Ibid., 1833).

Figure 9.3 Gay-Lussac's eudiometer (2).

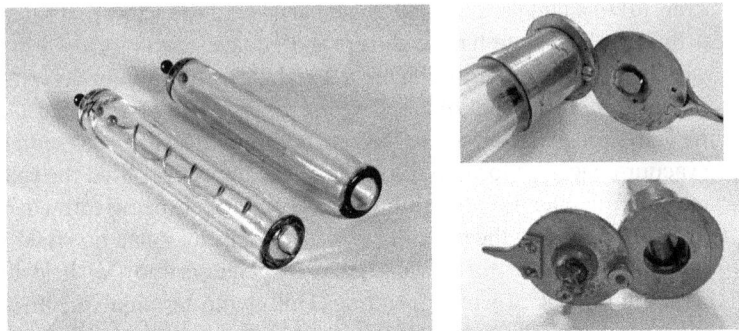

(Left) Eudiometrical tube with the spiral wire, belonging to the Collège Henri IV, Bergerac. (Right) Views of the conical opening closed by a valve, from a copy that belonged to the Lycée Bertran de Born, Périgeux. Photographs courtesy of Francis Gires.

Gay-Lussac recommended his renewed version of Volta's eudiometer to chemists on the basis of the instrument's practicality, which he had recognized for several years. This endorsement led to speculation that Gay-Lussac may have used this eudiometer in his outstanding research work on prussic acid (hydrogen cyanide), presented in a paper to the Institut de France in September 1815. Few compounds had been the object of more study than prussic acid, yet few were less well known. This research was conclusive in showing the qualitative and quantitative composition of prussic acid, but it can also be regarded as paradigmatic case, among others, of the significant role played by Volta's eudiometer in chemical laboratories during the first quarter of the nineteenth century.[25]

> Taking advantage of the hot days of August and the volatility of prussic acid, a glass jar was about two-thirds filled with oxygen over a mercurial trough at a temperature of between 86° and 95° and then filled completely with the vapour of prussic acid. When the temperature of the mercury was lowered to that of the ambient air, a determinate volume of the gaseous mixture was taken and washed in a solution of potash. The residue, when compared with the absorption that had taken place, gave the exact ratio of the oxygen to the prussic vapour. Then, without fear that the vapour of prussic acid might condense, a given volume of this gaseous mixture was introduced into Volta's eudiometer equipped with platinum wires,

---

[25] Gay-Lussac, 1815, pp.147- 148; 1816, pp. 354-355.

and an electric spark was triggered through the instrument. A highly vigorous combustion, accompanied by a white vapour from a little nitric acid and the water of water thus produced, and a diminution in volume occurred, which was ascertained by measuring the residue in a graduated tube. Once washed with a solution of potash, this residue underwent a further diminution as a result of the absorption of the carbonic acid gas formed. Lastly, this final residual gas was analyzed over water by hydrogen and was determined to be a mixture of nitrogen and oxygen, since this last gas had been employed in excess.

By his law of combining volumes, Gay-Lussac concluded that one volume of prussic acid vapour contained just as much carbon as would form its own volume of carbonic acid gas, as well as half a volume of nitrogen and half a volume of hydrogen. He was then confronted by the question of whether these elements were the only ones that entered into the composition of prussic acid and whether the proportions were exact. He thus compared the density of the prussic acid vapour with the sum of that of its elements,[26] according to which the density of the prussic acid vapour should be equal to the sum of that of the vapour of carbon and to half that of nitrogen and hydrogen.[27] The difference of one-hundredth between the calculated and experimentally determined densities of prussic acid vapour was estimated to be within the range of experimental error. Thus, the fact that one volume of prussic acid vapour contained one volume of the vapour of carbon, half a volume of nitrogen and half a volume of hydrogen, was assumed to have been demonstrated, and that no other substance entered into its composition.[28] A further significant

---

[26] Gay-Lussac was able to back up his ideas on the composition of compound gases by evidence from densities, i.e. the density of a compound gas was an additive property of its constituents (Crosland, 1978, p.102)

[27] Gay-Lussac applied his volumetric approach on matter to non-volatile solids, such as carbon. He had determined that, on combining with carbon, one volume of oxygen expanded to two volumes of carbon oxide, and therefore assumed that if a proportional amount of carbon could be vaporized, it would occupy two volumes (Rocke, 1984, p. 111). The density of the vapour of carbon was calculated by subtracting the density of oxygen from that of carbonic acid gas.

[28] Gay-Lussac, 1815, pp. 149-152; 1816, pp. 355-357. Additionally, the weight composition of prussic acid was also determined (44.39% carbon, 51.71% nitrogen, 3.90% hydrogen)

finding by Gay-Lussac in his research was the isolation of cyanogen. Actually, he believed he had isolated the radical of prussic acid.[29]

Proceeding in the same manner as in the case of prussic acid, Gay-Lussac was justified in concluding that a volume of cyanogen contained two volumes of the vapour of carbon and one volume of nitrogen.[30] Both aforementioned cases exemplify, firstly, although only partially, Gay-Lussac's first general method for organic analysis established mutually with Thenard,[31] and secondly, his volumetric approach to matter by expressing the composition of both substances in terms of 'volume

---

[29] In order to investigate the nature of the new compound, a mixture of cyanogen with about twice and a half its volume of oxygen was strongly detonated in Volta's eudiometer. Assuming operation on 100 parts of cyanogen, a diminution in volume of between approximately four to nine parts was found. When the residue was treated with potash, it diminished from 200 to 195 parts due to the loss of carbonic acid gas. When the new residue was analyzed over water by hydrogen, it yielded from 94 to 98 parts of nitrogen. The oxygen contained in this same residue, added to that in the carbonic acid gas, was equal (to within four or five per cent) to that which had been employed. (Gay-Lussac, 1815, p.181; 1816, p. 40.)

[30] Ibid., 1815, pp. 182-183; 1816, p. 41.

[31] Gay-Lussac & Thenard, 1811, Vol. 2, pp. 268-350. The method devised by Gay-Lussac and Thenard in 1811, based on oxidizing the organic substance with potassium chlorate, evolved between 1811 and 1815 thanks to the mutual collaboration between Gay-Lussac and Berzelius. Rather than indirectly using volumes, the latter decided to measure directly by weight the water and the carbonic acid gas produced. He proposed to absorb them in condensed-phase using calcium chloride and potassium hydroxide, respectively. Gay–Lussac, for his part, suggested using the more stable copper (II) oxide instead of potassium chlorate. Nevertheless, this method demanded experimenters skilled in measuring gases and also took up a great deal of time, at least two days for each analysis. An important limitation was the allotted maximum sample size, which could not be much greater than a tenth of gram, restricted by the maximum gas volumes that could be handled. A persistent problem was the difficulty of determining nitrogen accurately because of the complex mixtures of nitrogen compounds produced after the oxidation. Liebig's invention of his *Kaliapparat* in 1830 eliminated the separated volumetric absorption of water and carbonic acid gas, capturing them by means of the Liebig's combustion train and measuring both by weight in a single operation. Liebig's contributions to nitrogen determination were of little significance, but drove Duma's innovation that solved the problem in 1833 (Rocke, 2000, pp. 273-287; Usselman, 2003, pp. 74-88; Usselman *et al.*, 2005, pp. 3-4, 35-42). Liebig's *Kaliapparat* coupled with Duma's improvement made analysis of nitrogenous organic compounds a reliable method.

formulas'.[32] At that time, Volta's eudiometer had already become much more than an instrument for establishing the composition of atmospheric air or the inflammability of a gas; it had become an indispensable apparatus for determining the volumetric composition of gas compounds. Within the material instrumental framework of gas chemistry, Volta's eudiometer was the best practical apparatus that exemplified the volumetric approach to matter governed by Gay-Lussac's law of combining volumes and its sequel, the additive nature of the densities of the ingredients in a compound gas. Throughout the review of the analysis of organic compound in the 1830s, the need for making any eudiometrical measurement vanished. Nevertheless, in a quite different context, Volta's eudiometer would recover a transformative protagonism in the hands of the German chemist Robert Bunsen.

Bunsen. The birth of gasometry in the context of the cast iron industry

Robert-Wilhelm-Eberhard Bunsen (1811-1899) studied sciences at the University of Göttingen where he completed his doctorate. Aided by a grant from the Hanoverian government, he toured Europe from the beginning of 1830 to the fall of 1833. In September 1832 he arrived at Paris for a nine-month residence. There he worked in Gay-Lussac's laboratory and learnt from Dumas, Chevreul, Pelouze, Regnault, Reiset and others. In 1836, he succeeded Wöhler at the Kassel Technische Hochschule, where he began his celebrated researches into eudiometry and the cacodyl compounds. In 1839, he was appointed professor of chemistry at the University of Marburg and in 1852 succeeded Gmelin at the University of Heilderberg, where he remained until his retirement in 1889. The turbulent political atmosphere after the revolution of 1848 made it easy for Bunsen to leave Marburg, but it was in Heilderberg where sowed the seeds of a future dynamic academic chemical community.

In addition to the carbon-zinc battery (1841), the photometer (1844), the gas burner named after him (1855) and other pieces of laboratory equipment, Bunsen accomplished great work in many branches of chemistry. He concentrated mostly on physical and inorganic chemistry, in which he developed a variety of analytical techniques and highlighted the importance of accurate quantitative measurements. However, his researches into the cacodyl compounds, his only venture into organic

---

[32] Rocke, 1984, pp. 110-112. These 'volume formulas' could be transcribed into modern symbolic notation, for the sake of clarity, as $H_{1/2}CN_{1/2}$ for prussic acid and C_2N for cyanogen.

chemistry, elevated him to the highest position as an experimentalist chemist. After 1843 Bunsen excluded organic chemistry from his teaching and research. His aversion to that subject was apparently related to his strong general disinterest in theory.[33]

Between 1836 and 1846, he was involved in the investigation of the industrial production of cast iron in Germany. He examined the gases emitted from the Vickerhagen blast furnace operated by charcoal, and by analyzing the composition of these gases and applying scientific foundations he was able to optimise the production of cast iron. Lyon Playfair, who had made Bunsen's acquaintance at Marburg, was quick to appreciate the importance of these investigations. In his position as an organic chemist on the Geological Survey of the London Museum of Economic Geology, Playfair had been assigned to the analysis of mixtures of naturally occurring hydrocarbons, a requirement of a parliamentary commission on explosions in coal mines. At Playfair's suggestion, Bunsen visited England and undertook a similar series of experiments for the blast furnaces at Alfreton, which were fed with coke and coal. It was thus that a research model for the application of scientific investigation methods to the elucidation of industrial problems was initiated.

However, Bunsen's success in this field was a result of his research work on accurate methods of gas mixture analysis in the 1830s. In 1842, Hermann Kolbe drafted the first detailed description of Bunsen's gas analysis methods, published in 1842 in a handbook of pure and applied chemistry (*Handwörterbuch der reinen und angewandten chemie*) edited by Liebig, Poggendorff and Whöler. These methods came to the attention of the English public, albeit not in a comprehensive manner, through communication presented by Bunsen and Playfair to the meeting of the British Association held at Cambridge in June 1845. In October of the same year, Bunsen sent Kolbe as an assistant to Playfair to analyze mixtures of gases collected from coal mines, with the aim of providing means of preventing explosions. In 1857, Bunsen compiled his research on the analysis of gas mixtures carried out in relation to the improvement of cast iron production, which appeared in the book *Gasometrische Methoden*,[34] the only one he ever published. The English version of the book (*Gasometry*) was published in the same year, 1857.[35]

---

[33] Ibid., 1993, pp. 25, 47.

[34] A second and greatly enlarged edition appeared in 1877.

[35] Roscoe, 1900, pp. 519-520.

During the 1860s, the name of Bunsen became inextricably linked to that of Gustav Kirchhoff. In 1860, both scientists developed the spectroscope, a new and powerful analytical tool that led not only to the explanation of Fraunhofer's lines in the solar spectrum but also paved the way to finding the chemical composition of the sun and other stars. Bunsen took advantage of the application of spectral analysis to examine earthy materials. Spectral analysis led Bunsen and Kirchhoff to announce the discovery of caesium in 1860, and of rubidium one year later. Over the following years, several other elements were identified using spectroscopic methods.

Rectifying sources of error in Volta's eudiometer

When Hermann Kolbe wrote the entry *Eudiometer, Eudiometrie* for the second volume of the aforementioned chemical dictionary-style handbook, he was at that time one of Bunsen's assistants in Marburg. He was therefore in an excellent position to provide a further account of Bunsen's expertise in eudiometry. Kolbe expressed the current view that although eudiometry had lost its original significance, eudiometers continued to be used for analysing gas mixtures. However, eudiometry had a very limited application and was considered unreliable in comparison with the other branches of analytical chemistry, because the physical characteristics of gases made their handling difficult and stood in the way of their accurate measurement. Kolbe dismissed the eudiometric methods known until that time, except for Volta's eudiometer, whose sources of error could apparently be overcome by appropriate use of the instrument and by observing a few precautions.[36]

Bunsen's improvements to eudiometric analysis consisted not so much in the invention of a new device, but rather the elimination of the sources of error that affected Volta's method. In addition, Bunsen introduced new procedures for the separation and determination not only of oxygen, hydrogen, nitrogen and the carbonic acid gas but also of nearly all the permanent gases that could be collected over mercury. In what follows, the description of Bunsen's eudiometric device is centred in those aspects concerning the attempt to control or mitigate these sources of error. Kolbe's entry in the handbook of 1842 and the Bunsen-Playfair paper of 1846 are used as reference sources for the description below. Nevertheless, this account has been completed when necessary with Bunsen's text of 1857 in the interests of comprehension.

---

[36] Kolbe, 1842, pp. 1051-1053.

The glass tube was between 600 and 700 mm long, with an inner diameter of approximately 20 mm and a thickness of glass that did not exceed 1 mm. Two platinum wires with the thickness of horsehair were fused into the tube for triggering the electric spark through the gases. These wires were internally bent into the curve in the head of the tube so that the tips approached each other up to a distance of 3 mm (Figure 9.4). It was inadvisable to position the wires straight across the tube, since they were very likely to be bent on cleaning or when filling the instrument with mercury.

Since the effectiveness of the apparatus depended very much on how the platinum wires were fixed in the glass, great care had to be taken when fusing the wires into the glass to prevent the slightest breach from opening up between the glass and the wires, which could lead to the diffusion of the enclosed gases. For this reason, the corresponding process of implanting the platinum wires was described in full detail.[37]

Figure 9.4 The closed end of Bunsen's eudiometric tube showing the two platinum wires bent into the curve of the tube.

From *Handwörterbuch der reinen und angewandten chemie*, 1842, Vol. 2, p.1053, Fig. 88.

Bunsen's eudiometer differed from former eudiometers in that the graduation on the tube was independent of its capacity. Graduation of the tube on a volumetric scale was unadvisable, not only because it was more difficult to perform, but also because the volume of gas at the closed end of the tube would not correspond to the number etched on the divisions. A volumetric scale was by no means as exact as a length scale (i.e. millimetre scale), whose individual divisions were then calibrated with respect to the volume of the tube.

---

[37] Ibid., pp. 1053-1054; Bunsen, 1857, pp. 23-24.

For the purpose of graduating the eudiometric tube, Bunsen constructed a special partitioning machine (Figure 9.5). It consisted of a wooden board 2.5 m long and 3 cm thick with a groove (a) in which a model tube of hard glass (bb) graduated millimetrically was situated. This tube was held firmly in position by a straight brass strip fixed down by screws, but which allowed the curvature of the glass to rise slightly above the edge of the brass strip. The tube to be graduated (dd) was covered with a thin coating of wax containing a little turpentine and was fastened in a similar manner by two pieces of brass whose edges were separated by a few millimetres. The deeply engraved divisions on the model tube could easily be transferred onto the waxed surface of the tube (dd) by means of a rod that had a knife-edge at one end and a sharp point at the other. This was done by allowing the sharp point to pass lightly from one division to another on the model tube. In order to engrave the partitions marked on the wax onto the glass, gaseous hydrofluoric acid evolved from a mixture of fluorspar (the mineral form of calcium fluoride) and sulphuric acid was used.

Figure 9.5 Bunsen's copying machine for the graduation of the eudiometric tube.

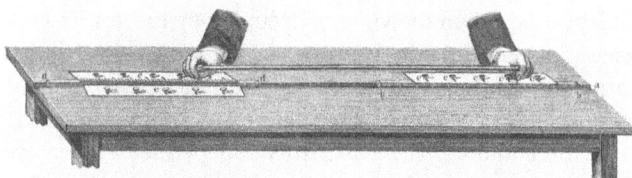

From Robert W.E. Bunsen, *Gasometry*, 1857, p. 26. Fig. 21.

The engraved graduation on the tube could not be used to indicate the capacity of the instrument because the inner diameter of the tube was not uniform. In order to calibrate the tube volumetrically, it was therefore necessary to pour the same quantity of mercury into various parts of the tube and then read off in each case the height to which the mercury rose on the graduated scale. In this manner it was possible to find the volume corresponding to each graduation on the tube. These equivalences were arranged into two columns (Table 7) in such a way that the linear divisions were given in the (Mm) column, while the next column gave the corresponding capacity of the tube according to a comparable standard.

Table 7
Correspondences between the millimetre scale and capacities in the Bunsen's eudiometric tube.

Mm.		Mm.		Mm.		Mm.	
10	9,2	22	21,3	34	33,3	46	45,4
11	10,2	23	22,3	35	34,3	47	46,4
12	11,2	24	23,3	36	35,3	48	47,4
13	12,2	25	24,3	37	36,3	49	48,4
14	13,2	26	25,3	38	37,3	50	49,3
15	14,2	27	26,3	39	38,3	51	50,3
16	15,2	28	27,3	40	39,4	52	51,3
17	16,3	29	28,3	41	40,4	53	52.3
18	17,3	30	29,3	42	41,4	54	53,3
19	18,3	31	30,3	43	42,4		etc.
20	19,3	32	31,3	44	43,4		
21	20,3	33	32,3	45	44,4		

From *Handwörterbuch der reinen und angewandten chemie*, 1842, Vol. 2, p.1058.

However, the volume found in the table still required a slight correction, because the height of the mercury in the tube was read at the highest point of the convexity of its meniscus, so the corresponding volume given by the table was less than the volume occupied by the gas. To correct this error, twice the space arising from the fault due to convexity had to be added to the volume taken from the table. That space was determined experimentally and provided the definitive volume reading.[38] The accuracy of the eudiometric measurements depended on the eye of the experimenter being in a perfectly horizontal plane with the surface of the mercury column.[39]

---

[38] Kolbe, 1842, pp. 1056-1059; Bunsen & Playfair, 1846, pp. 147-148, Bunsen, 1857, pp. 28-32.

[39] In order to obviate these parallax errors, Bunsen advised the use of a small moveable mirror placed on the opposite side of the tube. If the pupil of the experimenter's eye as seen through the tube in the mirror appeared to be halved by the mark corresponding to the convexity of the mercury, then the reading could be considered as correct. Instead of the mirror reading, Bunsen also made use of a horizontally oriented telescope, movable in the vertical direction and placed at a distance of between 6 and 11 feet away from the eudiometer. Apart from the fact that the readings could be carried out with much greater ease, it had the great advantage that the experimenter was situated at a safe distance from the tube and, unlike in the case of the mirror reading, did not have to fear any expansion of gas

A major requirement for the performance of accurate eudiometric analyses was to conduct them in places protected as much as possible from changes in atmospheric pressure and temperature, and at the same time with enough direct illumination by sunlight to enable exact measurements. For this purpose, it was desirable to conduct the analysis in a room with thick walls,[40] with no adjoining heated chambers and with large north-facing windows.

The experiments were conducted on a large bench or table (Figure 9.6) furnished with a tube (a) to siphon off the mercury spilt during the operation. Two upright supports (bb) with moveable arms (cc), which could be placed in any direction on the table, served as holders for the barometer (h) and the eudiometric tube (m). The analytical equipment was completed with a small wooden pneumatic trough filled with mercury (Figure 9.7) about 350 cm long and 80 cm wide. This trough had a glass window (H) on the side facing the observer through which the lower mercury level in the trough at the foot of the eudiometric tube could be read. The telescope (g) used for reading the height of the mercury in the barometric and eudiometric tubes stood on the floor a few feet away from the table.

Figure 9.6 Bunsen's laboratory setting for eudiometric analysis.

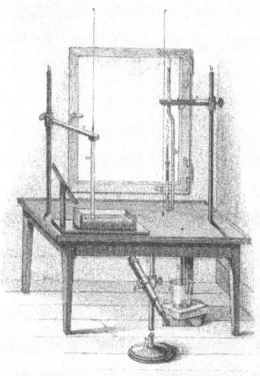

From Robert W. E. Bunsen, *Gasometry*, 1857, p. 22. Fig. 17.

---

when heating the tube due to the proximity of the experimenter's body. (Kolbe, 1842, pp. 1058; Bunsen & Playfair, 1846, pp. 148, Bunsen, 1857, p. 36.)

[40] The six-foot thick walls of Bunsen's laboratory building at Marburg provided outstanding temperature stability for eudiometric analysis (Rocke, 1993, p. 111).

Figure 9.7 Pneumatic trough with a glass window.

From *Handwörterbuch der reinen und angewandten chemie,* 1842, Vol. 2, p.1060.

In order for the changes in temperature of the mercury to coincide as closely as possible with those of the surrounding air, it was necessary that nobody entered the room for at least half an hour between each combustion and the subsequent reading, so that the gas, as well as the nearby thermometer, achieved exactly the ambient temperature. It was also necessary to allow between half an hour and two hours to elapse between each test.[41]

Another source of error arose from air bubbles that attached to the tube glass wall or around the platinum wires, for which particular precautions had to be taken when filling the eudiometer with mercury and transferring the gas sample. A preventive action consisted in cleaning and drying the tube with filtering paper after washing it out with water after every test. When practicable, a drop of water was transferred to the head of the clean tube by means of an iron wire, taking care not to wet the tube in the process. This quantity of water was more than sufficient to saturate the gas sample with water vapour at ordinary temperature. Once formed, air bubbles could be removed by touching them with a wire or by detaching a large air bubble and leaving a lot of tiny bubbles behind. Another precautionary measure consisted of filling the eudiometric tube with mercury by means of a funnel with a long nozzle, ending in a narrow opening at the lower end and placed at the bottom of the tube. The mercury flowed in a fine stream and gradually rose up the tube to form a perfectly mirror-like surface on the wall of the tube. To prevent air bubbles from forming when transferring the gas sample from a collecting tube to the eudiometer, the closed end of this former tube was broken by pressure against the bottom of the trough and underneath the open end of the eudiometer. In this way, the gas sample rose into the eudiometer, thereby displacing the mercury within.

---

[41] Kolbe, 1842, p. 1060; Bunsen, 1857, pp. 21-22, 33-34, 38.

Special care had also be taken to ensure that atmospheric air neither entered nor escaped during the combustion of the gas in the eudiometer. This was prevented from happening by pressing the open end of the tube during the explosion on a perfectly smooth sheet of rubber placed under the mercury in the pneumatic trough. However, it was necessary to take care that the rubber did not carry any air down with it that might find its way into the eudiometer by a pressure drop.[42]

The algebraic approach to eudiometric analysis

Bunsen dealt with the eudiometric analysis of mixed gases by dividing it into two parts: the separation of the absorbable gases by suitable absorbents, and the determination of the combustible gases by combustion with oxygen. In this latter case, Bunsen was able to prove that the eudiometric analyses of gases achieved a degree of accuracy that was unsurpassed by the most thorough analytical methods, and further showed that the presence of nitrogen had no disturbing influence on the analysis of explosive gas mixtures.[43] Apart from the detailed experimental procedures and the novelties in the laboratory equipment, Bunsen introduced an algebraic dimension into the eudiometric analysis of gas mixtures. This algebraization of the eudiometric analysis implied the planning of mathematical equations to calculate the composition of combustible gas mixtures.

As an example of calculation, Bunsen presented the case of the gas mixture resulting from the combustion of coke, brown coal and wood in a cast iron furnace. The only combustible constituents of such a mixture were carburetted hydrogen (methane), carbon oxide (carbon monoxide) and hydrogen. The composition of the mixture could be easily calculated

---

[42] Kolbe, 1842, pp. 1061; Bunsen & Playfair, 1846, pp. 148, Bunsen, 1857, pp. 34-35, 45-46. The observed volume of gas v, found in the table of capacity, was reduced by calculation to the volume V occupied in the dry state at 0 °C and under a pressure of 1 metre of mercury. This volume was found from the following equation:

$$V = \frac{(v + m)(b - b_1 - b_2)}{(1 + 0.00366 \cdot t)}$$

In which b represented the height of the barometer (atmospheric pressure), b_1 the height of the column of mercury rising from the level of the trough into the eudiometer, t the room temperature, b_2 the saturation pressure of water vapour for the temperature t and m the error of the meniscus (Bunsen & Playfair, 1846, p. 148; Bunsen, 1857, p. 38).

[43] Bunsen & Playfair, 1848, p. 152; Bunsen, 1857, p. 38.

by determining the volume of the oxygen consumed and that of the carbonic acid gas produced.

The reasoning behind the corresponding equations was based on the fact that hydrogen and carbon oxide consumed half of their volumes of oxygen. Carburetted hydrogen, however, required twice its own volume of oxygen. On the other hand, carburetted hydrogen and carbon oxide produced an equal volume of carbonic acid gas.[44] Then, for any given volume **A** of a gas mixture, consisting of x hydrogen, y carburetted hydrogen, and z carbon oxide, if **B** was the volume of oxygen necessary for the combustion and the **C** was the volume of carbonic acid gas produced, the following equations were obtained:[45]

$$x + y + z = A \qquad \tfrac{1}{2}\,x + 2\,y + \tfrac{1}{2}\,z = B \qquad y + z = C,$$

which gave the following:

$$x = A - C \qquad y = (2\,B - A) / 3 \qquad z = C - (2\,B - A) / 3.$$

The eudiometric analysis for the above-mentioned case yielded the following result:

Volume at 0 °C - 1 m Hg

Original gas sample	45.08 (1)
After addition of oxygen	113.91 (2)
After the first combustion	72.94 (3)
After absorption of carbonic acid gas	53.61 (4)
After the addition of hydrogen	159.48 (5)
After the second combustion	39.44 (6)

The first combustion was performed after the addition of oxygen to the original gas sample that also contained nitrogen. The second combustion was performed after, first, the absorption of the carbonic acid gas in the

---

[44] 1 vol. hydrogen + ½ vol. oxygen = 1 vol. water vapour

1 vol. carburetted hydrogen (methane) + 2 vol. oxygen = 1 vol. carbonic acid gas + 2 vol. water vapour

1 vol. carbon oxide + ½ vol. oxygen = 1 vol. carbonic acid gas

Then a mixture consisting of equal volumes (1) of hydrogen, carburetted hydrogen and carbon oxide (in total, 3 volumes) required for combustion ½ + 2 + ½ (in total, 3) volumes of oxygen and yielded: 1 + 1 (in total, 2) volumes of carbonic acid gas.

[45] Kolbe, 1842, p. 1066; Bunsen & Playfair, 1848, p. 152.

residue of the first explosion and, second, the addition of hydrogen.[46] All the oxygen contained in the volume (4) and not consumed in the first combustion, remained after the absorption of the carbonic acid gas and combined, in a second combustion, with the hydrogen added in excess to the volume (5). Then, volume (6) contained only the original nitrogen and the excess hydrogen. Consequently, the third part of the vanished volume [(5) – (6)] after the last combustion (159.48 – 39.44) / 3 = 40.01 was the volume of oxygen unburnt in the first combustion. So the difference [53.61 – 40.01 = 13.60] would be the volume of nitrogen in the original sample. The volume of the combustible gas mixture **A** was obtained by subtracting the nitrogen content from the initial volume (1). The net quantity of oxygen added to the initial volume was found by subtracting the initial volume (1) from the volume (2) after the addition of oxygen [113.91 – 45.08 = 68.83]. The volume of oxygen consumed in the first combustion **B** was obtained by subtracting the amount of unburnt oxygen from the quantity of oxygen added. The amount of carbonic acid gas formed in the first combustion was expressed by the difference in the volumes (3) and (4)

A = 45.08 – 13.60 = 34.48

B = 68.83 – 40.01 = 28.82

C = 72.94 - 53.61 = 19.33

After solving the corresponding algebraic system of equations, the following values of x, y and z were obtained:

Hydrogen (x)	12.15
Carburetted hydrogen (y)	8.72
Carbon oxide (z)	10.61
Nitrogen	13.60

This process of algebraization could be predictive in the sense that if the calculation gave such small values for some components as would fall within the limits of experimental error, it was wise to exclude these suspicious substances from the composition of the gas mixture. Additionally, if the analysis gave a large negative value for any constituent,

---

[46] Carbonic acid gas was determined by absorption with a potash-ball attached to a platinum wire. This potash-ball had to contain as little amount of water as would retain the impression of a fingernail, and also to be moistened externally with water before admission to the gas sample (Bunsen, 1857, pp. 80-81).

it meant that the gas sample contained other gases than those under consideration.[47]

Gasometrie was the work in which the whole of Bunsen's gasometrical researches were collected and updated. The chemist Henry Enfield Roscoe (1833-1915) was responsible for the English translation of this book. Roscoe became acquainted with Bunsen in the autumn of 1852, and they remained close friends for nearly fifty years. Roscoe commented about the book thus:[48]

'For originality of conception, for success in overcoming difficulties, for ingenuity in the construction of apparatus, and for accurate methods, this book as a record of experimental work is, I believe, unequalled.'

However, when Bunsen's work appeared in 1857, the French chemist-physicist Regnault and the physiologist Reiset had already published an impressive research paper on animal respiration that provided a novel complementary perspective on eudiometrical gas analyses.

Regnault. Gasometry in the context of the research on animal respiration

Henri Victor Regnault (1810-1878) entered the École des Mines in 1832, after having studied for two years in the École Polytechnique. During the years 1834-1835 he visited different countries to further enhance his academic studies. As a part of this educational tour, in 1835 he made a brief sojourn at the renowned laboratory of Liebig in Giessen, which was to provide him with access to the field of organic chemistry.

Back in France, in 1835 he was appointed to the chemistry laboratory of the École des Mines, where he acquired distinction as an organic chemist by conducting in this field research work that has failed to receive as much consideration by historians of science as his work in thermodynamics. In particular, he was involved in the argument between Liebig and Dumas about the composition of ethylene dichloride. By operating with pure constituents, Regnault was able to confirm Dumas' analyses and established the incomplete purification of the products as the source of error in Liebig's results.

---

[47] Kolbe, 1842, pp. 1068-1069; Bunsen, 1857, p. 103.

[48] Roscoe, 1900, p. 520.

During the first stage of his career (1835-1840) in organic chemistry, his publications followed one another without interruption. In this field, Dumas and Liebig again influenced the research on alkaloids conducted by Regnault, who had opposed these two leading chemists in regard to the problematic determination of the nitrogen content. At the end of this five-year period, Regnault was prompted by Dulong and Petit's findings to make some important remarks on the specific heat of simple or compound substances. In 1841, he became professor of general and experimental physics at the Collège de France, where he began his life's work on research into specific heats.[49] This refocusing of his career did not entail a renunciation of chemistry, for it was in 1848 that he began publishing his *Cours élémentaire de chimie*, while in 1852 he was appointed director of the porcelain works at Sèvres. At the same time, he supervised the experiments carried out to improve the gas lighting in Paris. Finally, it should be added that he was involved, as mentioned above, in significant work on animal respiration.[50]

Regnault's eudiometrical apparatus

For this research into animal respiration, Regnault worked together with the physiologist Jules de Reiset (1818-1896) on the design of a somewhat complex device that included a eudiometrical apparatus and was conceived specifically for assaying the exchange of respiratory gas between animals over long periods of time.[51]

> It consisted of two sections that could be assembled and separated at will: the measuring tube and the laboratory tube (Figure 9.8 and 9.9). The measuring tube (ab) with an internal diameter of 15-20 mm was graduated into millimetres and terminated at the top by a curved capillary tube (bcr'). The lower end of this tube was affixed to an iron component by two tubings (a,i) and provided with a three-way tap (R). A straight tube (ih) also graduated into millimetres, open at both ends and with the same diameter as the former tube, was sealed into the second tubing (i). It was therefore possible to establish a direct

---

[49] Fox, 1971, pp. 281-318.

[50] De Lapparent, 1895, Vol. 1, pp. 326-332; Poncet & Dahlberg, 2011; Reif-Acherman, 2012.

[51] Regnault & Reiset, 1849, pp. 310-333. Regnault's interest in assaying the exchange of respiratory gas in animals over a long periods of time using a complex apparatus had to do with his almost obsessive perfection in assessing all possible causes of error (Dörries, 2001, pp. 236, 239-240, 242)

connection between the two tubes (ab), (ih) or to connect only one of them with the outside. These two vertical tubes constituted a manometric apparatus enclosed in a cylindrical glass sleeve (pqp'q') filled with water and kept at a constant temperature, given by the thermometer (T), throughout the duration of the analysis. The water in the sleeve was stirred by means of the air blown through a tube dipped downwards. The measuring tube was crossed at the top by two opposite platinum wires (not depicted in the figure) sealed through the glass whose ends approached each other to a distance of a few millimetres inside the tube. The external ends of both wires were fixed with a little wax on the edge of the sleeve. The manometric apparatus was screwed into an iron support (ZZ').

The laboratory tube (gf) was open at the bottom and terminated at the top by a curved capillary tube (fer). This tube was submerged in an iron trough (U) containing mercury that stood on a shelf and could be raised at will along the vertical support (ZZ') by means of the rack (vw), which engaged with a toothed pinion (o) set in motion by means of the crank (B). The ratchet (k) enabled the rack, and consequently the trough (U), to be stopped at any position. The laboratory tube was held in a fixed vertical position by means of a clamp (u). The ends of the capillary tubes that terminated in the laboratory and the measuring tubes were affixed into two small steel taps (r, r').

Figure 9.8 Regnault's eudiometrical apparatus.

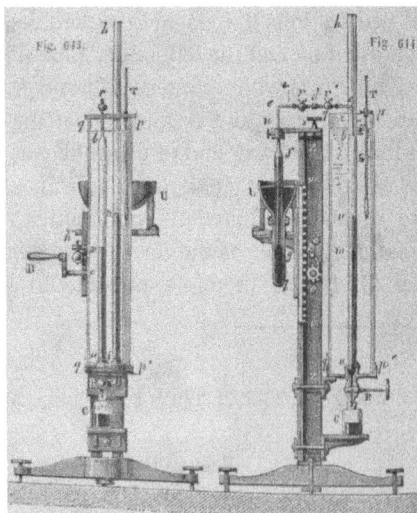

Geometric projection of its front face *(Fig. 643)* and a vertical cross-section of this front face *(Fig. 644)*. From Henri Victor Regnault, *Cours élémenatire de chimie*, 1850, Vol. 4, pp. 75-76.

Figure 9.9 Regnault's eudiometrical apparatus.

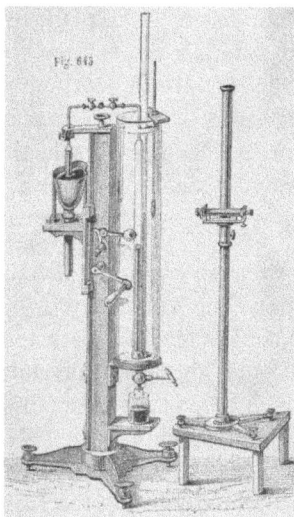

An overall perspective of the apparatus and *(Fig. 645)*. From Henri Victor Regnault, *Cours élémenatire de chimie*, 1850, Vol. 4, pp. 75-76.

The description of how the apparatus functions is based on Regnault's assumption of the analysis of a mixture of atmospheric air and carbonic acid gas.

> The measuring tube (ab) was completely filled with mercury poured through the tube (ih). The laboratory tube (gf) was also filled with mercury and completely immersed in the trough (U) with the tap (r) opened. The air was then sucked through the mouth using a glass tube connected to the tap (r) by a rubber tube; when mercury began to emerge the tap (r) was shut off. The gas sample to be analysed was collected in a small bell jar and transferred to the laboratory tube (gf). Then, the trough (U) was raised while mercury was poured from the measuring tube through the tap (R). The gas was then passed from the laboratory tube to the measuring tube. The tap (r) was closed when the end of the mercury column was made up to a mark ( ) drawn on the horizontal branch (er) at a small distance from the tap (r). The mercury level was then brought to a given division (m) of the measuring tube, and the difference in height between the two columns of mercury (h) was read immediately on the scale of the tube (ih).

Let **t** be the temperature of the water that was kept constant throughout the duration of the analysis; **f** the saturation pressure of water vapor at this temperature [the gas sample collected in the measuring tube was always saturated with moisture because the walls of the tube (ab) were moistened with a small quantity of water]; **V** the volume of the gas sample; **H** the barometric pressure. Then, **H + h - f** was the pressure of the dry gas sample. In order to avoid errors of parallax, readings were made with the aid of a level viewer (Figure 9.9, Right, right side) to facilitate indications of one-tenth of a millimetre. Although this degree of accuracy was regarded as sufficient, it could be improved by measuring with a cathetometer.

> To ensure that the gas sample was free of carbonic acid gas, a drop of a concentrated solution of potash was brought up into the laboratory tube by means of a curved pipette. This gas sample then passed from the measuring tube into the laboratory tube, and the solution of potash completely moistened the walls of the tube. After waiting for a few minutes to allow the absorbing action of the potash, the gas from the laboratory tube was again transferred to the measuring tube by raising the trough (U) and pouring the mercury from the tap (R). These operations could be repeated several times if it was deemed to be suitable, but usually after the second operation the carbonic acid gas was completely absorbed. Afterwards, the mercury level was brought to (m) in the tube (ab), and the difference in height *h'* between the two columns of mercury (ab) and (ih) was measured and the barometric pressure *H'* was recorded.

Thus, the pressure of the dry gas sample free of carbonic acid gas was *H'* + *h'* - *f*. Consequently, *(H + h – f) – (H' + h' – f) = (H – H' + h - h')* was the drop in pressure caused by the absorption of carbonic acid gas, and

$$\frac{H - H' + h - h'}{H' + h' - f}$$

represented the proportion of carbonic acid gas contained in the gas sample assumed to be dry.

> To determine the proportion of oxygen that existed in the dry gas sample, the laboratory tube was detached and washed several times with water. It was then dried with absorbent paper and placed in connection with a vacuum pump for a few moments. When it was completely filled with mercury it was again adjusted to the measuring tube. Once the trough (U) had been brought to its maximum point, the mercury retained by the tap (R) was made to flow, the taps (r, r') were carefully opened, and mercury passed from the laboratory tube into the measuring tube (ar'). The tap (r') was closed when the top of the mercury column reached a second marker

( ) drawn on the vertical branch (bc), and the mercury in the measuring tube was again brought back to the level (m).

The difference in height *h"* between the two columns of mercury and the barometric pressure *H"* were determined. *H" + h" − f* was therefore the pressure of the dry gas sample. The quantity of this gas was a little less than that obtained immediately after the absorption of the carbonic acid gas, because a small amount, approximately $^1/_{3000}$, was lost when detaching the laboratory tube from the measuring tube. This small loss had no influence on the result of the analysis since parameters were determined again.

Once the laboratory tube had again been detached from the measuring tube, hydrogen was introduced in order to burn it with oxygen. The difference in height *h'''* between the two columns of mercury was determined, as well as the barometric pressure *H'''*. Thus, *H''' + h''' − f* was the pressure of the mixture of hydrogen with the gas sample to be analysed. As it took some time for the gases to mix perfectly, combustion by the electric spark could not be carried out immediately.

Once the gas mixture had been detonated, the pressure of the dry residual gas measured after making up the mercury in (m) was: *H'''' + h'''' − f.* Therefore, *(H''' + h''' − f) − (H'''' + h'''' − f) = (H''' − H'''' + h''' − h'''')* was the pressure of the dry gas mixture of hydrogen and oxygen that combined after detonation (i.e., the pressure of the gas mixture lost during combustion).

Since *1/3 (H''' − H'''' + h''' − h'''')* was the pressure of the dry oxygen in the gas sample that combined with hydrogen, and *H" + h" − f* was the pressure of the dry gas sample, then

$$1/3 \frac{H''' - H'''' + h''' - h''''}{H'' + h'' - f}$$

was the proportion of oxygen in the carbonic acid-free gas sample.

According to Regnault, the operations carried out with his apparatus were very simple and the experimenter could perform them without assistance. The analysis could be performed in less than three-quarters of an hour, most of this time being taken up by the absorption of the carbonic acid gas and the cleaning of the tube after this operation. Actually, an analysis of carbonic acid-free air was carried out in less than twenty minutes. It is worth remarking that the composition of the gas mixture was determined by measuring pressures rather than volumes.[52] It was not necessary to gauge the capacity, which was always a delicate operation. The volume of the gas sample was constantly the same, and only pressures of dry gases were determined, which made it unnecessary to avoid measuring errors due to the presence of moisture in the gas sample. Gas pressures were measured by a direct reading on the graduated tubes, whose divisions corresponded to the columns of mercury.[53]

A worldwide survey on the composition of atmospheric air

Regnault and Reiset checked their device by analyzing free acid carbonic atmospheric air and obtained a mean value of 20.943 % of oxygen, 0.028 being the highest difference. This gave a greater precision than had yet been achieved by the eudiometrical methods available at that time. Then they went on to conduct a series of trials in order to identify the causes of error in the eudiometric analyses and to seek the means of avoiding them. Thus, they first determined the limits of the explosive capacity of the mixtures of hydrogen and oxygen in which one of both gases predominated, as well as the greatest variations in the composition of these mixtures within which the eudiometrical analysis remained accurate.[54] The second series of trials was devoted to the analysis of diverse gaseous mixtures including binary, ternary and quaternary combinations of oxygen, hydrogen, nitrogen, carbon oxide, methane and ethene. Following Bunsen's example, Regnault also planned the calculation of the composition of the different cases on the basis of systems of algebraic equations.[55] Finally, the main trials were devoted to the analysis of the composition of the air exhaled by different groups of animals (mammals, birds, reptiles and insects) in relation to their feeding regime.[56]

---

[52] Regnault's apparatus could be employed differently: instead of keeping the volume of the gas constant and measuring pressures, the pressure could be kept constant and volumes measured. The manipulations were the same. Nevertheless, the pressures method was more accurate because it only required the graduation of the measuring tube in millimetres, while the volumes method required a rigorous gauging of this same tube.

[53] Regnault & Reiset, 1849, pp. 333-341; Regnault, 1850-1854, Vol. 4, pp. 75-81; Ogier, 1885, pp. 175-180.

[54] Regnault & Reiset, 1849, pp. 341-356.

[55] Ibid., pp. 361-386; Regnault, 1850-1854, Vol. 4, pp. 87-105.

[56] Regnault & Reiset, 1849, pp. 386-496.

The many gas analyses carried out by Regnault and Reiset during their research on animal respiration prompted Regnault to apply the same method to the analysis of atmospheric air, in order to resolve an issue that still gave rise to doubt: to determine whether atmospheric air maintained a constant composition throughout the year, and whether this composition was identical in all parts of the world. On February 7th, 1848, while Regnault was still occupied in his research with Reiset, he presented before the Académie des Sciences in Paris a plan for the first worldwide cooperative investigation into the composition of air.[57]

According to Regnault's plan, atmospheric air samples were to be collected in a large number of suitably chosen localities from all around the world, on the 1st and 15th of each month at about the hour of true midday in each place throughout a whole year. These air samples were to be sent to the Collége de France, where they would be analyzed under exactly the same conditions with the same eudiometer employed in his joint research with Reiset, and then compared with the air collected in Paris. Regnault himself carried out many of these analyses, while young assistants working in his laboratory conducted the others with the same eudiometer.

Regnault devised a simple experimental protocol for collecting air samples and preserving them without alteration, which would enable people unfamiliar with scientific experiments to apply this protocol. The devices had to be inexpensive and provided with adequate protection during transport to prevent breakage. The protocol was described in a small printed note that was sent to each of the persons who were willing to assist Regnault in his enterprise. The various operations that had to be carried out when collecting the air samples were illustrated faithfully in this printed note (Figure 9.10).

Figure 9.10 Three steps in Regnault's protocol for collecting air samples.

A glass tube ending in two open points covered to prevent breakage was employed to collect an air sample.

---

[57] *Comptes rendus hebdomadaires des séances de l'Académie des Sciences*, 1848, 26, pp. 156-157.

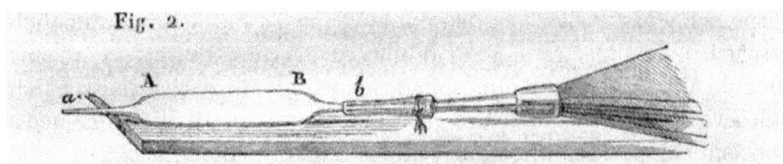

One of the terminal points was connected to a bellows, and the tube was filled
with ambient air.

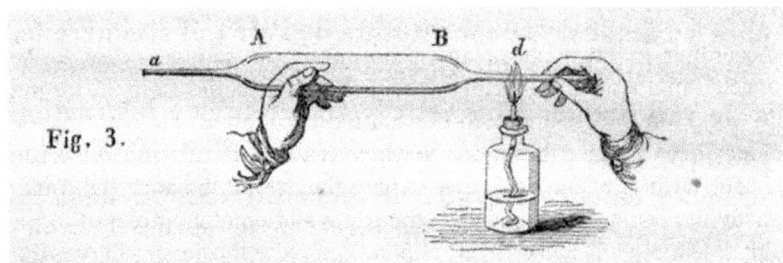

The tube was sealed airtight by closing both ends with the flame of an alcohol lamp.
From *Annales de chimie et de physique*, 1852, Vol. 36, p. 388.

Unfortunately, this ambitious scientific survey was partly disrupted by
the political events of 1848. Most of the tubes that had been sent to their
destination and those at the Ministry of Foreign Affairs were lost.
Nevertheless, Regnault continued to analyze the atmospheric air of Paris
and that of a large number of air samples collected at different locations
across France, Switzerland, Berlin, Madrid and the Mediterranean Sea, as
well as from the most distant places (the Atlantic, East Indian and Arctic
Oceans and the Caribbean Sea) by travellers and naval officers. Regnault
first concluded that atmospheric air usually exhibited sensible but very
small variations in composition, since the percentage of oxygen generally
varied only between 20.9 and 21.0, although in certain cases, most
frequently in warm countries, the percentage of oxygen could fall to 20.3.
Secondly, the average percentage of oxygen in the air of Paris during the
year 1848 was 20.96.[58]

The agenda of gasometry in the second half of the nineteenth century

Gasometrie entailed a remodelling of gas analysis in which Bunsen gave a
holistic interpretation of the subject. From the instrumental point of view,
Volta's simple graduated combustion tube remained at the core of

---

[58] Regnault, 1852.

Bunsen's experimental device, surrounded by other already known auxiliary instruments with the novelty of the absorption tube. The analysis of a gas sample commenced with the absorption of those gases that were easily decomposed or that entered easily into combination. Then the residual unabsorbed gas mixture, which usually contained inflammable constituents and nitrogen, was transferred to the combustion tube in order to be exploded. Absorption was performed in a small graduated tube, and in order to bring the solid absorbents into contact with the gas, they were moulded into small balls that were fastened on to the end of platinum wires.[59] This provided a means for obviating the inaccuracies inseparable from the introduction of liquid solvents. In almost all cases it was impossible to make use of liquid absorbents, as the gases were then dissolved. It was precisely to eliminate the disadvantages arising from the introduction of solids and liquids reagents into graduated tubes that Louis-Michel-François Doyère conceived a new eudiometric method based on a type of curved pipette with two balls (Doyère's pipette).[60] The gas sample was introduced into this pipette by suction and then constantly agitated with the required absorbents in order to achieve complete absorption.[61] In 1875, the Russian botanist and physiologist Kliment Timiryazev published a work on the assimilation of light by plants in which he presented a modification of Doyere's pipette for measuring extremely small quantities of gas during his researches on the decomposition of carbonic acid gas by the green parts of plants.[62]

Regnault and Bunsen's contributions gave rise to much of the agenda of gasometry for the following decades up to the closing years of the nineteenth century. After the publication of their works, instrumental gasometry underwent a continuous evolution to remedy the deficiencies inevitably found in any nascent analytical method. Bunsen's method was

---

[59] The most common solid absorbents were potash (potassium hydroxide) for carbonic acid gas, manganese dioxide moistened with phosphoric acid for hydrogen sulphide and sulphurous acid, alkaline pyrogallol (1,2,3-trihydroxibenzene) for oxygen and copper (I) chloride for carbon oxide. The alkaline solution of pyrogallol turns brown on the absorption of oxygen, and solutions of CuCl in hydrochloric acid or ammonia absorb carbon monoxide to form colourless complexes such as the dimer $[CuCl(CO)]_2$.

[60] Doyère submitted his new method and apparatus to the Paris Academy of Sciences in 1847: «Note sur une nouvelle method pour l'analyse des gaz», *Comptes rendues hebdomadaires des séances de l'Académie des Sciences*, 25, pp. 928-931.

[61] Doyère, 1850; Ogier, 1885, pp. 199-201; Dennis, 1913, pp. 99-101.

[62] Timiryazev, 1877; Ogier, 1885, pp. 208-210.

mainly criticized for the length of time it took, while Regnault's method, which rectified this defect, was underestimated because of the considerable loss of accuracy in the process. The principal criticism was that considerable changes in volume were numerically underrepresented when compared with those obtained by Bunsen's method.

In an attempt to remedy these deficiencies, while at the same time combining the advantages of the two methods, E. Frankland and W.J. Ward constructed an apparatus by taking Regnault's device as a model with the incorporation of some mechanical modifications. According to both authors, their apparatus was self-correcting because the temperature was kept constant throughout the entire analysis, while the atmospheric pressure did not exert any influence on the volumes and the pressure of water vapour was internally balanced.[63]

Edward Frankland (1825–1899) was apprenticed to a Lancaster pharmacist in about 1840, and in 1845 he went to study at the London Museum of Economic Geology under Playfair's guidance, where he was introduced to Kolbe and made spectacular progress in systematic chemical analysis. In 1847, Frankland and Kolbe went to work under Bunsen at Marburg for three months, and then in the autumn of 1848 he entered Marburg University, where he continued his studies on the action of zinc on the alkyl iodides, which later led to the synthesis of the organometallic compounds. Frankland gained his doctorate from Marburg in 1849 before going to work under Liebig in Giessen. One of the reasons for Frankland's indebtedness to Bunsen was the training he received in new techniques of gas analysis. In a letter to John Tyndall in 1857, Frankland extolled Bunsen's contribution to gas analysis: [64]

'[Bunsen´s] improvements in the construction of eudiometers enabled any operator to construct these instruments for himself with facility, and possessing an accuracy of measurement, which could not be exceeded in the workshop of the philosophical instrument maker. These improvements, together with his new methods of applying absorbents to gaseous mixtures, have imparted to eudiometrical determinations a degree of accuracy never exceeded and rarely equalled in other departments of analytical chemistry. It is also perhaps worthy of notice that the subsequent improvements in gaseous analysis have been chiefly if not entirely made by the pupils of Bunsen.'

---

[63] Frankland & Ward, 1854; Ogier, 1885, pp. 180-183.

[64] Quoted from Russell (1996, p. 74)

The last sentence clarifies Frankland's involvement in the improvements made to the last generation of eudiometers. The complicated nature of Frankland and Ward's apparatus appears to have limited its use to some extent. The issue was how to compensate for the effects caused by the changes in temperature and pressure during the eudiometrical trial. In order to tackle this drawback, Alexander-William Williamson and William J. Russell introduced a standard quantity of air into a pressure-tube over mercury in the trough used in Bunsen's apparatus. The procedure consisted of marking off the height of the mercury in this tube at ambient temperature and pressure, and then at any other temperature or pressure, of raising or lowering the eudiometrical tube in the trough in order to bring the enclosed air exactly to its original volume. Thanks to this invention, they were able to obtain very accurate results with considerably less difficulty than by Bunsen's method, with the additional advantage of not having to perform any calculations.[65]

Nevertheless, this improvement proved to be inadequate when it came to solid absorbents, which were often inconvenient to use. Thus, Williamson and Russell devised a further modification for their instrument that enabled the action of liquid reagents. They believed that neither Regnault nor Frankland's eudiometrical methods enjoyed general use due mainly to the fact that these apparatuses were of fragile nature. In order to make the device robust enough for routine laboratory use, Williamson and Russell designed a new apparatus without stopcocks or other delicate mechanisms.[66] A persistent inconvenience of all these types of apparatuses was the need for a separate tube for the reagent, which rendered them much more complicated than Bunsen's apparatus. To overcome this need for a laboratory tube, as well as the slow action of solid reagents, Russell described a method by which the liquid reagent could be introduced and removed from the eudiometer itself.[67] The following year, Herbert McLeod presented a new modification of Frankland and Ward's version of Regnault's eudiometer,[68] while in 1879 Joseph W. Thomas devised an apparatus of greater sensitivity than the Frankland and McLeod eudiometers to enable the manipulation of small volumes of gas.[69] Later, in 1888, William Marcet introduced a new form of eudiometer

---

[65] Williamson & Russell, 1858.

[66] Williamson & Russell, 1864; Ogier, 1885, pp. 169-173.

[67] Russell, 1868.

[68] McLeod, 1869; Ogier, 1885, pp. 183-193.

[69] Thomas, 1879; Ogier, 1885, pp. 196-198.

that he claimed to be reliable in its functioning, simple in construction and easy to handle. He used a slightly modified version of the instrument for his researches in determining the proportions of oxygen transformed into carbonic acid gas and of oxygen retained in the blood in human respiration.[70]

In France, in 1875, Louis Orsat addressed the problem of the analysis of industrial gasses due to the difficulties of using Bunsen or Regnault's eudiometric methods in metallurgical factories. Orsat designed an apparatus to enable the rapid analysis of the gases produced by combustion in blast furnaces, with sufficient precision for industrial purposes (Figure 9.11). Orsat's device was based on the absorbent properties of potash, pyrogallol and copper (I) chloride.[71] Along this same line was the contribution by J.J. Coquillion, whose eudiometer was principally applied to the measurement of gases released from industrial furnaces in order to detect firedamp from marsh gas present in the mines and to determine the quality of lighting gas. A salient feature of this apparatus was the burner of the gas sample, which consisted of an inverted glass jar with a palladium or platinum spiral wire connected to an electric battery. This mechanism was based on the burning without explosion that occurred when an inflammable gas mixture surrounding a platinum or palladium wire was brought to red heat by an electric current.[72]

Over the last three decades of the nineteenth century, Germany experienced a rapid and spectacular economic development that greatly benefited the production of chemicals. It deeply affected the production of acids and alkalis, the growth of the dyestuffs industry and the expansion in other branches of manufacturing, particularly textiles, glass, soap, paper, metals, pharmaceuticals and fertilizers. Additionally, tar distillation and related industries such as the recovery of ammonia from gas-works fluids

---

[70] Marcet, 1888; 1891.

[71] Orsat, 1875; Ogier, 1885, pp. 211-216; Dennis, 1913, pp. 78-89. The Orsat gas analyser still remains as a piece of laboratory equipment, relatively simple to use, to examine gas samples of fossil fuels for its oxygen, carbon monoxide and carbon dioxide content.

[72] Coquillion, 1877; 1881. In 1817, Humphry Davy published a paper in which he described how, during his research work that led to the miner's safety lamp, he had observed that a hot platinum wire introduced into a mixture of coal gas and air immediately became incandescent.

and the production of organic intermediates were developed.[73] In Germany, as well as in France, gas-consuming and gas-producing industries also lacked easy monitoring by chemical analysis; they required a fast and simple method of analysis for gas mixtures that also provided the desired precision.

Figure 9.11 The Orsat gas analyzer.

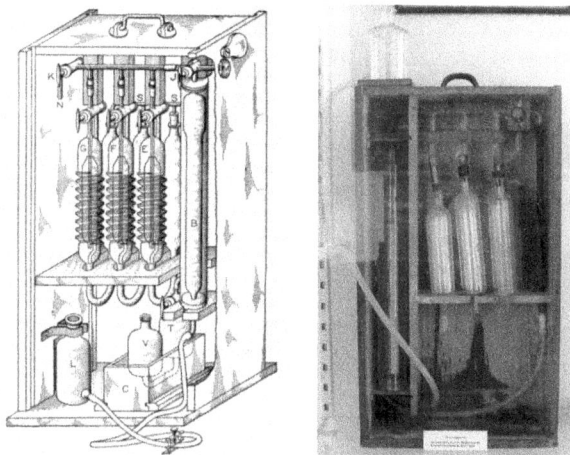

(Left) Apparatus modified by Dennis. From Louis Munroe Dennis, *Gas Analysis*, 1913, p. 86. Fig. 48. (Right) Apparatus used in beet sugar production. Work released into the public domain. Current location, Zucker-Museum, Berlin.

The talented chemist Walther-Matthias Hempel (1851-1916) possessed great manual dexterity and expertise in the art of glass blowing and attained the objective of adapting gas analysis to technical needs. Hempel studied chemistry at the Royal Polytechnical School in Dresden for three years, and from 1871 to 1872 learned organic chemistry at the Friedrich-Wilhelms University in Berlin. In 1872, he went to Robert Bunsen in Heidelberg, where he passed his doctoral examination in the same year. He then returned to Dresden to work as an assistant at the chemical centre for public health, and from 1876 until his official professional accreditation in 1878 he was an assistant at the chemical laboratory of the renamed Royal Saxon Polytechnical College. His accreditation thesis *Neue Methoden zur Analyse der Gase* (New Method for the Analysis of Gases) was published in 1880 and represented the continuation of Bunsen's work on gasometrical methods. In this thesis, Hempel referred to some new

---

[73] Haber, 1958, pp. 121-128.

types of apparatus: the gas burette, intended for rapid analyses of gases insoluble in water; the gas-absorbing pipettes for solid and liquid reagents (Figure 9.12); the explosion pipette; the hydrogen-producing pipette and the U-shaped tube containing a palladium sponge[74] for the determination of hydrogen.[75]

Figure 9.12 The Hempel simple absorption pipette for liquid reagents.

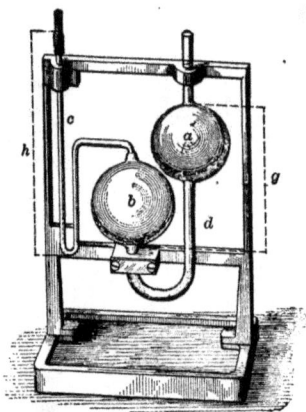

From Louis Munroe Dennis, *Gas Analysis*, 1913, p. 54. Fig. 34.

In 1890, after more than ten years experience, Hempel gave an account of his vast and varied work on gas analysis in the treatise *Gasanalytische Methoden*, in which he described a number of apparatus designed both for technical purposes and for purely scientific investigations. The book ran to three more updated German editions (1889, 1899 and 1913), and various English translations served to confirm the leading role played by Hempel in the field of gas analysis.[76]

---

[74] In 1866, Thomas Graham had discovered that finely divided palladium readily absorbed hydrogen at room temperatures. While this property, which implies the formation of palladium hydrides, is common to many transition metals, palladium has a uniquely high absorption capacity.

[75] Ogier, 1885, pp. 219-226; Dennis, 1913, pp. 51-70, 90-98, 146-147.

[76] Hempel, 1892.

Summary

Bunsen's *Gasometrie* marked a turning point in gas analysis in the mid-nineteenth century. He felt obliged to reinvestigate the entire subject of eudiometry, the initial purpose of which - measuring the goodness of air - had already been superseded. During the second half of that century, chemists were engaged in devising new laboratory instruments for gas analysis in which the explosion pipette was the only remaining vestige of the solitary and original eudiometrical tube. A number of features were involved in that process of transformation, among the most notable being those concerning experimental procedures, the overcoming of operating difficulties, the need for safety measures, the innovations in laboratory equipment, and the research and external contexts that drove this remarkable development.

The early analysis of inflammable gas mixtures by rapid combustion with oxygen employing Volta's eudiometer revealed the danger that the eudiometrical tube might break when burning highly explosive mixtures. For safety reasons, Henry devised in 1819 an apparatus with a eudiometrical tube that enabled slow combustion with extremely ignitable samples. A significant drawback arising as a result of powerful detonations when operating in closed mode over water was the increase in residual gas due to the vacuum created after ignition. Faced with the choice of operating in open mode, which entailed the risk of losing gas, Gay-Lussac designed in 1817 a version of Volta's eudiometer that was equipped with an anti-vacuum valve. Apart from deficiencies in the laboratory equipment, eudiometrical methods did not always provide reliable measurements.

Bunsen was convinced that the problem with the eudiometrical methods resided mainly in the detection of their sources of error, which could be corrected by constructing more precise instruments and using them more appropriately. Bunsen's endeavours in making eudiometry a respectable subdiscipline of analytical chemistry began in the 1830s and culminated in 1857 with the publication of his *Gasometrische Methoden*. Regarding the improvements in the laboratory equipment, Bunsen introduced novelties into the explosion tube that proved to be crucial. The platinum wires were fused into the glass walls and bent into its curved closed end to prevent the gas from leaking and the wires from bending. By attaching solid absorbent balls to the platinum wires, the explosion tube was thereby adapted to ensure the capture of certain non-combustible gases such as carbonic acid. In this way, he eliminated the inaccuracies arising from the introduction of liquid solvents into the tube. The tube was graduated on a millimetre scale independently of volume, with subsequent calibration of the actual volume of each division of this independent graduation. For this purpose, he invented a portioning machine to engrave the scale on the glass tube. However, further correction of the capacity figures due to the fault of the convexity of the mercury meniscus was also required. To carry out his readings Bunsen employed a telescope placed some distance from the tube in order to prevent gas expansion resulting from the experimenter's body heat, as well as to avoid parallax errors.

A virtue of Kolbe's and Bunsen's accounts was the detailed description of some physical gestures made when manipulating the eudiometrical device. The performance of such gestures consisted of tacit knowledge that was usually taken for granted or kept secret and therefore not written down.[77] In this regard, it is worth mentioning the disclosure of manipulative procedures to prevent atmospheric air from entering during combustion, the formation of air bubbles attached to the inner glass walls or to ensure that these walls were free from any adhering impurity, and also to bring a drop of water drop to the head of the tube to obtain the sample saturated with water vapour.

Mathematical procedures for calculating the composition of gas mixtures after ignition with a known proportion of oxygen had been widely used in early attempts at employing Volta's eudiometer as a gas sample analyser. It was possible to work out these calculations on the basis of a few assumptions: that the formation of each volume of carbonic acid gas consumed the same volume of oxygen, and that two volumes of hydrogen consumed one of oxygen to form water. Berthollet, Dalton, Henry, Saussure and Gay-Lussac were particularly active in this field. Bunsen provided a new dimension to gas analysis by means of his algebraic approach. By formulating and solving a series of mathematical equations, he was able not only to calculate the composition of complex gas mixtures but also to endow his approach with some predictive value. This was because the solutions to the equations could suggest either the exclusion or the inclusion of components in the unknown gas sample. Bunsen's algebraic approach served to reinforce the apparent degree of accuracy attained by his eudiometrical methods. However, a crucial requirement for a successful practical application of these methods was to ensure that the laboratory was not subject to changes in atmospheric pressure and temperature. Thus, in the case of temperature control, Bunsen advised that entrance to the laboratory should be strictly controlled, as should the time elapsing between two consecutive analyses.

---

[77] Sibum, 1995, p. 76, note 8.

In order to address the issue of temperature control, Regnault enclosed the eudiometer in a cylinder filled with water. His eudiometer also made it easier to perform the various operations by dividing the apparatus into two tubes; one for absorption and other for measurement. The former was particularly suitable for eliminating the carbonic acid from the gas sample. The whole apparatus was designed to prevent external air from entering air into the system by internal gas transference between both tubes. Additionally, Regnault adopted the method on every occasion of determining the gaseous volumes after they had been saturated with water vapour. Nevertheless, these advantages were in some respects diminished by the experimental procedure of keeping the volume of the gas constant and measuring the pressure exerted by that constant gas volume, a case in point being the numerical underrepresentation of large changes in volume. Although Regnault was convinced that his eudiometer was straightforward to operate and required minimal assistance, things were actually not as simple as he believed. While it appears that the instrument was capable of providing accurate results in experienced hands, its operation was rather complicated and required apprenticeship and training. The apparatus was awkward to transport, very fragile, and prone to accidents causing damage that was difficult to repair.[78] After Bunsen and Regnault's contributions, the agenda of gas analysis was devoted to surmounting the shortcomings of both methods and to facilitating the use of solid and liquid reagents with the eudiometrical instrumentation.

Gas leakage was already a constant source of annoyance in the early days of eudiometers. Stopcocks were necessary components in the Regnault-type apparatus employed in the second half of the nineteenth century. They were usually connected to glass or metallic tubing by means of short rubber tubing. In many cases, steel or brass stopcocks did not provide a perfectly tight fit and occasionally allowed gas to leak out or let in atmospheric air. When stopcocks made of finely ground glass became available, they could be employed with only a very slight risk of leakage provided they were properly lubricated.[79] At that time, vulcanized rubber tubing became commonplace in laboratory equipment.[80] It was used extensively for keeping the tubes in place and made the whole device more flexible, thereby guarding against fracture, facilitating laboratory operations such as aligning mercury levels or transferring gases from one

---

[78] Ogier, 1885, p. 175.

[79] Gas leakage is still a critical factor in modern gas chromatography and gas chromatography-mass spectra systems.

[80] Smeaton, 2000, p. 231.

tube to another. To fix the rubber tubing in place, the ends of the tubes were rounded in the flame and brought close together in the section of rubber tubing, which was fastened around the glass tubes with copper wire ligatures. Long rubber connections were not advisable, however, because of the porosity of rubber and also because the air in the tube adhered to the walls.[81]

Volta's eudiometer not only proved its efficacy in the analysis of inflammable gas mixtures but also in the early elucidation of the chemical composition of some organic compounds. Dalton's analysis of methane and ethene in 1804 provided him with a new case of multiple proportions, in addition to the oxides of nitrogen. In 1810, Saussure determined the composition of ethene with greater accuracy, while Gay-Lussac's discovery of the nature and composition of prussic acid and cyanogen in 1815 greatly contributed to underpinning Volta's eudiometer as the vehicle for the materialization of his volumetric approach to matter. These are but a few of the significant examples of the early use of Volta's eudiometer in academic research, which was later taken up in other fields such as animal respiration, plant physiology and Regnault's international survey on the composition of atmospheric air. However, this context of academic research is insufficient for explaining the renewed momentum of gas analysis during the second half of the nineteenth century.

Considerations of a wider industrial and economic context are required to reach a better understanding of that extraordinary growth. In Great Britain at the beginning of the nineteenth century, the need to control the quality of gas from industrial coal distillation for use in artificial illumination prompted Henry to embark on the creation of a new eudiometrical device. Bunsen's concept of an innovative eudiometrical methodology for gas analysis emerged in the mid-nineteenth century as a result of the need to optimize the production of cast iron in Germany as well in England. In the last quarter of the nineteenth century, the release and consumption of gaseous fluids by industries required an easy monitoring system by eudiometrical methods for their large-scale processes. It was in Germany, in the context of a growing economic development that propelled the chemical industry, where Hempel succeeded in adapting gas analysis to industrial requirements with the introduction of new methods and apparatuses. Thus, it was in this context of large-scale chemical production that eudiometry, that is, gas analysis, underwent its definitive evolution.

---

[81] Frankland *et al.*, 1876, Vol. 2, pp. 13-21; Dennis, 1913, pp. 90, 99, 114-115.

10.

Concluding remarks

Preliminary observations

As an experimental device, all eudiometers were based on the fact that the respirable part of atmospheric air could be extracted from an air sample by the action of a particular substance. These absorbent substances could be solid materials (phosphorous, iron filings with sulphur and potassium sulphide), aqueous solutions (iron sulphate impregnated with nitrous gas and alkaline or calcium sulphides) and gaseous substances such as nitrous gas and hydrogen. Actually, the first eudiometrical device, in which nitrous gas was used, was designed to replace mice in the determination of the goodness (i.e., respirability) of an air sample. The purpose of this development was not only to achieve greater precision but also to remove the inconvenience of maintaining a stock of mice in the appropriate conditions.[1] In general, the different kinds of eudiometers were conceived as volumetric approaches to the quantification of matter, and only a few involved direct gravimetric determinations. Together with gasometers, and with the priceless support of the chemical balance, eudiometers contributed to the transformation of pneumatic chemistry into a precise quantitative science in the second half of the eighteenth century onwards.

The practice of field eudiometry conducted outside the laboratories stimulated the construction of portable eudiometers, which had to be easy to handle and transport as well as packed in the manner of an apothecary chest. In one way or another, the different types of eudiometers were in active use during the last decade of the eighteenth century and were often employed in parallel to compensate for their limitations. In the early nineteenth century, Volta's eudiometer enjoyed a renewed life as a gas mixture analyser, but it was thanks to Bunsen that in the middle of that century eudiometry acquired the status of a reputable subdiscipline of analytical chemistry.

---

[1] Levere coined the expression 'mouse-free route' as a brief explanation of the meaning of this replacement (Levere, 2000, p. 110)

The evolution of eudiometry has revealed some rhetorical topics used to persuade the scientific community of personal contributions. Thus, Landriani addressed the general topic of testimony by referring to Priestley's authority when he presented his nitrous air eudiometer as an improved version of his own concept of a seemingly non-existent Priestley eudiometer. The persuasive force of images was employed by Fontana in the imposing iconography of his eight eudiometers, while Volta also invoked Priestley's authority when presenting his prototype of the inflammable air eudiometer.[2] Nevertheless, voices of authority may sometimes be counterproductive, as in the case of calls on Gay-Lussac's authority, which served to strengthen the criticisms aimed at Martí's eudiometrical test.

Experimental procedures, material equipment and resemblances among eudiometers

The nitrous gas eudiometer

The evolution of the nitrous gas eudiometer can be fittingly illustrated by the attempts to establish a standardized experimental procedure. This turned out to be no easy task, since the achievement of a systematic experimental performance involved critical issues such as the test endpoint and the training of eudiometrists. The determination of the test endpoint was a troublesome factor in the experimental procedure, because it was associated to a number of other unresolved factors, such as the reaction ratio of nitrous gas and oxygen at the point of saturation, the dosing of the air sample, and the time at which the contraction in volume should be observed. Additionally, other issues such as the source of the water used in the test, the quality of the nitrous gas and the time spent on different operations (endpoint reading, submerging of the air measure and shaking of the eudiometrical tube) contributed to compromising the reliability of the test.

Material constraints in the experimental equipment also conditioned the running of the test. These were basically the use of acid-resistant materials (ivory or glass) and gas or liquid leakage. Developers of the nitrous gas eudiometer such as Fontana, Ingenhousz and Magellan incorporated accessory utensils into the instrument to improve its performance: a magnifying glass for the readings; a brass ring for parallax

---

[2] Volta had also called on Priestley's authority and partnership for his electrical investigations and researches on inflammable airs.

errors; a measuring device for sampling replicability, and a stabilizer for the verticality of the tube. In particular, Ingenhousz was highly committed to making improvements to the test reliability, and to tackle this challenge he designed abridged procedures for reducing and controlling the execution time of the different operations, as well as conducting tests with highly oxygenated air samples. In spite of the disputable points surrounding the nitrous gas test, it appears that Ingenhousz eventually managed to achieve a reasonably standardized test. However, the aforementioned difficulties involved in the nitrous gas test, together with the coexistence of other reliable eudiometrical tests, finally eclipsed its utility at the end of the eighteenth century.

Humboldt's observations on the size of the recipient and the order of addition of the reagents in the test results may have influenced Dalton's reconstructed version of the nitrous gas test. Dalton praised the test in accordance with Priestley's approach to chemical instrumentation, and in fact restored the apparatus' former simplicity. In addition, he pointed out the influence on the size of the recipient, in terms of narrowness or width of the surface of the water, and the shaking of the gas mixture on the contraction in volume. Gay-Lussac regarded the nitrous gas test as the most unreliable from the outset. However, guided by Dalton's conclusions on the influence of the size of the eudiometrical tube on the test outcomes, he was subsequently able to redevise a later version of the nitrous gas eudiometer.

Volta's eudiometer

The perception of Volta's eudiometer at the beginning of the nineteenth century was that it was both less accurate and less sensitive than the nitrous gas eudiometer, given its poor efficacy regarding air samples with low ratios of oxygen and the difficulty in obtaining hydrogen with an appropriate degree of purity. On the other hand, the instrument was quickly appreciated for its portability and for providing a more reliable test endpoint than the nitrous gas test. Volta endorsed a modular conception of the instrument, although its development was a collaborative process of many actors in addition to him.

There is no doubt that Gay-Lussac and Humboldt's contributions strengthened the analytical capability of the instrument. They succeeded in applying a procedure to overcome the weakness arising from the low sensitivity of the instrument for poorly oxygenated samples, and in making the test highly expeditious and accurate. Even more decisive was their achievement in establishing the volumetric composition of water and in confirming the general composition of common air, which provided the instrument with an invaluable impetus. For its part, Volta's eudiometer constituted the materialization of Gay-Lussac's volumetric conception of matter. Early analysis of highly explosive gas mixtures once again revealed an acknowledged weakness in Volta's eudiometer, the breakability of its glass explosion tube. Iron armour and devices for performing the test in low combustion mode were invented to remedy this major drawback.

In the mid-1800s, Bunsen provided Volta's eudiometer with a decisive boost by implementing rigorous experimental procedures and novel improvements in the equipment, specifically in the explosion tube. His aim was to eliminate as far as possible the many sources of error inherent in the eudiometrical test. With regard to procedures, Bunsen was especially concerned with aspects of gestural knowledge, the control of changes in temperature and pressure in the laboratory itself, and the establishment of a mathematical approach to estimate the composition of inflammable gas mixtures, which provided eudiometry with an added predictive value in terms of gas analyses.

Together with Bunsen's approach to eudiometry, Regnault's eudiometer, based on the measurement of changes in pressures rather than volumes, also deserves recognition in the development of Volta's eudiometer at that time. However, Regnault's eudiometrical method required quite experienced eudiometrists for its execution. In the last decade of the nineteenth century, Hempel brought the field of eudiometry up to date with a kit of specific gas burettes, and solid and liquid reagents began to enter into use in gas analysis. But in spite of all these developments, gas leakage remained the most serious and persistent problem for eudiometry and constituted a major source of inaccuracy. The emergence of rubber tubing as a habitual material in laboratory equipment ensured safer connections between metallic stopcocks and glass or metallic tubing. Furthermore, the use of lubricated ground glass stopcocks went a long way towards alleviating gas leakage.

Phosphorous eudiometer

The phosphorous eudiometer emerged in the context of portable eudiometry as a response to the poor portability of the nitrous gas and Volta's eudiometers. Achard and Reboul's first versions of this type of eudiometer had their continuation in the very simple instruments designed by Giobert and Spallanzani. In contrast with the nitrous gas and Volta's eudiometers, these later versions of the phosphorous eudiometer may be regarded as eudiometry's return to instrument simplicity. This eudiometrical test, based on the fast combustion of phosphorous, was also implemented in a laboratory bench device developed by Séguin. These phosphorous eudiometers were much more affordable than the nitrous gas and Volta's eudiometers. The test proved to be accurate and expeditious, although lacking in reliability when employed with low oxygenated samples. The nature of the test entailed some material limitations regarding the breakability of the glass tube, as in the case of Volta's eudiometer, because of the sudden increase in temperature on the inside or by bringing the igniting flame too close to the outside of the tube.

The discovery of the phenomenon of the slow combustion of phosphorous led to a new kind of eudiometer, first developed by Berthollet. This new eudiometrical test had a precise endpoint but was nevertheless subject to controversy. The final reading of the test required a correction factor and the experimenter had to take care to avoid the spontaneous burning of the piece of phosphorous. A fully functioning phosphorous eudiometer, and the Volta-type eudiometer, both required resistant glass tubes, fine instrument makers and the efficient transmission of tacit knowledge.

In general, eudiometrical tests were not used alone, but jointly with one another. For a complete picture of the eudiometrical scenario, it is necessary to add to the abovementioned eudiometers other devices such as those based on the use of alkaline and calcium sulphides, a mixture of iron filings with sulphur and a solution of iron sulphate impregnated with nitrous gas. They were all simple laboratory devices providing reasonable accuracy, ease of handling and great affordability. The alkaline sulphide and the iron filings mixed with sulphur tests took much longer to perform than the calcium sulphide test, while on the other hand the solution of iron sulphate with nitrous gas proved to be as accurate as the slow combustion of phosphorous and the alkaline sulphides tests.

Resemblances

There existed resemblances among some eudiometers in so far as they had a similar appearance in some respects, but not in others. In particular, two basic points of resemblance were the equipment design and the action of the eudiometrical reagents. The inverted retort ideated by Chaussier and Guyton for their experiments with phosphorous was first taken up by Guyton himself for his potassium sulphide eudiometer, and later redesigned in its simplest form (a tube bent at a right angle) in the phosphorous eudiometers of Giobert and Spallanzani. Chemists sometimes prefer to modify existing apparatus rather than inventing new ones.

Martí's and Davy's eudiometers not only resembled each other in that both used a simple graduated glass eudiometrical tube but also in the action of the corresponding eudiometrical reagents. Martí's solution of calcium sulphide impregnated with nitrogen and Davy's iron sulphate solution impregnated with nitrous gas were both employed to absorb oxygen without absorbing nitrogen. These two latter eudiometrical reagents entered into the picture again when Henry suggested the use of Hope's eudiometer to replace the original solution of calcium sulphide with another of iron sulphate impregnated with nitrous gas. Such cases as these are relevant for establishing the existence of a transfer of material and procedural characteristics between different kinds of eudiometers and therefore explain their resemblances.

Theoretical accommodation

Eudiometers were highly adaptable to theoretical frameworks in the sense that their design could be tailored according to theoretical principles. It may be said that up to end of the eighteenth century, the disputes arising from the differences between the theoretical frameworks of phlogiston-centred chemistry and oxygen-centred chemistry resided to a great extent, directly or indirectly, in the most significant changes taking place in eudiometry. Such adaptations primarily involved the conceptualization, design and experimental procedures of the eudiometrical devices. Dalton's law of the multiple proportions of combination also played its role in the development of eudiometry at the beginning of the nineteenth century.

A controversial issue was that surrounding the performance of eudiometrical tests either over mercury or over water, a question that should not be addressed on theoretical grounds only. In some cases, the high difference in cost between mercury and water may have favoured the choice of the latter. In the case of the nitrous gas test, the use of mercury was not a practical alternative to the use of water because of the danger of mercury reacting with the nitrous gas. On the other hand, from a theoretical point of view, the contraction in volume observed in this test was initially explained by the argument that phlogiston disengaged from the nitrous acid of the nitrous gas and united with the acid principle of the common air sample, whereas water absorbed the nitrous acid released. Thus, water provided a faster and more complete reduction in volume than mercury. The emergence of oxygen-centred chemistry did nothing to change the practical outcome of the test. The contraction in volume occurred because oxygen from the common air sample united with the nitrous gas to form the corresponding acid, which dissolved in water.

Hence, the concept of the nitrous gas eudiometer proved to be immune to those theoretical disputes.

A crucial step in the development of oxygen-centred chemistry was taken in 1783 when Lavoisier and Laplace identified water as the product of the explosion of a mixture of hydrogen and oxygen triggered by an electrical spark. This discovery affected the issue of "water vs. mercury" in the case of Volta's eudiometer. The inflammable air test was performed over water on the condition that the ignition of a mixture of inflammable (hydrogen) and dephlogisticated gas (oxygen) was understood as transference of phlogiston from hydrogen to oxygen. However, once water vapour had been identified in the inflammable air test as the residual product, mercury began to be used in preference to water.

The belief that during the fast combustion of phosphorous certain hypothetical fixed air was given off from the air sample prompted Achard to perform the eudiometrical test over water in his first phosphorous eudiometer of 1780. But the suspicion that the phlogiston released by phosphorous could mix with water and cause irregularities made him change his mind, in his phosphorous eudiometer of 1786 Achard replaced water with mercury. Reboul used mercury instead of water in his own phosphorous eudiometer of 1788 because the idea that certain fixed air was removed from the common air during burning had already been abandoned. Thus, first of all, the controversy about the use of water or mercury ceased to be a theoretical dispute, and secondly, the phosphorous eudiometer proved to be unaffected by the shift from phlogiston to oxygen-centred chemistry. Giobert, in 1784, and Spallanzani, in 1796, used water in place of mercury in their extremely simple phosphorous eudiometers in order to improve the apparatus portability as well to make it more affordable. This change obliged them to replace the ordinary vertical eudiometrical tube with a tube bent at a right angle.

Eudiometers not only had to contend with theoretical innovations; eudiometers also existed whose conception and redesign took place in the mainstream of these innovations. For example, Séguin's phosphorous eudiometer of 1790 was equally indebted to Lavoisier's early experiments on the combustion of phosphorous, which were at the core of the oxygen-centred chemistry, and to his latest ideas on animal respiration. In this latter case, the crucial factor that launched the development of Séguin's eudiometer was Lavoisier's belief that carbon dioxide was not the only product of animal respiration, but also water. Furthermore, the emergence of Spallanzani's phosphorous eudiometer mentioned above can in part be placed within the context of the theoretical debates surrounding the rejection of Göttling's theories, as an alternative to Lavoisier's oxygen-centred chemistry, concerning the interpretation of the chemiluminescence of phosphorous during its slow combustion. This phenomenon also formed the basis of Berthollet's phosphorous eudiometer, which can be regarded as a spin-off research outcome of his refutation of Göttling's ideas. However, Berthollet's theoretical approach to eudiometry should be placed in the wider context of his view of chemistry and physics as allied disciplines. Finally, at the beginning of the nineteenth century, Dalton revisited the nitrous gas test to justify the formulation of his law of multiple combining proportions, which later made it possible to calculate the composition of inflammable gas mixtures using Volta-type explosion tubes.

Eudiometers in context

While eudiometry was still more concerned with the salubrity rather than with the respirability of air, there were eudiometrists who envisioned the need of introducing an overall eudiometrical control of atmospheric air. In this regard, in the eighteenth century, Ingenhousz was pondering a project consisting of a network of stations for monitoring the influence of vegetation on the salubrity of common air, and Humboldt for his part was also planning a similar project. This ambition to acquire a general overview of the composition of atmospheric was eventually realized in the nineteenth century thanks to the international survey organized by Regnault.

The determination of air respirability in crowded spaces as well as in open locations was far from being the only area of concern of eudiometry; its range of applications extended to other fields, the foremost of which was to monitor research work in different areas such as animal and plant physiology, chemical composition and the chemical industry. If eudiometers were able to determine the respirability of an atmospheric air sample, they could also play a role in determining the respirability of air samples collected in experiments on animal respiration, and it was in this field of research that eudiometers became an indispensable instrument.

Lavoisier used the nitrous air test for his experiments on animal respiration in 1775, which he took up again in 1783. Endeavours in this same field were decisive for the development of Séguin's phosphorous eudiometer in 1790. Lavoisier and Séguin also made use of the Volta's eudiometer in their investigations into animal respiration. Between 1779 and 1783, Ingenhousz enlarged the scope of action of the nitrous air test to include plant physiology, with special attention to the phenomena of respiration and photosynthesis. He also relied upon that test to assess the therapeutic value of coastal airs and oxygen for respiratory diseases. In 1796, Spallanzani conceived his phosphorous eudiometer in circumstances that were partially determined by his researches on animal and plant respiration, while in 1803 Hope's eudiometer was particularly involved in checking the absorption of nitrogen in human respiration. Pepys and Allen employed their polyvalent gas mixture analyser in the eudiometrical mode for the research work they carried out on plant and human respiration in 1807. Finally, the Regnault's Volta-type eudiometer was used extensively in the remarkable research on animal respiration conducted by Regnault himself and Reiset in 1848.

The development of eudiometry was also associated with some significant aspects of research into chemical composition. Lavoisier and Humboldt, among others, addressed the question of the saturation ratio between nitrous gas and oxygen during their work with the corresponding eudiometrical test. Dalton took up the nitrous gas test for the formulation of his law on the multiple proportions of combination. Similarly, Gay-Lussac's reshaping of the nitrous gas eudiometer was a result of his first attempt to establish a theory of the composition of the oxides of nitrogen. Lastly, it should be pointed out that Dalton, Saussure and Gay-Lussac also employed Volta's eudiometer, a key instrument in the early elucidation of the composition of some organic substances.

Despite its many sources of error, the nitrous gas test fascinated men of science from its very beginnings. It is also the eudiometrical means that has attracted the most attention from the perspectives of history, philosophy and sociology of science, the eventual conclusion being that the nitrous gas eudiometer was a failed instrument.[3] However, Golinski's analysis on why the nitrous gas eudiometer failed to live up to its original expectations, and Boantza's study illustrating how it actually failed, are subject to two limitations.[4] Firstly, both focus only on the original aim of the test; that is, for measuring the relative salubrity of an air sample. Secondly, neither considers the test over a period of time much longer than that of the eighteenth century. As mentioned above, it was in the last quarter of the eighteenth century when the nitrous gas test proved to be essential for research on animal and plant physiology. Furthermore, in the early nineteenth century the test played an important part in the conception of Dalton's law of the multiple proportions of combination, as well as in Gay-Lussac's endeavours to establish the composition of the oxides of nitrogen. Given this much broader perspective, the contention

---

[3] See chapter 1, note 54.

[4] Golinski, 1992, pp. 117-127; Boantza, 2013.

that the nitrous gas eudiometer was a failed instrument is rather more difficult to sustain.

In Great Britain, at the beginning of the nineteenth century, artificial illumination by gas lighting provided an economic context within which early eudiometrical gas analysis equipment for hydrocarbon mixtures was developed. This was the prelude to the emergence of gasometry from the second half of the nineteenth century onwards, which was largely underpinned by the favourable industrial and economic conditions in Germany. Bunsen's eudiometrical innovations in the production of cast iron and Hempel's adaption of gas analysis to industrial demands are two significant examples of that influence.

Eudiometers and the chemical revolution

The chemical revolution was a historical landmark stemming from the overthrow of the phlogiston theory by Lavoisier's oxygen theory. In the years leading up to this event, between 1772 and 1779, Lavoisier carried out a series of experiments on the gain in weight of several substances on combustion, and it was during this same period that the first eudiometers were developed. It is therefore apposite to ask what the role of these instruments was in the chemical revolution, and indeed if there was such a role.

In addressing this question, it is mandatory to state that the chemical revolution can be regarded from other perspectives other than that provided by the sole explanation of combustion and calcination based on the oxygen theory. The following two complementary approaches to the chemical revolution may help to shed light on this question. First of all, by considering Lavoisier's theory of respiration as an extension of his theory of combustion, and secondly, by taking into account that Lavoisier's commitment to an exhaustive quantitative experimental method was inexorably linked to the use of precise instruments.[5]

---

[5] In addition to these two perspectives, it is worth mentioning other alternative approaches that provide a better understanding of the chemical revolution, a number of which have paved the way to an exploration of the novelties introduced by Lavoisier, which were developed or completed during the nineteenth century. The oxygen theory may be regarded as part of a general theory of the gaseous state as well as a particular theory of acidity. The chemical revolution can be viewed as a reinterpretation of chemical composition with the definition of simple substance and its accompanying table of simple substances. The reform of chemical nomenclature provides a complementary linguistic approach to the chemical

With regard to the apparatus involved in establishing and demonstrating the principles of oxygen-centred chemistry, Holmes distinguished between two classes of instruments. First, Lavoisier's more complicated instruments (the ice calorimeter, the gasometer, the precision balance and the large round-bottomed flask for the synthesis of water), designed to address problems requiring new solutions. Second, the relatively simple devices (retorts, furnaces, receivers, pneumatic vessels and modifications or combinations of Hales apparatus) that helped Lavoisier to formulate his quantitative methodology.[6] Eudiometers did not fall into either of these two classes of instruments. Incidentally, Priestley conceived the first eudiometrical test (i.e. the nitrous gas test) a few months before Lavoisier performed his first experiments on the combustion of phosphorous.

Lavoisier used the nitrous gas test in his first experiments on animal respiration between April and October 1776. These experiments led Lavoisier to conclude that respiration was nothing more than a slow combustion of carbon with the consumption of oxygen and the release of carbon dioxide. His interest in respiration did not figure on his agenda again until February – March of 1783, when jointly with Laplace he conducted his first experiments using the calorimeter. In the initial phase of this second series of experiments on respiration, the two chemists used Fontana-Ingenhousz's nitrous gas eudiometer to determine the proportion of impurities in a sample of vital air.[7]

In June 1783, Lavoisier and Laplace conducted their brilliant experiment on the synthesis of water from hydrogen and oxygen. This was a key step in Lavoisier's work on the composition of water (1781-1783) that culminated in the large-scale experiments of 1785, thanks to which he discovered that the hydrogen released in the action of dilute acids on metals came from the water. It is quite likely that Volta had shown his eudiometer to Lavoisier during his visit to Paris in April 1782, and in fact Lavoisier subsequently used Volta's eudiometer to repeat the experiments on the synthesis of water. Furthermore, in 1805, Gay-Lussac and Humboldt definitively determined the volumetric composition of water by means of this eudiometer. Thus, it is clear that the nitrous gas and Volta's

---

revolution. This broad view of Lavoisier's achievements also covers his treatment of vegetable and animal substances (Crosland, 1980, pp. 406-407, 414, 409; 1994, pp. 14-15, 18-19, 182; 2009, pp.104-106)

[6] Holmes, 2000b, pp.76, 137-138.

[7] Ibid., 1985, p. 170.

eudiometers were actively present on the experimental scene during the first stage of the chemical revolution.[8]

Lavoisier took up his experiments on respiration again with the assistance of Séguin in 1790. The fact that Lavoisier was already convinced that the product of respiration was not only fixed air, but also water, prompted him to use eudiometrical methods to determine the amount of oxygen in the air both before and after being respired. Séguin adopted a fresh approach to the experimental apparatus that Lavoisier had developed in his previous researches on the combustion of phosphorus to devise a new version of the phosphorous eudiometer for their joint work on respiration.[9]

Firstly, in summary, the nitrous gas and Volta's eudiometers proved to be valuable complementary instruments in the formulation of Lavoisier's oxygen theory. Secondly, Lavoisier's early experiments on the combustion of phosphorous were not only important for establishing his oxygen theory, but also as an inspiration for the construction of Séguin's phosphorous eudiometer for experiments on animal respiration. Thus it is possible to say that, to some extent at least, eudiometers played an unacknowledged role in the chemical revolution.

The didactic second life of some eudiometers

Different nineteenth-century versions of Volta-type eudiometrical tubes operating over mercury were used for research, the most commonly employed of which were: the ordinary tube with brass or iron wires inserted straight across its head; Bunsen's tube with platinum wires bent into the head; Gay-Lussac's tube with a spiral wire, and Regnault's measuring tube. While they could also be used for teaching purposes, Guyton's Volta-type eudiometer, operating over water, was preferred for

---

[8] The chemical revolution may divided into two major stages. The first stage depended on the interpretation of a series of quantitative experiments on combustion and calcination (1772-1779) and on the composition of water (1781-1783, 1785), from which Lavoisier was able to arrive at a definitive formulation of the oxygen theory. After that, in 1785, he felt confident enough to commence his assault on the phlogiston theory. In the second, more conceptual and linguistic stage (1787-1789), Lavoisier was more concerned with disseminating and reorganising chemistry. The new periodical *Annales de chimie* and his textbook *Traité élémentaire de chimie* became the necessary means by which the new chemistry was propagated (Crosland, 2009, pp. 99-101, 113).

[9] However, in addition to the phosphorous eudiometer, the alkaline sulphide, the nitrous air and Volta's eudiometers were also used in these researches.

teaching demonstrations because it was more suitable and easier to use.[10] This eudiometer was just one among many other apparatus that enjoyed a second life as didactic instruments after they had exhausted their potential usefulness for research. The occasional appearance of research apparatus as teaching aids is often a sign that such instruments were no longer in active use for research.[11]

Another similar case was that of the phosphorous eudiometer, which underwent a reincarnation as a didactic instrument in chemical treatises. These works featured eudiometrical devices employed in the fast and the slow combustion of phosphorous; they had not been redesigned for further research, but rather for demonstrating the composition of common air effectively, and thus could be used to illustrate examples in chemistry lessons. Instruments based on the slow combustion of phosphorous were derived from Berthollet's original eudiometer (Figure 10.1, left), while those based on the fast combustion of phosphorous resembled Giobert's and Spallanzani's eudiometers (Figure 10.1, right). These latter instruments were in fact modified using simpler and user-friendly materials.

Figure 10.1 Phosphorous eudiometers from French textbooks of chemistry (1).

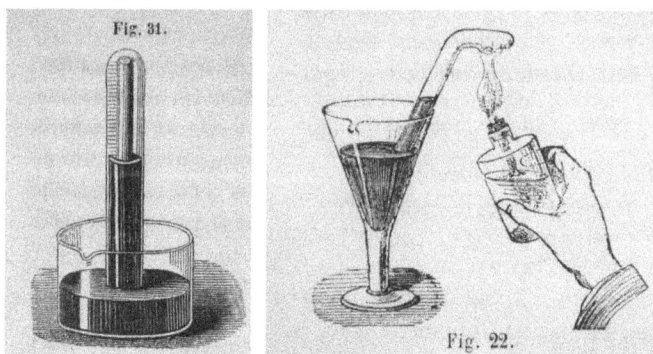

(Left) Eudiometrical device based on the slow combustion of phosphorous (over mercury). From Auguste Cahours, *Leçons de chimie générale élémentaire*, 1855, p. 98, Fig. 31. (Right) Eudiometrical device based on the fast combustion of phosphorous (over mercury). From Charles-Adolphe Wurtz, *Leçons élémentaires de chimie modern*, 1867, p. 60, Fig. 22.

---

[10] See Figure 2.10. Regnault, 1850, Vol. 1, p. 114; Pelouze & Fremy, 1854, Vol. 1, pp. 244-245; Salleron, 1861, pp. 83-84.

[11] Golinski, 1992, pp. 126-127; Brenni, 2012, pp. 192-195.

The test based on the slow combustion of phosphorous was praised for giving more exact results than that based on fast combustion. It was not much quicker, however, since it took from two to three hours to complete. This test reached its endpoint when the stick of phosphorous no longer glowed in the dark or when the volume of the residual gas underwent no further reduction. In general, authors recommended performing the test over mercury, although in a few cases the figure in the textbook depicted the test conducted over water (Figure 10. 2, left).[12]

Figure 10.2 Phosphorous eudiometers from French textbooks of chemistry (2).

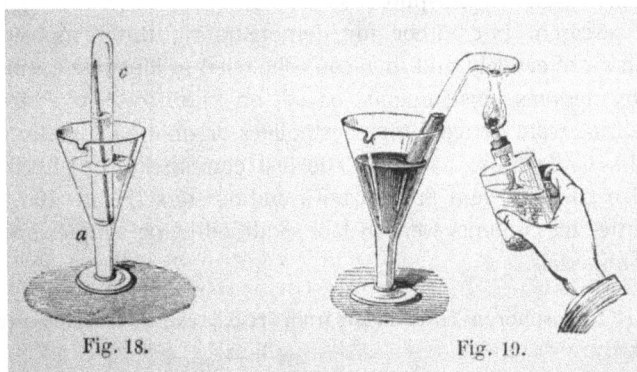

Fig. 18.   Fig. 19.

(Left) Eudiometrical device based on slow combustion (over water) and (Right) eudiometrical device based on the fast combustion of phosphorous (over mercury). From Edmond Frémy and Auguste Terreil, *Le guide du chimiste*, 1885, p. 119, Figs. 18, 19.

The test based on the fast combustion of phosphorous was not always accurate, but it was easier to perform and sufficient for approximate tests. It was necessary to heat the piece of phosphorus thoroughly; otherwise it could evaporate without catching fire. The tube would then be filled with an explosive gas mixture of oxygen, nitrogen and phosphorus, which would break the tube and would expel the inflamed phosphorus.[13] However, there was a lack of consensus among different authors about performing both tests over mercury or water (Figures 10.3 and 10.4).

---

[12] Nevertheless, the inner wall of the tube and the surface of the stick of phosphorous had to be wet in order to saturate the air sample with water, otherwise the combustion came to a halt.

[13] Cahours, 1860, Vol.1, p. 123.

Figure 10.3 Phosphorous eudiometers from French textbooks of chemistry (3).

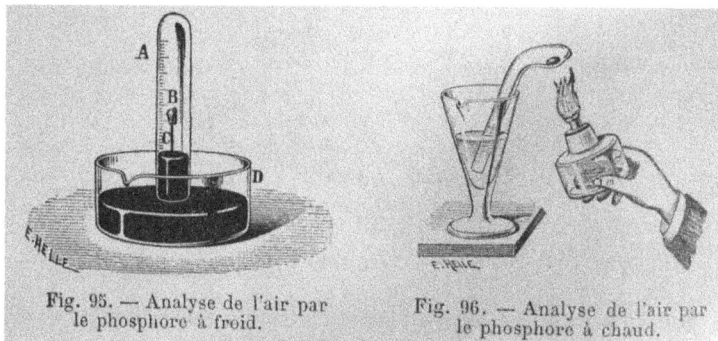

Fig. 95. — Analyse de l'air par le phosphore à froid.

Fig. 96. — Analyse de l'air par le phosphore à chaud.

(Left) Eudiometrical device based on the slow combustion (over mercury) and (Right) eudiometrical device based on the fast combustion of phosphorous (over water). From Louis Serres, *Cours de chimie*, 1901, p. 116, Figs. 95, 96.

Figure 10.4 Phosphorous eudiometers from French textbooks of chemistry (4).

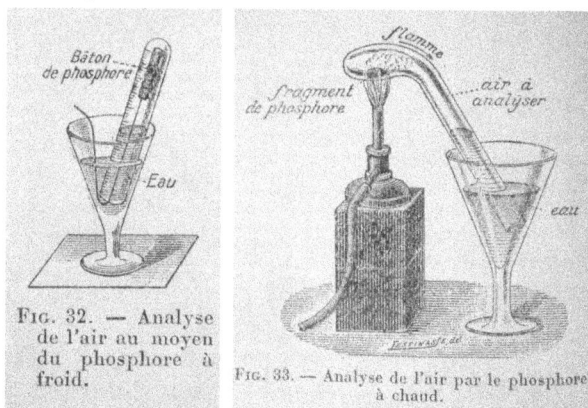

FIG. 32. — Analyse de l'air au moyen du phosphore à froid.

FIG. 33. — Analyse de l'air par le phosphore à chaud.

Eudiometrical devices based on the slow combustion (Left) and on the fast combustion of phosphorous (Right) both over water. From E. Henry, *Cours de chimie*, 1912, p. 44, Figs. 32, 33.

Appendix

I. Fontana's fourth nitrous air *macchina*

This experimental device (Figure 11.1) consisted of a wooden cabinet varnished on both the inside and outside, *Fig. I*, competently assembled with the sections tightly joined together. The cabinet was divided into three chambers (A, B, C), two of which (A, C) were identical to each other. The third chamber (B) was as big as the other two together but with a vault attached to the top. The two identical chambers (A, C) had two apertures, each with its stopper, two of which (a, b) were located on the top of the cabinet and the other two on the sides. The three chambers were separated by thick, inwardly slanting walls, while the front consisted of three large panes of glass assembled on wooden frames so that the mastic adhesive prevented any air from entering. The two chambers (A, C) were directly connected to the middle chamber (B) by two large, square openings (c, d) in the common walls. Both chambers were closed with two wooden hinged sections, twice as heavy the top, where strong but flexible threads were attached. These two threads acted as valves which, when falling under their own weight, perfectly closed off the openings because they were inclined at the same angle as the common walls. Two silk - or preferably aloe - threads (e, e) attached to each valve joined at (B) and came together through the funnel (o, p). The chamber (B) was open at the top and attached to a box (x) pierced in the base by a screw with a male thread. This box was covered with a skin that fitted perfectly into the convex wall of the chamber (B) preventing any air intake. Both the chamber (B) and the box (x) held a single volume that extended into the funnel. The box (x) was perforated at the top, *Fig. II*, allowing the nozzle of the funnel, *Fig. III*, to pass through and almost touch the bottom of the middle chamber. The stopper (m) provided a direct connection through the slot (r) made in the nozzle to the chamber (B), *Figs. I* and *III*.

The chamber (B) was filled with mercury through the funnel (oB), while the other two chambers (A, C) were filled with nitrous air and respirable air, respectively, ensuring that no original air remained in either chamber. The next operation consisted of pouring mercury from a bottle containing a weighted amount of mercury into the funnel (o) while the threads (p) were pulled upwards. Consequently, both valves (c, d) opened, and mercury flowed from the middle chamber into the lateral chambers until the levels of the liquid were the same. In this way, respirable and nitrous air were completely mixed, and the three chambers remained connected. Both airs were mutually annihilated on mixing, and mercury entered the middle chamber through the funnel until the interior air (i.e. air pressure) was equal to the external one. Finally, after replacing the stopper (m) and removing the funnel, the remaining mercury was poured back into the bottle, which was then weighed again to determine the quantity of mercury that had been introduced into the box. This amount of mercury had to be proportional to the goodness of respirable air used in the experiment.

Figure 11.1 The fourth "macchina" of Felice Fontana.

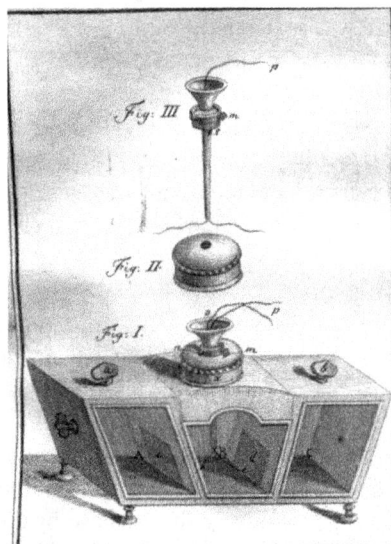

From Felice Fontana, *Descrizione ed usi di alcuni stromenti per misurare la salubritat dell'aria* **(Firenze, 1775), plate 4.**

II. Gérardin's nitrous air eudiometer

Gérardin's device (Figure 11.2) consisted of two main parts: a cylindrical flask and an inverted siphon. The graduated cylindrical glass flask (*Fig. 5*, c) had a capacity of 3.5 *septiers*[1]. At the top of this flask there were two short glass necks, each tightly connected to a retort-shaped vial (a, b) with a capacity of half a *septier*. Between each vial and its respective neck there was a tap (R, R), and the top end of each vial was closed off by its

---

[1] The *septier* was an old French unit of capacity roughly equivalent to a half a litre.

respective taps or screw caps (*bouchons à vis*) (m, m). On the side of the flask, a graduated glass tube, the lower end sealed with a ground glass stopper, descended from beneath the necks. This tube was as long as the flask and had the same capacity as each vial. The flask was affixed to a tripod by means of an iron ring (*Fig. 6*, p).

The inverted siphon consisted of two components: a glass bottle (*Fig. 5*, S) with a capacity of 1 *chopine*[2] filled with water for the purpose of receiving nitrous air. The top of the bottle had a neck with a glass stopper (m). The second component was a recipient (SS) with a capacity of 3 *poissons*[3] for containing nitrous air. At the bottom there was a tap (R) – like the two in the cylindrical flask – for controlling the flow of nitrous air.[4] The top of this recipient ended with a tap (u) that was connected to the neck of one vial by means of a T-shaped tube with a glass stopper (x). At the intersection point (y) of those two components pieces there was a glass tap or a screw cap.

The first step in the procedure was to open the tap (m) of the vial (a) to fill the cylindrical flask (c) and the two vials (a,b) with water. Then the tap (R) of the same vial (a) was opened to replace water by atmospheric air. After closing the taps (m, R), the stopper (x) was removed and substituted by a glass funnel. The next step was to refill the T-shaped tube with water and replace the stopper (x). Then the taps (m, R) of the other vial (b) were opened, and after that the tap (u) of the recipient (SS) was also opened. The nitrous air flowed into the vial (b), while water ran out through the tap (R). Once all the taps (m, R) of the vial (b) were closed and the T-shaped tube disconnected, the cylindrical flask was fixed in the iron ring of the tripod and positioned upside down. The nitrous and atmospheric air then mixed and the residue rose up to the graduated part of the flask. Meanwhile, water in the lateral tube went down in accordance with the diminution in volume of the air mixture. When the contraction in volume was completed, the graduation in the flask indicated the diminution in volume of both airs, and therefore the degree of respirability of the atmospheric air.

---

[2] The Paris *chopine* was roughly equivalent to a half a litre.

[3] The Paris *poisson* or *posson* was roughly equivalent to a 0.12 L

[4] Nothing was mentioned in relation to the obtaining of nitrous air and, therefore, to the operations involved with the device (SS, S) on the table (*Fig. 5*).

Figure 11.2 Gérardin's eudiometer.

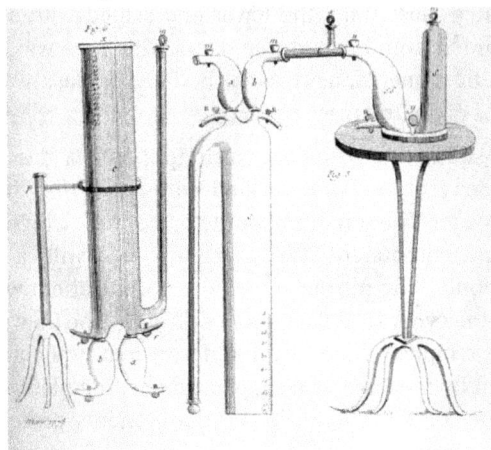

From Observations sur la physique, sur l'histoire naturelle et sur les arts (1778), plate 2.

III. Lavoisier's experimental results and calculations for the determination of the vital air content and the purity of the nitrous air

Determination of the vital air content

For the first addition, 100 parts (**a**) of a sample of vital air were added to 300 parts (**b**) of a sample of nitrous air. Once the contraction had been observed, the residual air mixture occupied 131 parts (**c**). Hence, 100 parts of the sample contained 98.72 parts (**x**) of pure vital air and 1.28 parts of irrespirable air. For the second addition, 100 parts (**a**) of a sample of common air were added to 300 parts (**b**) of a sample of nitrous air. Once observed the contraction in volume, 100 more parts of common air were added. The same operation was repeated for the third time. Table 8 shows the corresponding results.

Table 8 Experimental results of the determination of the vital air content of a sample of atmospheric air.

Addition	Common air a	Nitrous air b	Total a+b	Residue c
1a	100	300	400	331
2a	100		500	367
3a	100		600	394

From these data, the parts of vital air could be calculated as shown in Table 9.

Table 9 Calculations of the vital air content of a sample of atmospheric air.

Addition	ΔV	Parts of vital air (0.367· ΔV)
1[a]	400 – 331 = 69	25.3/1 = 25.3
2[a]	500 – 367 = 133	48.8/2 = 24.4[5] (25.0)
3[a]	600 – 394 = 206	75.6/3 = 25.2

Determination of the purity of the nitrous air

After adding five consecutive doses of 100 parts of nitrous air to a sample of 300 parts of vital air, the following results were obtained (Table 10).

Table 10 Experimental results of the determination of the purity of the nitrous air used as a reagent.

Addition	Vital air a	Nitrous air b	Total a+b	Residue c
1[a]	300	100	400	251
2[a]		100	500	201
3[a]		100	600	153
4[a]		100	700	93
5[a]		100	800	62

From these data the parts of pure nitrous air could be calculated (Table 11).

Table 11 Calculations of the purity of the nitrous air used as a reagent.

Addition	ΔV	Parts of nitrous air (0.633 · ΔV)
1[a]	400 – 251 = 149	94.3/1 = 94.3
2[a]	500 – 201 = 299	189.27/2 = 94.6
3[a]	600 – 153 = 447	282.95/3 = 94.3 (94.8)[b]
4[a]	700 – 93 = 607	384.23/4 = 96.1 (96.0)
5[a]	800 – 62 = 738	467.15/5 = 93.4

[5] This is a discordant result with that of Lavoisier's paper, in which he assigned the value 25.0. The data of the final residue for this value should have been 364. However, it may have been a misprint.

[6] This is actually the only significantly discordant value with that of Lavoisier's paper. It may be due to a misprint of the digit "3", which was replaced by "8".

IV. The nitrous air eudiometer of Adam Wilhelm Hauch

The apparatus consisted of a glass tube (Figure 11.3, *Fig. 1*, K) 18 inches long and a diameter of 6 lines with an attached brass scale. The lower end of the tube was fitted to the upper neck of a round flask (B), whose lower neck was fitted to the bottom of the cylindrical glass vessel (A) by an inner cylindrical tube. This vessel had a length of 4.5 inches, a diameter of 2 inches and was open at the top. A valve placed inside the inner tube connected it to the outer vessel. The bottom of this vessel had two holes and a shutter (h) sliding over them in such a way that either the round flask (B) or the vessel (A) could be connected to a glass cylinder (C) (see also *Fig. 2*). This cylinder was the upper piece of the mobile part (CEDFHI) of the eudiometer that could be shifted from right to left and vice-versa. It was coupled through a brass Y-shaped connector (EDF) to the glass bottles (H) and (I). This connector had a shut-off valve (G) interconnecting the three tubings. Six shutters were distributed throughout the eudiometrical device (k, I, g, f, IH, d). Each shutter was similar to the flat slider of the measure used in Fontana-Ingenhousz's eudiometer, except that in Hauch's eudiometer it served as a shut-off valve for air as well as for water.[7]

The test was first conducted by filling the bottle (H) with common air and the bottle (I) with nitrous air. The entire eudiometrical device - except the two bottles - was then filled with water. An amount of water was also poured into the vessel (A) up to a certain level (m). By manipulating all the shutters and the shut-off valve (G) correctly, the common and the nitrous air were eventually mixed together in the round flask (B). The contraction in volume of the air mixture induced a pressure drop in the eudiometer and, as a consequence, the valve placed inside the inner tube of the vessel (A) opened and water started flowing from the latter vessel into the eudiometrical device until the inner and the atmospheric pressure balanced. Finally, the residual air mixture rose up through the tube (K) and its volume was determined.

V. Achard's portable nitrous air eudiometer

This eudiometer consisted of a U-shaped glass tube (abcdef) of 4 lines in diameter (Figure 11.4, left), curved in such a way that the branch (ac) was equal in length to the (ed) branch, each branch being 7 inches long. A glass globe (g) of 15 lines in diameter was coupled to the middle part (cd) of the tube. This globe was joined directly through its neck (h) to the interior of the tube. The whole device was filled with water before starting

---

[7] Levere, 1999, p. 15.

a test. Crystal stoppers (L, m) were fitted snugly into the two openings (ab, ef) of the tube, which was attached to a small board (DGFE) 7 inches long and 24 lines wide. A small pendulum (HI) was attached to this board to keep it in a vertical position.

Figure 11.3 Hauch's nitrous air eudiometer.

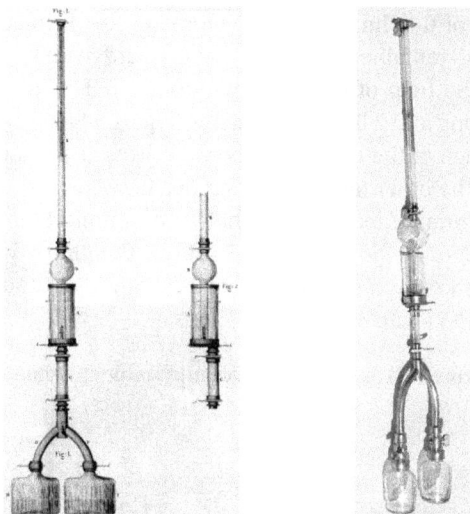

(Left) From Adam Wilhelm Hauch, *Nye Samling af det Kongelige Danske Videnskabernes Selskabs skrifter* (1793), Vol. 5, plate p. 134. (Right) Copy at the Sorø Akademi. Photography courtesy of Jørgen From Andersen.

To determine the purity of an air sample, the lower neck of a glass globe (n) 12 lines in diameter (Figure 11.4, right) was fitted into the opening of the crystal bottle (X), which was equipped with a tap (YZ). This bottle could hold about six or eight times the capacity of the other globe and contained nitrous air. After opening the tap (st) of the globe, it was filled with water through the upper opening (u), replacing any air that might remain between both taps. The opening (u) was then closed off with a crystal stopper (Z), and the tap (YZ) of bottle X was opened. When the water had descended into the bottle and the nitrous air filled the globe (n), both taps were closed.

The next step consisted in removing the stopper (m) of the left opening (ef) of the tube and fitting the neck (opqr) of the globe (n) into it. This being done, the globe (i) of 12 lines in diameter was filled with water that was then drained out by holding the globe upside down so that it filled with the air to be examined. The stopper (L) of the right opening (ab) of the tube was then removed, and the neck (K) of the globe (i) was fitted correctly into that opening (ab). Afterwards, the entire device was reversed, so that the air contained in the globes (i) and (n) entered the globe (g), a procedure that was necessary to accelerate the air mixture. The instrument was then inclined so that the air mixture was transferred to the globe (n) and to the left branch (fed) of the tube. After that, the globe (i) was removed from the tube, and the water level descended into the branch (ac) as a result of the diminution in volume of the air mixture. When the water level had remained stationary, the board was kept in a vertical position with the help of the small pendulum (HI). Finally, a solid tube (FG) 7 inches long and 2 lines in diameter (Figure 11.4, left) was immersed in the (ac) branch of the tube until the water was at the same level in this branch and in the other (ed). The difference between the length of the air column in the branch (ed) and the height of an air column in the same tube, whose volume was the same as that of the globe (i), indicated the diminution in the volume of the air mixture, and therefore the degree of salubrity of the air sample.

Figure 11.4 Achard's portable nitrous air eudiometer.

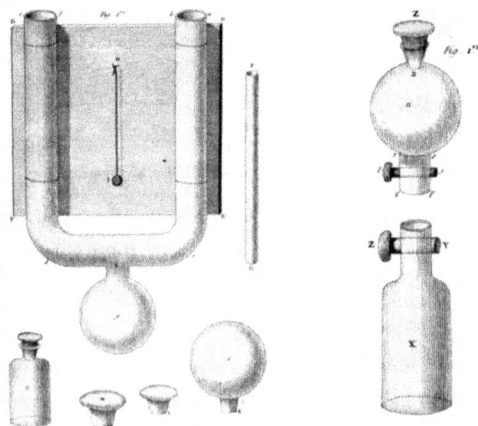

From *Observations sur la physique, sur l'histoire naturelle et sur les arts* (1784), plates 1 and 2, pp. 84, 86[8].

---

[8] A simplified drawing of this eudiometer was also published in the *Nouveaux Mémoires de l'Académie Royale des Sciences et Belles-Lettres à Berlin* (1780), plate 5, p. 161.

VI. Parrot's eudiometer (oxygenmeter)

Parrot's eudiometer (Figure 11.5) consisted of a glass absorption tube (*fig. 3*, AB). The upper part (AC) was wide and short with a lubricated screw thread (Dd) at the top to provide an airtight seal. This part ended in the neck (C) and continued with the open-ended narrow tube (CB) that was graduated in 350 divisions.[9] The absorption tube was placed within a glass tubular container (*fig. 4*, cb), which was opened at the top and closed at the bottom. Its length was slightly higher than that of the graduated scale, and its inner diameter was 3 inches larger than that of the absorption tube. This container was held to a wooden board (pqo) by two metal circles (rr) and hanged on a nail.

After noting the temperature and the atmospheric pressure, the tubular recipient was half filled with mercury.[10] Then, a stick of phosphorous was introduced (*fig. 4*) through the upper part (ac) of the absorption tube, which was closed with the screw afterwards (d).[11] To fill the absorption tube with an air sample, the tube was first filled with mercury and then emptied in the place which air wanted to be analysed. If the air sample was stored in a vessel, then the usual procedure with the pneumatic trough with mercury was used. After that, the tube was put to its original upright position within the tubular container with mercury (cb).

The duration of the assay ranged from 7 to 8 hours, at a temperature of 15 -19 ℃, until the glow of phosphorous disappeared. In the meantime, mercury should be added within the tubular container to prevent the mercury level in this recipient from being lower than in the absorption tube (CB). Alternatively, an additional weight could be put over the screw. At the same time, care should be taken to keep the hands and the face

---

[9] The number of divisions had to represent the net volume of the absorption tube. For this reason, before drawing the scale it was necessary to determine the volume of the stick of phosphorus to be used. This stick displaced an equal volume of air of the absorption tube and, therefore, its volume had to be deducted from the total volume of the tube. The volume of the stick of phosphorus was calculated from its weight. Then, after filling the absorption tube with mercury up to (aa) and weighing the mercury used, the entire volume of the tube could be calculated. Subtracting the volume of the stick of phosphorous from this total volume, the net absorption volume of the tube was determined.

[10] Parrot noticed that using water instead of mercury involved some uncertainties such as the absorption of air or any humidity remaining in the stick of phosphorus and in the inner walls of the tubes.

[11] It was needed that the neck (C) was narrow enough to prevent the stick from passing to the tube (CB) (*Fig. 3*)

away from the tube. Finally, to determine the height of the mercury column, the absorption tube was moved until the mercury column within the tube (CB) levelled with the mercury within the tubular container (cb). Then, the level of mercury could be read on the graduated scale to know the proportion of oxygen to the thousandth.[12]

Figure 11.5 Parrot's eudiometer (oxygenmeter).

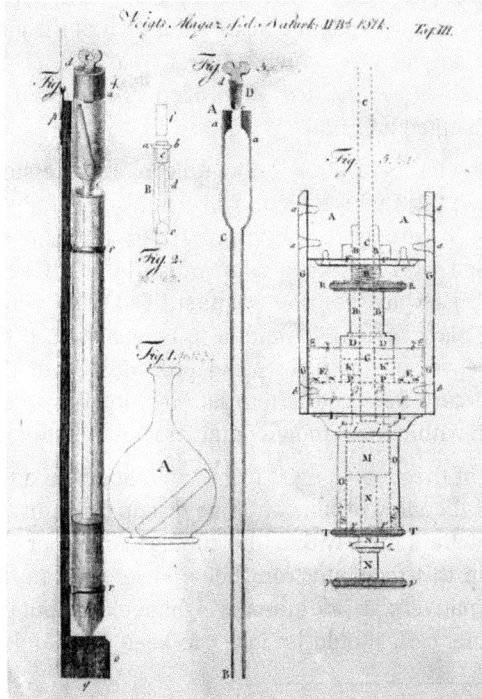

From *Magazin für den neusten Zustand de Naturkunde*, 1800, Vol. 2, plate 3.

---

[12] Thermometric correction: 0.0046 for each degree Reaumur. Barometric correction: 0.00225 for each line of variation in the barometer.

Bibliography

Abbreviations

LM Lavoisier, A.L. 1805, *Mémoires de chimie*, Paris: unpublished, 2 vols.; 2004, *Mémoires de physique et de chimie*, Bristol: Thoemmes Continuum, 2 vols.

During the year 1792 Lavoisier had decided to publish a new collection of *Mémoires de physique et de chimie*. The process of publication was truncated on 28 November 1793 when the National Convention ordered to arrest Lavoisier. The *Mémoires* were never actually published. Madame Lavoisier decided in 1805 to bind the surviving printed material of 1793 and to donate copies to institutions and friends. For the complex origins and circulation of the *Mémoires* see Beretta (2001b).

LO Lavoisier, A.L. 1864-1893, *Oeuvres*, Paris: Imprimerie Impériale, 6 vols.

VO Volta, A. 1918, *Le Opere di Alessandro Volta, Edizione Nazionale*, Milano: Ulrico Hoepli, 7 vols.

Bibliography of primary sources

Accum, F.C. 1803, *A System of Theoretical and Practical Chemistry*, London: Accum and Kearsley; Edinburgh: Bell and Bradfute; Glasgow: Brash and Reid; Dublin: Archer and Colbert, 2 vols.

Achard, F.C. 1780 (pour l'année 1778), «Mémoire sur la mesure de la salubrité, renferment la description de deux nouveaux eudiometres», *Nouveaux Mémoires de l'Académie Royale des Sciences et Belles-Lettres à Berlin, Classe de philosophie expérimentale*, 91-100.

_____. 1783 (pour l'année 1781), «De l'effet des parfums sur l'air», *Nouveaux Mémoires de l'Académie Royale des Sciences et Belles-Lettres à Berlin*, 33-40;1785, *Observations sur la physique, sur l'histoire naturelle et sur les arts*, 26, part 1, Février, 81-87.

_____.1784, «Mémoire sur la mesure de la salubrité, renferment la description de deux nouveaux eudiomètres», *Observations sur la physique, sur l'histoire naturelle et sur les arts*, 24, Part 1, Janvier, 33-40.

_____.1786a (pour l'année 1784), «Recherches faites dans la vue de découvrir une méthode exacte pour mesurer les quantités relatives de phlogistique contenues dans une sorte d'air donné, de façon que les degrés de phlogistication de l'air soient réduits par cette méthode à des rapports justes & numériques», *Nouveaux Mémoires de l'Académie Royale des Sciences et Belles-Lettres à Berlin*, 27-43.

_____. 1786b (pour l'année 1784), «Détermination de la salubrité de l'air atmospherique, dans différens endroits compris dans l'étendue de 16 milles», *Nouveaux Mémoires de l'Académie Royale des Sciences et Belles-Lettres à Berlin*, 44-57.

Allen, W.; Pepys, W.H. 1807, «On the Quantity of Carbon in Carbonic Acid, and on the Nature of the Diamond», *Philosophical Transactions*, 97, 267-292; 1807-1808, *The Philosophical Magazine*, 29, 216-227, 315-324.

Allen, W.; Pepys, W.H. 1808, «On the Changes Produced in Atmospheric Air, and Oxygen Gas, by Respiration», *Philosophical Transactions*, 98, 249-281; *The Philosophical Magazine*, 32, 242-267.

Berthelot, M. 1890, *La révolution chimique. Lavoisier*, Paris: Félix Alcan.

Berthollet, A.B. 1809, «Mémoire sur l'analyse de l'ammoniaque», *Mémoires de physique et de chimie de la Société d'Arcueil*, Vol. 2, 268-294, Paris: Bernard, 3 vols.

Berthollet, C.L. 1788a (pour l'année 1785), «Analyse de l'alcali volatil», *Histoire de l'Acadèmie Royale des Sciences*, 316-326.

_____ . 1788b (pour l'année 1785), «Suite des recherches sur la nature des substances animales, et sur leurs rapports avec les substances végétales», *Histoire de l'Acadèmie Royale des Sciences*, 331-349.

_____ . 1795, «Dixième leçon de chimie», *Séances des Écoles Normales, recueillies par des sténographes et revues par les professeurs*, Vol. 5, 67-88. Paris: Imprimerie du Cercle Social, an IV, 8 vols.

_____ . 1796, «Observations sur les propriétés eudiomètriques du phosphore», *Journal de l'École Polytechnique*, 1, 3, 274-278.

_____ . 1800a, «Observations eudiomètriques», *Annales de chimie*, 34, 73-85.

_____ .1800b, «Observations eudiomètriques», *Mémoires sur l'Égypte publiés pendant les campagnes du general Bonaparte, dans les années VI et VII*, Paris: Didot l'ainé, 284-294.

_____ . 1801, «Observations sur l'action que le sulphate de fer exerce sur le gaz nitreux», *Annales de chimie*, 30 Messidor an IX, 39, 3-17.

_____ . 1803a, «Observations sur le charbon et le gaz hydrogène carboné», *Mémoires de la classe des sciences mathématiques et physiques de l'Institut National de France*, 4, 1er série, an VI-X, 269-318, 319-324, 325-333.

_____ . 1803b, *Essai de statique chimique*, Paris: Fermin Didot, 2 vols.

_____ .1809, «Nouvelles observations sur les gaz inflamables désignés par les noms d'hydrogène carburé et d'hydrogène oxicarburé», *Mémoires de physique et de chimie de la Société d'Arcueil*, Vol. 2, 68-93. Paris: Bernard, 3 vols.

Biot, J.B. 1807a, «Extrait d'une lettre de Mr.Biot à Mr. Berthollet», *Annales de Chimie*, 61, 271-281.

_____ . 1807b, «Extract of a Letter from Mr. Biot to Mr. Berthollet», *A Journal of Natural Philosophy, Chemistry and the Arts*, 22, 126.

_____ . 1808, «Einige Bemerkungen aber die Absorption von Gasarten durch Waffer, und über die Eudiometrie, von Herrn Antonio de Marty in Tarragona; aus einem Briefe Biot's an Berthollet», *Annalen der Physik*, 28, 417-426.

Bouillon-Lagrange, E.J.B. 1798-1799, an VII, *Manuel d'un cours de chimie*, Paris: Bernard, 2 vols.

_____ . 1801, an IX, *Manuel d'un cours de chimie*, seconde édition, Paris: Bernard, 3 vols.

Bunsen, R. 1857, *Gasometry*, London: Walton and Maberly.

Bunsen, R.; Playfair L. 1846, «Report on the Gases Evolved from Iron Furnaces, with Reference to the Theory of the Smelting Iron», In *Report of the Fifteenth Meeting of the British Association for the Advancement of Science, 1845*, London: John Murray, 142-186.

Cahours, A. 1855, *Leçons de chimie générale élémentaire*, deuxième edition, Paris: Mallet-Bachelier, 2 vols.

_____ . 1860, *Traité de chimie élémentaire*, deuxième edition, Paris: Mallet-Bachelier, 3 vols.

Cavallo, T. 1781, *A Treatise on the Nature and Properties of Air and Other Permanent Elastic Fluids. To Which is Prefixed an Introduction to Chymistry*, London: printed for the author.

Cavendish, H. 1766, «The Three Papers Containing Experiments on Factitious Airs», *Philosophical Transactions*, 56, 141-183.

_____ . 1783, «An Account of a New Eudiometer», *Philosophical Transactions*, 73, 1, 106-136.

_____ . 1785, «Experiments on Air», *Philosophical Transactions*, 75, 2, 372-285.

Coquillion, J.J. 1877, «Sur les appareils grisoumètres qui peuvent servir à doser l'hydrogène protocarboné dans les mines», *Comptes Rendus Hebdomadaires des Séances de l'Académie des Sciences*, 84, 458-549.

_____ . 1881, *Analyse des gaz. Description et usage des eudiomètres à fil de platine*, Paris: J. Baudry.

Dalton, J. 1805a, «Experimental Enquiry into the Proportions of the Several Gases or Elastic fluids Constituting the Atmosphere», *Memoirs of the Literary and Philosophical Society of Manchester*, 2nd series, 1, 244-258.

_____ . 1805b, «On the Absorption of Gases by Water and Other Liquids», *Memoirs of the Literary and Philosophical Society of Manchester*, 2nd series, 1, 271-287.

_____ . 1808, 1810, *A New System of Chemical Philosophy*, Manchester: R. Bickerstaff, 2 parts.

_____ . 1819, «Memoir on Sulphuric Ether», *Memoirs of the Literary and Philosophical Society of Manchester*, 2nd series, 3, 446-482.

_____ .1837, «Sequel to an Essay on the Consitution of the Atmosphere, Published in the Philosophical Transactions for 1826; With Some Account of the Sulphurets of Lime», *Philosophical Transactions*, 127, 347-363

Davy, H. 1800a, «On the Nitrous Oxide or Gaseous Oxide of Azote; on Certain Facts Relating to Heat and Light; and on the Discovery of the Decomposition of the Carbonate and Sulphate of Ammoniac», *A Journal of Natural Philosophy, Chemistry and the Arts*, 3, February, 515-518.

_____ .1800b, *Researches chemical and philosophical, chiefly concerning nitrous oxide*, London: J. Johnson; Bristol: Biggs and Cottle.

_____ .1801, «An Account of a New Eudiometer», *The Philosophical Magazine*, 10 (June - September), 2, 56-58.

_____ . 1836-1840, *The Collected Works of Sir Humphry Davy*, edited by his brother John Davy, London: Smit, Elder and Co. Cornhill, 9 vols.

Davy, J. 1836, *Memoirs of the Life of Sir Humphry Davy*, London: Longman and others, 2 vols.

De Lapparent, A. 1895, *École Polytechnique, Livre du Centenaire, 1794-1894*, Paris: Gauthier-Villars et fils, 3 vols.

De Luc, J.A. 1772, *Recherches sur les modifications de l'atmosphere*, Genève: s. n., 2 vols.

Dennis, L.M. 1913, *Gas Analysis*, New York and London: The Macmillan Company.

Desormes, C.B.; Clément, N. 1801, «Mémoire sur la réduction de l'oxide blanc de zinc par le charbon, et sur le gaz oxide de carbone qui s'en dégage», *Annales de chimie*, 39, 26-64.

Doyère, L.M.F. 1850, «Études sur la respiration. Premier mémoire. Procédés et observations eudiométriques», *Annales de chimie et de physique*, 3e sèrie, 28, 5-56.

Fontana, F, 1775a, *Ricerche fisiche sopra l'aria fissa*, Firenze: Gaetano Cambiagi; «Recherches physiques sur l'aire fixe», *Observations sur la physique, sur l'histoire naturelle et sur les arts*, 1775, 6, octubre, 280 – 289.

_____ . 1775b, *Descrizione ed Usi di Alcuni Stromenti per Misurare la Salubrità dell'Aria*, Firenze: Gaetano Cambiagi.

_____ . 1779a, «Experiments and Observations on the Inflammable Air Breathed by Various Animals», *Philosophical Transactions*, 69, 337-361.

_____ . 1779b, «Account of the Airs Extracted from Different Kinds of Waters; With Thoughts on the Salubrity of Air at Different Places», *Philosophical Transactions*, 69, 432-453.

_____ . 1779-1780, «Science de l'air», In *Studi su Felice Fontana*, edited by F. Abbri, Cosenza: Edizioni Brenner, 23-52.

Fourcroy, A. F. 1792, «Appareil», *Encyclopédie Méthodique. Chimie, Pharmacie et Métallurgie*, Vol. 2, 350-371, Paris: Chez Panckoucke; Liège: Chez Plomteux, 6 vols.

_____ . 1796, «Chimie», *Encyclopédie Méthodique. Chimie, Pharmacie et Métallurgie*, Vol. 3, 262-781, Paris: Chez H. Agasse, 6 vols.

_____ .1800, *Système des connaissances chimiques, et leurs applications aux phénomènes de la nature et de l'art*, Paris: Baudoin, 10 vols.

_____ . 1805, «Eudiomètre», «Eudiomètrie», «Laboratoire», *Encyclopédie Méthodique. Chimie et Métallurgie*, Vol. 4, 276-279, 279-283, 566-581, Paris: Chez H. Agasse, 6 vols.

Fourcroy, A.F.; Vauquelin, N.L.; Séguin, A. 1791, «Mémoire sur la combustion du gaz hydrogène dans des vaisseaux clos», *Annales de Chimie*, 8, 230-308; 9, 30-50.

Frankland, E. et al. 1876, «Dr. Frankland's Address: On Eudiometers», In *Conferences Held in Connection with the Special Loan Collection of Scientific Apparatus, 1876*, London: Chapman and Hall, 2 vols.

Frankland, E.; Ward, W.J. 1854, «On an Improved Apparatus for the Analysis of Gases», *Quarterly Journal of the Chemical Society*, 6, 197-205.

Frémy, E.; Terreil, A. 1885, *Le guide du chimiste*, Paris: G. Maison.

Gay-Lussac, J. L. 1804, «Relation d'un voyage aérostatique», *Journal de Physique, de Chimie, d'Histoire Naturelle et des Arts*, 59, 454-462.

_____ . 1809a, «Mémoire sur la combinaison des substances gazeuses, les unes avec les autres», *Mémoires de Physique et de Chimie de la Société d'Arcueil*, Vol. 2, 206-234, Paris: Mad. V^e. Bernard, 3 vols.

_____ . 1809b, «Mémoire sur la vapeur nitreuse, et sur le gaz nitreux consideré comme moyen eudiométrique», *Mémoires de Physique et de Chimie de la Société d'Arcueil*, Vol. 2, 235-251; Paris: Mad. V^e. Bernard, 3 vols.

_____ . 1815, «Recherches sur l'acide prussique», *Annales de chimie*, 95, 136-231.

_____ . 1816, «Experiments on Prussic Acid», *Annals of Philosophy*, 7, 350-364; 8, 37-52, 108-115.

_____ . 1817, «Description d'un eudiomètre de Volta», *Annales de chimie et de physique*, 4, 188-189.

_____ . 1833, «Nouvelle simplification de l'eudiométre de Volta», *Annales de chimie et de physique*, 66, 443-444.

Gay-Lussac, J. L.; Thenard, L. J. 1811, *Recherches physico-chimiques*, Paris: Deterville, 2 vols.

Gérardin, L.R. 1778, «Observations sur les eudiomètres», *Observations sur la physique, sur l'histoire naturelle et sur les arts*, March, 11, 248-254.

Giobert, G.A. 1789, «Des eaux sulfureuses et thermales de Vaudier», *Journal de Physique, de Chimie, d'Histoire Naturelle*, Messidor an VI, 4, 47, 197-201.

_____ . 1793, *Des eaux sulphureuses et thermales de Vaudier, avec des observations physiques, économiques et chimiques sur la vallée de Gesse et des remarques sur l'analyse des eaux sulphureuses en général*, Turin: Imprimiere de Jaques Fea.

Göttling, J.F. 1794, *Beytrag zur Berichtigung Antiphlogistischen der Chemie auf Versuche Gegründet*, Weimar: Hoffmanns Wittwe und Erben.

_____ . 1795, «Ueber das Leuchten des Phosphors in Stickluft», *Neues Journal der Physik*, 1, 1-15.

Guyton de Morveau, B. L. 1789, *Encyclopédie Méthodique. Chymie, Pharmacie et Métallurgie*, Vol.1, partie II, Paris: Panckoucke; Liège: Plomteux.

_____ . 1795, «Description et usage d'un eudiomètre à sulfure de potasse», *Journal de l'ÉcolePolytechnique*, cahier 2, 166-168.

Hales, S. 1727, *Vegetable Staticks: or, an account of some statical experiments on the sap in vegetables,..* , London: W. and J. Innys; T. Woodward.

_____ . 1738, *Statical Essays Containing Vegetable Staticks*, London: W. and J. Innys & T. Woodward, 2 vols.

Hauch, A.W. 1793, «Beskrivelse af en nye LuftprØver eller Eudiometer», *Nye samling af det Kongelige Danske Videnskabernes Selskabs skrifter*, 5, 537-544.

Hempel, W. M. 1892, *Methods of Gas Analysis*, London and New York: Macmillan (translated from the second German edition of 1889)

Henderson, A. 1804, «Experiments and Observations on the Change which the Air of the Atmosphere Undergoes by Respiration, Particularly with

Regard to the Absorption of Nitrogen», *A Journal of Natural Philosophy, Chemistry and the Arts*, 8, 40-45.

Henry, W. 1801, 1803, *An Epitome of Chemistry*, London: J. Johnson.

_____ . 1805, «Experiments on the Gases Obtained by the Destructive Distillation of Wood, Pit, Pit-Coal, Oil, Wax, &c. With a View to the Theory of their Combustion, when Employed as Sources of Artificial Light, and Including Observations on Hydro-Carburets in general and the Carbonic Oxide», *A Journal of Natural Philosophy, Chemistry and the Arts*, 11, 66-74.

_____ . 1808, *An Epitome of Chemistry*, New York: Collins and Perkins. (First American from the fourth English edition of 1806).

_____ . 1808b, «Description of an Apparatus for the Analysis of the Compound Inflammable Gases by Slow Combustion; With Experiments on the Gas from Coal, Explaining its Application», *Philosophical Transactions*, 98, 282-303.

_____ . 1810, *The Elements of Experimental Chemistry*, London: J. Johnson and Co., 2 vols.

_____ . 1818, *The Elements of Experimental Chemistry*, London: Baldwin, Cradock and Joy and R. Hunter, 2 vols.

_____ . 1819, «Experiments on the Gas from Coal Chiefly with a View to its Practical Application», *Memoirs of the Literary and Philosophical Society of Manchester*, 3, 391-429.

Hope, T.C. 1803, «Account of a Simple Eudiometric Apparatus Constructed and Used by T.C. Hope», *A Journal of Natural Philosophy, Chemistry and the Arts*, 6, 61-62, 210-212, plates IV, XII.

Humboldt, A.F. 1798a, «Mémoire sur la combinaison ternaire du phosphore, de l'azote et de l'oxygène, ou sur l'existence des phosphures d'azote oxidés», *Annales de Chimie*, 30 Messidor an VI, 27, 141-160.

_____ . 1798b, «Mémoire sur sur le gaz nitreux et ses combinaisons avec l'oxigéne», *Bulletin des Sciences par la Société Philomatique*, Thermidor an VI, 17, 132-134.

_____ . 1798c, «Lettre de Fréderic von Humboldt à Garnerin l'aîné sur l'analyse de l'air atmosphérique, pris à l'hauter de 669 toises avec aérostat», *Journal de Physique, de Chimie, d'Histoire Naturelle*, Fructidor an VI, 4, 202-203.

_____ . 1798d, «Expériences sur le gaz nitreux et ses combinaisons avec l'oxigene», *Annales de Chimie*, 30 Vendemiaire an VII, 28, 123-180.

_____ . 1798e, «Extrait d'une lettre de Humboldt au D. Ingenhousz sur la propriété des terres simples de décomposer l'air atmosphèrique», *Journal de Physique, de Chimie, d'Histoire Naturelle*, Brumaire an VII, 4, 377-378.

_____ .1799a, «Lettre de Humboldt à J.C. De la Métherie sur l'absorption de l'oxigene par les terres simples», *Journal de Physique, de Chimie, d'Histoire Naturelle et des Arts*, Nivôse an VII, 48, 132-134.

_____ . 1799b, «Mémoire sur l'absorption de l'oxigène par les terres simples, et son influence dans la culture du sol», *Annales de Chimie*, 30 Nivôse an VII, 29, 125-160.

_____ . 1799c, *Versuche über die Chemische Zerlegung des Luftkreises und über einige andere Gegenstände der Naturlehre*, Brauschweig: Friedrich Vieweg.

_____ . 1811, «Sur la respiration des cocodriles», In *Recueil d'observations de zoologie et d'anatomie comparée faites dans l'océan atlàntique, dans l'intérieur du noveau continent et dans la mer du sud pendant les années 1799, 1800, 1801, 1802 et 1803*, edited by A. F. Humboldt and A. Bonpland, Paris: Schoell and Dufour, 1, 253-259.

Humboldt, A.F.; Bonpland, A. 1814-1825, *Voyage aux regions équinoxiales du noveau continent fait en 1799, 1800, 1801, 1802, 1803 et 1804*, Paris: Librairie Greque-Latine, Allemande, 3 vols. English translation: 1814-1829, *Personal Narrative of Travels to the Equinoctial Regions of the Continent During the Year 1799-1804*, London: Longman, Hurst, Rees, Orme, Brown, Murray and Colburn, 7 vols.

Humboldt, A.F.; Gay-Lussac, L.J. 1805, «Expériences sur les moyens eudiométriques et sur la proportion des principes constituans de l'atmosphère», *Journal de Physique, de Chimie, d'Histoire Naturelle et des Arts*, Pluviose an XIII, 60, 129-168.

Humboldt, A.F.; Vauquelin, N.L. 1798, «Notice sur la cause et les effets de la dissolubilité du gaz nitreux dans la solution du sulfate de ferre», *Annales de Chimie*, 30 Vendemiaire an VII, 28, 181-188.

Ingenhousz, J. 1776, «Easy Methods of Measuring the Diminution of Bulk, Taking Place upon the Mixture of Common Air and Nitrous Air; Together with Experiments on Platina», *Philosophical Transactions*, I, 66, 257-267.

_____ . 1779, *Experiments upon Vegetables, Discovering their Great Power of Purifying the Common Air in the Sunshine, and of Injuring it in the Shade and at Night. To which is Joined a new Method of Examining the Accurate Degree of Salubrity of the Atmosphere*, London: P. Elmsly and H. Payne.

_____ . 1780a, «On the Degree of Salubrity of the Common Air at Sea, compared with that of the Sea-Shore, and that of Places far Removed from the Sea», *Philosophical Transactions*, 70, 354-377.

_____ . 1782, «Some Farther Considerations on the Influence of the Vegetable Kingdom on the Animal Creation», *Philosophical Transactions*, 72, 426-439.

_____ . 1784, «Sur la vertu de l'eau imprégnée d'aire fixe, de différents acides, et de plusieurs autres substances, pour en obtenir, par le moyen des plantes et de la lumière du soleil, de l'air déphlogistiqué», *Observations sur la physique, sur l'histoire naturelle et sur les arts*, 24, Mai, 337-348.

_____ . 1785a, «Sur la construction et l'usage de l'eudiomètre de M. Fontana, et sur quelques propriétés particulières de l'air nitreux», *Observations sur la physique, sur l'histoire naturelle et sur les arts*, 26, Mai, 339-359.

_____ . 1785b, *Nouvelles expériences et observations sur divers objets de physique*, Paris: P. Théophile Barrois le jeune

_____ . 1787, *Expériences sur les végétaux, spécialment sur la propriété qu'ils possèdent à un haut degré, soit d'améliorer l'air quand ils sont au soleil, soit de le corrompre la nuit, ou lorqu'ils sont à l'ombre; auxquelles on a joint une méthode nouvelle de juger du dregré de salubrité de l'atmosphère.* Vol. 1. Paris: Chez Théofile Barrois le jeune.

_____ . 1789, *Expériences sur les végétaux, spécialment sur la propriété qu'ils possèdent à un haut degré, soit d'améliorer l'air quand ils sont au soleil, soit de le corrompre la nuit, ou lorqu'ils sont à l'ombre; auxquelles on a joint une méthode nouvelle de juger du dregré de salubrité de l'atmosphère*, Vol. 2. Paris: Chez Théofile Barrois le jeune.

Julia de Fontenelle, J.S.E. 1823, *Recherches històriques, chimiques et médicales sur l'air marécageux*, Paris: Chez Gabon et Compagnie.

Kolbe, H. 1842, «Eudiometer, Eudiometrie», *Handwörterbuch der Reinen und Angewandten Chemie*, Vol. 2, 1050-1073, Braunschweig: Friedrich Vieweg und Sohn.

Landriani, M. 1775a, «Lettre de M. Le Chevalier Marsilio Landriani à l'aiteur de ce recueil», *Observations sur la physique, sur l'histoire naturelle et sur les arts*, octubre, 6, 315-316.

_____ . 1775b, *Ricerche fisiche intorno allà salubrità dell'aria*, Milano: s.n.

Lavoisier, A.L. 1774, *Opuscules physiques et chymiques*, Paris: Durand, Didot et Esprit.

_____ . 1780, (pour l'année 1777), «Expériences sur la combinaison de l'alun avec les matières charbonneuses et sur les altérations qui arrivent à l'air dans lequel on fait brûler du pyrophore», *Histoire de l'Acadèmie Royale des Sciences*, 363-372.

_____ . 1789, *Traité élémentaire de chimie*, Paris: Cuchet, 2 vols.

Lavoisier, A.L.; Séguin, A. 1814, «Second mémoire sur la respiration», *Annales de chimie*, 91, 318-334.

Le Roy, J.B. 1778, «Lettre relative aux expériences sur l'air inflamable des marais, découvert par M. Volta», *Observations sur la physique, sur l'histoire naturelle et sur les arts*, 11, Part 1, Mai, 401-403.

Libes, A. 1813, *Histoire philosophique des progrés de la physique. Tableau des progrès de la physique depuis la naissance de la chimie pneumatique jusqu'à nos jours*, Paris: Coucier, 4 vols.

Macquer, P.J. 1766, *Dictionnaire de chimie, contenant la théorie et la pratique de cette science, son application à la physique, à l'histoire naturelle, à la médicine, et à l'economie animale*, seconde édition, Paris: Chez Lacombe, 2 vols.

_____ . 1778, *Dictionnaire de chimie, contenant la théorie et la pratique de cette science, son application à la physique, à l'histoire naturelle, à la médicine, et aux arts dépendans de la chimie*, seconde édition, 2 vols. Paris: Chez Théophile Barrois.

Magellan, J. H. 1777, *Description of a Apparatus for Making Mineral Waters, like those of Pyrmont, Spa seltzer, &c, Together with the Description of Some New Eudiometers*, etc., London: W. Parker, J. Johnson and W. Brown.

_____ . 1780, «Extrait d'une lettre de M. Magellan», *Observations sur la physique, sur l'histoire naturelle et sur les arts*, 16, Part 2, Juillet, 74.

_____ . 1783, *Description of a Apparatus for Making the Best Mineral Waters, like those of Pyrmont, Spa seltzer, Aix-La-Chapelle &c, Together with the Description of Two New Eudiometers, etc. The Third Edition [...] with an Examination of the Strictures of Mr. T. Cavallo, F.R.S. upon these Eudiometers*, London: Printed for the author.

Marcet, W. 1888, «A New Form of Eudiometer», *Proceedings of the Royal Society of London*, 44, 383-387.

_____ . 1891, «Researches on the Absorption of Oxygen», *Proceedings of the Royal Society of London*, 50, 58-75.

Martí-Franquès, A. 1795, «Memoria sobre los varios métodos de medir la cantidad de ayre vital de la atmosfera», *Memorial literario, instructivo y curioso de la Corte de Madrid*, 261-275, 347-360, 389-404.

_____ . 1801a, «Mémoire sur la quantité de l'air vital de l'atmosphère et sur les différentes méthodes de la mesurer», *Journal de Physique, de Chimie et d'Histoire Naturelle*, 52, 173-185.

_____ . 1801b, «Memoir on the Quantity of Vital Air in the Atmosphere, and the Different Methods of measuring it», *The Philosophical Magazine*, 9, 250-262.

_____ . 1805, «Antonio de Martí's eudiometrische Untersuchungen, ausgezogen vom Herausgeber», *Annalen der Physik*, 19, 4, 389-393.

Mayow, J. 1674, *Tractus Quinque Medico-Physici. Quorum Primus Agit de Sal-nitro et Spiritu Nitro-aereo*, Oxon: Theatro Sheldoniano. (English translation: 1907, Edinburgh: The Alembic Club Reprint, No. 17.)

McLeod, H. 1869, «On a New Forma of Apparatus for Gas Analysis», *Journal of the Chemical Society*, 6, 313-323.

Monge, G., 1786 (pour l'année 1783), «Mémoire sur le résultat de l'inflammation du gaz inflammable & de l'air déphlogistiqué, dans des vaisseaux clos», *Histoire de l'Académie Royale des Sciences*, 78-88.

Ogier, J. 1885, «Analyse des gaz», In *Encyclopédie chimique*, directed by E. Fremy, Vol. 4. Paris: Vve. Ch. Dunod, Analyse chimique, 10 vols.

Orsat, L. 1875, «Analyse industrielle des gaz», *Annales des mines*, 7e sèrie, 8, 485-506.

Parrot, G.F. 1800, «Ueber die eudiometrischen Eigenschaften des Phosphors, nebst Beschreibung eines richtigen Phosphor-Eudiometers», *Magazin für den neusten Zustand de Naturkunde (Johann Heinrich Voight Magazin)*, 2, 154-185.

Pelletier, B. 1798, *Mémoires et observations de chimie*, Paris: Croullebois, Fuchs, Barrois and Huzard, 2 vols.

Pelouze, J; Fremy, E. 1854, *Traité de chimie générale, comprenant les applications de cette science à l'analyse chimique, à l'industrie, à l'agriculture et à l'histoire naturelle*, Paris: Librairie de Victor Mason, deuxième édtion, 3 vols.

Pepys, W. H. 1807, «A New Eudiometer, Accompanied with Experiments, Elucidating Its Applications», *Philosophical Transactions*, 97, 247-259; 1807-1808, *The Philosophical Magazine*, 29, 116-126, plate IV.

Priestley, J. 1772, «Observations on Different Kind of Airs», *Philosophical Transactions*, 62, 147-252.

_____ . 1775, 1776, 1777, *Experiments and Observations of Different Kinds of Air*, London: J. Johnson.

_____ . 1779, 1781, 1786, *Experiments and Observations Relating to various Branches of Natural Philosophy, with a Continuation of the Observations on Air*, Birmingham: J. Johnson.

_____ . 1790, *Experiments and Observations on Different Kinds of and other Branches of Natural Philosophy, Connected with the Subject*, Birmingham: T. Pearson and J. Johnson, 3 vols.

Pringle, J. 1783, «A Discourse on the Different Kinds of Airs. Delivered at the Anniversary Meeting of the Royal Society, November 30, 1773», In *Six Discourses Delivered by Sir John Pringle Bart. Before the Royal Society. On Occasion of Six Annual Assignments of Sir Godfrey Copley's Medal*, London: Straham and Cadell.

Reboul, H. 1788, «Description d'un eudiomètre atmospherique», *Mémoires de l'Académie Royale des Sciences, Inscriptions et Belles Lettres de Toulouse*, 3, 378-383.

_____ . 1792, «Description d'un eudiomètre atmospherique», *Annales de chimie*, 13, 38-46.

Regnault, H. V.; Reiset, J. 1849, «Recherches chimiques sur la respiration des animaux des diverses classes», *Annales de chimie et de physique*, 26, 299-519.

Regnault, H. V. 1850-1854, *Cours élémentaire de chimie*, Paris: Victor Masson, Langlois et Leclercq, 4 vols.

_____ . 1852, «Recherches sur la composition de l'air atmosphérique», *Annales de chimie et de physique*, 36, 385-405.

Roscoe, H.E. 1900, «Bunsen Memorial Lecture», *Journal of the Chemical Society, Transactions*, 77, 513-554.

Russell, W. J. 1868, «On Gas Analysis», *Journal of the Chemical Society*, 21, 128-141.

Salleron, J. 1861, *Notice sur les instruments de précision, construïts par J. Salleron*, Paris: Chez Bonaventure et Ducessois.

Saussure, H.B. 1779, *Voyages dans les Alpes, précédés d'un essai sur l'histoire naturelle des environs de Geneve*, Vol. 1. Neuchatel: Samuel Fauche, 4 vols.

_____ . 1803, *Recherches chimiques sur la végétation*, Paris: Chez la Ve. Nyon.

Saussure, N.T. 1798, «Lettre de Saussure fils à J.C. De la Métherie pour prouver que les terres pures n'absorbent pas l'oxigen», *Journal de Physique, de Chimie, d'Histoire Naturelle*, Messidor an VI, 4, 470-471.

_____ . 1809, «Observations sur la combustion de plusieurs espèces de charbon et sur le gaz hydrogène», *Annales de chimie*, 71, 254-324.

_____ . 1810, «Observations on the Combustion of Several Sorts of Charcoal, and on Hidrogen Gas», *A Journal of Natural Philiosophy, Chemistry and the Arts*, 26, 161-175, 300-309.

_____ . 1811, «Analyse du gaz oléfiant», *Annales de chimie*, 78, 57-68.

_____ . 1812, «Analysis of Olefiant Gas», *A Journal of Natural Philiosophy, Chemistry and the Arts*, 31, 69-74.

Savèrien, A.J. 1753, *Dictionnaire universal de mathématique et de physique*, Paris: Jacques Rollin and Charles-Antoine Jombert, 2 vols.

Scheele, C.W. 1777, *Chemische Abhandlungen von der Luft und dem Feur*, Upsala und Leipzig: Verglegt von Magn. Swederus, Buchländler.

_____ . 1779, «Rón om rena Luftens mángd fom dageligen uti vár Luft-krets ár nárvarande», *Kongl. Vetenskaps Academiens Handlingar*, 40, January-February-March, 50-55.

_____ . 1780, *Chemical Observations and Experiments on Air and Fire*, London: J. Johnson.

_____ . 1781, *Traité chimique de l'air et du feu*, Paris: Rue et Hôtel Serpente.

_____ . 1782, «Expériences sur la quantité d'air pur qui se trouve dans notre atmosphère», *Observations sur la physique, sur l'histoire naturelle et sur les arts*, 19, 79-82.

_____ . 1785a, *Supplément au Traité chimique de l'air et du feu*, Paris: Rue et Hôtel Serpente.

_____ . 1785b, *Mémoires de Chymie de C. W. Scheele*, Vol. 2. Dijon: Chez l'Éditeur; Paris: Barrois et Cuchet, 2 vols.

_____ . 1901, *The Chemical Essays of Charles-William Scheele. Translated from the Transactions of the Academy of Sciences of Stockholm*, London: Scott, Greenwood and Co.

Scherer, J.A. 1782, *Eudiometria sive methodus aeris atmosphaerici puritatem salubritatemne examinandi*, Vienne: Typis Iosephi Nobilis de Kurzbeck.

Séguin, A. 1791, «Mémoire sur l'eudiomètre», *Annales de Chimie*, 9, 293-303; «Combustion du phosphore, employé comme moyen eudiométrique», LM, 2, 143-153.

_____ . 1814, «Mémoire sur la salubrité et l'insalubrité de l'air atmosphérique dans ses divers degrés de pureté», *Annales de Chimie*, 89, 251-272.

Séguin, A.; Lavoisier, A.L. 1793 (pour l'année 1789), «Première mémoire sur la respiration des animaux», *Histoire de l'Académie des Sciences*, 566-584; LO, 2, 688-703.

Senebier, J. 1782, *Mémoires physico-chymiques sur l'influence de la lumière solaire pour modifier les êtres de trois règnes de la Nature, et surtout ceux du règne végétal*, Genève: Chez Barthelemi Chirol, 3 vols.

_____ . 1783, *Recherches sur l'influence de la lumière solaire pour métamorphoser l'air fixe en air pur par la végétation*, Geneve: Chez Barthelemi Chirol.

_____ . 1788, *Expériences sur l'action de la lumière solaire dans la végétation*, Genève: Chez Barde, Manget & Compagnie; Paris: Chez Buisson.

Serres, L. 1901, *Cours de chimie à l'usage des candidats aux Écoles nationales des Arts et Métiers*, Paris: Librairie Polytechnique Ch. Béranger éditeur.

Servières, C. U. 1777, «Description d'un instrument pour mesurer la salubrité de l'air», *Observations sur la physique, sur l'histoire naturelle et sur les arts*, 10, octobre, 321-322.

Sigaud de La Fond, J.A. 1779, *Essai sur différentes especes d'air qu'on désigne sous le nom d'air fixe*, Paris: P. Fr. Gueffier.

Spallanzani, L. 1796a, *Chimico Esame degli esperimenti del Sig. Göttling Professore a Jena sopra la luce del fosforo di Kunkel osservata nell'aria commune*, Modena: Presso la Società Tipografica.

_____ . 1796b, «Descrizione ed uso dell'Eudiometro del sig. Giobert. Tratto dal chimico esame degli sperimenti del sig. Gotling», *Scelta di opuscoli interessanti scelti sulle scienze e sulle arti*, 19, 352-360.

_____ . 1803a, *Memorie su la respirazione*, Milano: Presso Angelo Nobile, 2 vols. [2010, Edizione nazionale delle opere di Lazzaro Spallanzani, Parte quinta, Vol. 5, 55-158, Modena: Mucchi Editore, 6 parts.]

_____ . 1803b, *Mémoires sur la respiration*, Genève: J.J. Paschoud. [2010, Edizione nazionale delle opere di Lazzaro Spallanzani, Parte quinta, Vol. 5, 159-302, Modena: Mucchi Editore, 6 parts.]

Thomas, J. W. 1879, «On Some Aspects in the Analysis of Combustible Gases and in the Construction of Apparatus», *Journal of the Chemical Society, Transactions*, 35, 213-224.

Timiryazev, K. 1877, «Recherches sur la décomposition de l'acide carbonique dans le spectre solaire, par la partie verte des végétaux», *Annales de chimie et de physique*, 5e sèrie, 12, 355-396.

Traill, T.S. 1849, «Memoir of Dr. Thomas Charles Hope, late Professor of Chemistry in the University of Edinburgh», *Transactions of the Royal Society of Edinburgh*, 16, 419-434.

Vauquelin, N.L. 1792, «Observations chimiques et physioluguiques sur la respiration des insectes et des vers», *Annales de Chimie*, 12, 273-291.

Venel, G.F. 1751, «Chymie», *Encyclopédie, ou Dictionnaire raisonné des sciences, des arts et des métiers, 1751-1772*, Vol. 3, 408-437, Paris : Chez Briasson, David, Le Breton & Durand, 17 vols.

_____ . 1766, «Instrument», *Encyclopédie, ou Dictionnaire raisonné des sciences, des arts et des métiers, 1751-1772*, Vol. 8, 802-804, Paris : Chez Briasson, David, Le Breton & Durand, 17 vols.

Volta, A. 1777a, *Lettere del Signor Don Alessandro Volta sull'aria infiammabile native delle paludi*, Milano: Giuseppe Marelli.

_____ . 1777b, «Lettera II del Sig. D. Alessandro Volta al Sig. Marchese Francesco Castelli», *Scelta di opuscoli interessanti sulle scienze e sulle arti*, 30, 97-109.

_____ . 1777c, «Lettera del Sig. Don Alessandro Volta al Sig. Dottore Giuseppe Priestley», *Scelta di oposculi interessanti sulle scienze e sulle arti*, 34, 65-82.

_____ . 1778a, *Lettres de Mr. Alexandre Volta sur l'air inflamable des marais, auxquelles on a ajouté trois lettres du même auteur*, tirées du Journal de Milan, Strasbourgh: J. H. Heitz.

_____ . 1778b, «Précis des lettres de M. Alexandre Volta sur l'air inflamable des marais», *Observations sur la physique, sur l'histoire naturelle et sur les arts*, 11, Part 1, Février, 152-158.

_____ . 1778c, «Première lettre adressée à M. Priestley sur la inflammation de l'air inflammable mêlé à l'air commun dans des vaisseaux fermés, et sur les phénomènes que présentent sa décomposition et la diminution qu'il produit dans l'air respirable avec lequel on le mêle», *Observations sur la physique, sur l'histoire naturelle et sur les arts*, 12, Part 2, Novembre, 365-373.

_____ . 1779, «Seconde lettre adressée à M. Priestley sur la inflammation de l'air inflammable mêlé à l'air commun dans des vaisseaux clos, et sur les phénomènes que présentent sa décomposition et la diminution qu'il produit dans l'air respirable avec lequel on le mêle», *Observations sur la physique, sur l'histoire naturelle et sur les arts*, 13, Part 1, Avril, 278-303.

_____ . 1783, «Aria nitrosa», «Eudiometro», *Dizionario di chimica del Sig. Pietro Giuseppe Macquer*, Vol. 2, 197-282; Vol. 4, 117-137, Pavia: Stamperia del R.I. Monastero di S. Salvatore, 11 vols.

_____ . 1790, 1791, «Descrizione dell'eudiometro ad aria infiammabile il quale serve inoltre di apparato universale per l'accensione al chiuso delle arie infiammabili d'ogni sorta mescolate in diverse proporzioni con aria respirabile più o meno pura, e per l'analisi di quelle e di queste», *Annali di chimica ovvero raccolta di memorie sulle scienze, arti e manifatture ad essa relative*, 1790: 1, 171-231; 1791: 2, 161-209; 3, 36-45.

White, W. 1778, «Experiments upon Air, and the Effects of Different Kinds of Effluvia upon It; Made at York», *Philosophical Transactions*, 68, 194-220.

Williamson, A.W.; Russell, W.J. 1858, «Note on the Measurement of Gases in Analysis», *Proceedings of the Royal Society of London*, 9, 218-222.

_____ . 1864, «On a New Method of Gas Analysis», *Journal of the Chemical Society*, 17, 238-257.

Young, A. 1792, *Travels During The Years 1787, 1788 and 1789. Undertaken more particularly with a view of ascertaining the cultivation, wealth, resources and national prosperity of the Kingdom of France*, London: J. Rackham.

Bibliography of secondary sources

Abbri, F. 1996, «Spallanzani e la "química nuova"», *La «Mal-laria» di Lazzaro Spallanzani e la respirabilità dell'aria nel Settecento*, Firenze: Leo S. Olschki Editore, 3-15.

Anderson, R. G. W. 1985, «Instruments and Apparatus», In *Recent Developments in the History of Chemistry*, edited by C.A. Russell, London: The Royal Society of Chemistry, 217-237.

_____ . 2015, «Thomas Charles Hope and the Limiting Legacy of Joseph Black», In *Cradle of Chemistry. The Early Years of Chemistry at the University of Edinburgh*, edited by R. G. W. Anderson, Edinburgh: John Donald, 147-162.

Beale, N.; Beale, E. 2011, *Echoes of Ingen Housz, the long lost story of the genius who rescued the Habsburgs from smallpox and became the father of photosynthesis*, Salisbury: The Hobnob Press.

Belloni, F. 1960, «L'eudiometro del Landriani», *Actes du Symposium International d'Histoire des Sciences, Florence-Vinci, 8-10 Octobre 1960*, Vinci: s.n., 130-151.

Benedict, F.G. 1912, *The Composition of the Atmosphere with Special Reference to its Oxygen Content*, Washington, DC: The Carnegie Institution of Washington.

Bensaude-Vincent, B. 1992, «The Balance: Between Chemistry and Politics», *The Eighteenth Century*, 33, 3, 217-237.

Bensaude-Vincent, B. 2000, «"The Chemist's Balance for Fluids": Hydrometers and Their Multiple Identities, 1770-1810», In *Instruments and Experimentation in the History of Chemistry*, edited by L.H. Holmes and T.H. Levere, Cambridge, MA: The MIT Press, London, 153-184.

Bensaude-Vincent, B.; Blondel, C. 2008, «A Science Full of Shocks, Sparks and Smells», In *Science and Spectacle in the European Enlightenment*, edited by B. Bensaude-Vincent and C. Blondel, London and New York: Routledge, Taylor and Francis Group, 1-10.

Beretta, M. 1989, «Gli scienziati italiani e la rivoluzione chimica», *Nuncius*, 4, 2, 119-146.

_____ . 1995, *Introduzione a «Ricerche fisiche intorno alla salubrità dell'aria» (Marsilio Landraini)*, Firenze: Giunti, 5-51.

_____ . 2000, «Pneumatics vs. "Aerial Medicine": Salubrity and Respirability of Air at the End of the Eighteenth Century», In *Nuova Voltiana. Studies on Volta and his Times*, edited by F. Bevilacqua and L. Fregonese, Vol. 2, 49-71. Pavia: Università degli Studi di Pavia; Milano: Editore Ulrico Hoepli, 5 vols.,

_____ . 2001a, «From Nollet to Volta: Lavoisier and Electricity», *Revue d'Histoire des Sciences*, 54, 1, 29-52.

_____ . 2001b, «Lavoisier and his Last Printed Work: the *Mémoires de physique et de chimie*, 1805», *Annals of Science*, 58, 327-356.

_____ . 2001c, *Imaging a Career in Science. The Iconography of Antoine Laurent Lavoisier*, Canton, MA: Science History Publications.

_____ . 2002, *Storia materiale della scienza. Dal libro ai laboratori*, Milano: Paravia Bruno Mondadori Editori.

_____ . 2014, «Between the Workshop and the Laboratory: Lavoisier's Network of Instrument Makers», *Chemical Knowledge in the Early Modern World, Osiris*, 29, 197-214.

Boantza, V.D. 2013a, *Matter and Method in the Long Chemical Revolution. Laws of Another Order*, Surrey and Burlington: Ashgate Publishing Limited.

_____ . 2013b, «The Rise and Fall of Nitrous Air Eudiometry: Enlightenment Ideals, Embodied Skills, and the Conflicts of Experimental Philosophy», *History of Science*, 51, 377-412.

Boklund, U. 1968, *Carl Wilhelm Scheele: His Work and Life*, Stockholm: Roos Boktryckeri AB, 2 vols. bound as one.

Bouchard, G. 1938, *Guyton-Morveau, chimiste et conventionnel (1737-1816)*, Paris: Librairie Acadèmiques Perrin.

Bradley, J. 1992, *Before and After Cannizzaro. A Philosophical Commentary on the Development of the Atomic and Molecular Theories*, North Ferriby: J. Bradley.

Brenni, P. 2012, «The Evolution of Teaching Instruments and their Use Between 1800 and 1930», *Science & Education*, 21, 2, 191-226.

Bret, P. 2017, «Du laboratoire de l'Académie de Dijon à celui de l'École Polytechnique, trente-six ans d'expérience d'enseignement de la chimie», *Bulletin de la SABIX*, 60, 9-36.

Bruhns, K. 1873, *Life of Alexander von Humboldt*, London: Longmans and Green and Co., 2 vols. (English translation)

Capuano, F.; Cavalchi, B. 1998, «Spallanzani e la respirabilità dell'aria nel tardo '700. Strumenti e misure della chmica pneumatica», Scandiano: Casa Spallanzani.

_____ . 2010, «Chimica pneumatica e fisiològica negli studi di Spallanzani sulla respirazione», *Edizione nazionale delle opere di Lazzaro Spallanzani*, Parte quinta, Vol. 5, 303-338, Modena: Mucchi Editore, 6 parts.

Cerruti, L. 1998, «Chemicals as Instruments. A Language Game», *HYLE – An International Journal for the Philosophy of Chemistry*, 4, 39-61.

Ciardi, M. 2010, «Per una riconstruzione delle ricerche e delle memorie di Lazzaro Spallanzani sulla respirazione animale e vegetale», *Edizione nazionale delle opere di Lazzaro Spallanzani*, Parte quinta, Vol. 5, 9-34, Modena: Mucchi Editore, 6 parts.

Clow, A.; Clow, N.L. 1952/1992, *The Chemical Revolution. A Contribution to Social Technology*, Philadelphia, PA: Gordon and Breach Science Publishers.

Crosland, M. 1967, *The Society of Arcueil. A View of French Science at the Time of Napoleon I*, Cambridge, MA: Harvard University Press,

_____ . 1978, *Gay-Lussac, Scientist and Bourgeois*, Cambridge: Cambridge University Press.

_____ . 1980, «Chemistry and the Chemical Revolution», In *The Ferment of Knowledge, Studies in the Historiography of Eighteenth-Century Chemistry*, edited by G.S. Rousseau and Roy Porter, Cambridge: Cambridge University Press, 389-416.

_____ . 1983, «A Practical Perspective on Joseph Priestley as a Pneumatic Chemist», *British Journal for the History of Science*, 16, 3, 223-238.

_____ . 1994, *In the Shadow of Lavoisier: The Annales de Chimie*, Oxford: British Society for the History of Science, The Alden Press.

_____ . 2000, «"Slippery Substances": Some Practical and Conceptual Problems in the Understanding of Gases in the Pre-Lavoisier Era», In *Instruments and Experimentation in the History of Chemistry*, edited by L.H. Holmes and T.H. Levere, Cambridge, MA and London: The MIT Press, London, 79-104.

_____ . 2009, «Lavoisier's Achievement; More than a Chemical Revolution», *Ambix*, 56, 2, 93-114.

Crosland, M.; Smith, C. 1978, «The Transmission of Physics from France to Britain: 1800-1840», *Historical Studies in the Physical Sciences*, 9, 1-61.

Daumas, M. 1955, *Lavoisier. Théoricien et expérimentateur*, Paris: Presses Universitaires de France.

Davoli V. 1996, «Lazzaro Spallanzani e gli studi eudiometrici», *La «Mal-laria» di Lazzaro Spallanzani e la respirabilità dell'aria nel Settecento*, Firenze: Leo S. Olschki Editore, 51-55.

Dettelbach, M. 1999, «The Face of Nature: Precise Measurement, Mapping, and Sensibility in the Work of Alexander von Humboldt», *Studies in History and Philosophy of Biological and Biomedical Sciences*, 30, 4, 473-504.

Dörries, M. 2001, «Purity and Objectivity in Nineteenth-Century Metrology and Literature», *Perspective on Science*, 9, 2, 233-250.

Doyle, W.P. 1982, «Thomas Charles Hope M.D., F.R.S.E., F.R.S. 1766-1844», *Scottish Men of Science*, Edinburgh: Scotland's Cultural Heritage.

Farrar, K. R. 1963, «A Note on a Eudiometer Supposed to Have Belonged to Henry Cavendish», *British Journal for the History of Science*, 1, 4, 375-380.

_____ . 1968, «Dalton's Scientific Apparatus», In *John Dalton & The Progress of Science*, edited by D. S. L. Cardwell, Manchester: Manchester University Press; New York, NY: Barnes & Noble Inc., 159-186.

Fors, H. 2008, «Stepping Through Science's Door: C.W. Scheele, from Pharmacist's Apprentice to Man of Science», *Ambix*, 55, 1, 29-49.

Fox, R. 1971, *The Caloric Theory of Gases. From Lavoisier to Regnault*, Oxford: Clarendon Press.

Gee, B. 1989, «Amusement Chests and Portable Laboratories: Practical Alternatives to the Regular Laboratory», In *The Development of the Laboratory: Essays on the Place of Experiment in Industrial Civilization*, edited by F. A. J. L. James, Houndmills, Basingstoke, Hampshire and London: Macmillan Press.

Gigli, A. 2002, «Volta's Teaching in Como and Pavia: Moments of Academic Life under All Flags», In *Nuova Voltiana. Studies on Volta and his Times*, edited by F. Bevilacqua and L. Fregonese, Vol. 4, 53-99. Pavia: Università degli Studi di Pavia; Milano, Editore Ulrico Hoepli, 5 vols.

Gooding, D. 1989, «History in the Laboratory: Can We Tell What Really Went on?», In *The Development of the Laboratory. Essays on the Place of Experiment in Industrial Civilization*, edited by F. A. J. L. James, Houndmills, Basingstoke, Hampshire and London: Macmillan Press, 63-82.

Golinski, J.V. 1989, «A Noble Spectacle. Phosphorus and the Public Cultures of Science in the Early Royal Society», *Isis*, 80, 1, 11-39.

_____ . 1992, *Science as Public Culture. Chemistry and Enlightenment in Britain, 1760-1820*, Cambridge: Cambridge University Press.

_____ . 2016, *The Experimental Self. Humphry Davy and the Making of a Man of Science*, Chicago and London: The University of Chicago Press.

Goupil, M. 1977, *Le chimiste Claude-Louis Berthollet, 148-1822. Sa vie, son oeuvre*, Paris: J. Vrin.

Grao, J.P. 2013, *Henri Reboul. L'aube du pyrénéisme*, Pau: MonHélios.

Grapí, P. 1997; Izquierdo; M., «Berthollet's conception of chemical change in its context», *Ambix*, 44, 3, 113-130.

_____ . 2014, «Berthollet's Revolutionary Course of Chemistry at the *École Normale* of the Year III. Pedagogical Experience and Scientific Innovation», *Proceedings of the 5th International Conference of the European Society for the History of Science, Athens, 1-3 November 2012*, Athens, National Hellenic Research Foundation & Institute of Historical Research, 541-548.

Grau, J.; Bonet, J. (eds.). 2011, *Antoni de Martí i Franquès. La química de l'aire*, Tarragona: Publicacions URV.

Guerlac, H. 1961/1990, *Lavoisier-The Crucial Year. The Background and Origin of His First Experiments on Combustion in 1772*, New York, NY: Cornell University Press.

Haber, L.F. 1958, *The Chemical Industry During the Nineteenth Century. A study of the Economic Aspect of Applied Chemistry in Europe and North America*, Oxford: The Clarendon Press.

Hackmann, W.D. 1989, «Scientific Instruments: Models of Brass and Aids to Discovery», In *The Uses of Experiment. Studies in the Natural Sciences*, edited by D. Gooding, T. Pinch and S. Schaffer, Cambridge: Cambridge University Press, 31-66.

Hankins, T.L.; Silverman, R. J. 1995, *Instruments and Imagination*, Princeton, NJ: Princeton Universitry Press.

Hannaway, O.; Hannaway, C. 1977, «La fermeture du cimetière des Innocents», *Dix-huitième siècle*, 9, 181-191.

Heering, P. 2005, «Weighing the Heat: The Replication of the Experiments with the Ice-calorimeter of Lavoisier and Laplace», In *Lavoisier in Perspective*, edited by M. Beretta, Munchen: Deutsches Museum, 27-42.

Heilbron, J.L. 1990, «Introductory Essay», In *The Quantifying Spirit in the Eighteenth Century*, edited by T. Frägsmyr, J.L. Heilbron and R.E. Reider, Berkeley, Los Angeles, Oxford: University of California Press, 1-24.

Hein, W.H. 1987, «The Young Alexander von Humboldt and Scientific Pharmacy», In *Alexander von Humboldt. Life and Work*, edited by W. H. Hein, Ingelheim am Rhein: C.H. Boehringer Sohn, 152-166.

Henning, J. 2003, «Bunsen, Kirchhoff, Steinheil and the Elaboration of Analytical Spectroscopy», *Nuncius*, year XVIII, 741-754.

Henry, W.C.H. 1854, *Memoirs of the Life and Scientific Researches of John Dalton*, London: The Cavendish Society.

Höettecke, D. 2000, «How and What Can We Learn From Replicating Historical Experiments? A Case Study », *Science & Education*, 9, 343-362.

Holmes, F.L. 1789, *Eighteenth-Century Chemistry as an Investigative Enterprise*, Berkely, CA: University of California Press.

_____ . 1985, *Lavoisier and the Chemistry of Life. An Exploration of Scientific Creativity*, Madison, WC: The University of Wisconsin Press.

_____ . 1998, *Antoine Lavoisier-The Next Crucial Year or the Sources of His Quantitative Method in Chemistry*, Princeton, NJ: Princeton University Press.

_____ . 2000a, «Phlogiston in the Air», In *Nuova Voltiana. Studies on Volta and his Times*, edited by F. Bevilacqua and L. Fregonese, Vol. 2, 73-113. Pavia: Università degli Studi di Pavia; Milano: Editore Ulrico Hoepli, 5 vols.

_____ . 2000b, «The Evolution of Lavoisier's Chemical Apparatus», In *Instruments and Experimentation in the History of Chemistry*, edited by L.H. Holmes and T.H. Levere, Cambridge, MA and London: The MIT Press, 137-152.

Homburg, E. 1999, «The Rise of Analytical Chemistry and its Consequences for the Development of the German Chemical Profession (1780 – 1860)», *Ambix*, 46, 1, 1-32.

Home, R.W., 2000, «Volta's English Connections», In *Nuova Voltiana. Studies on Volta and his Times*, edited by F. Bevilacqua and L. Fregonese, Vol. 1, 115-132. Pavia: Università degli Studi di Pavia; Milano: Editore Ulrico Hoepli, 5 vols.

Jacobsen, A.S. 2000, «A.W. Hauch's Role in the Introduction of Antiphlogistic Chemistry into Denmark», *Ambix*, 47, 2, 71-95.

Jungnickel, C., McCormmach, R., 1999, *Cavendish. The Experimental Life*, Lewisburg, PA: Bucknell University Press.

Klein, U. 2012, «The Prussian Mining Official Alexander von Humboldt», *Annals of Science*, 69, 1, 27-68.

_____ . 2015, *Humboldts Preu en. Wissenschaft und Technik im Aufbruch*, Darmstadt: Wissenschaftliche Buchgesellschaft.

Klein, U.; Lefèvre, W. 2007, *Materials in Eighteenth-Century Science. A Historical Ontology*, Cambridge, MA and London: The MIT Press.

Knight, D. 2009, «Chemists get down to Earth», *Geological Society, London, Special Publications*, 317, 1, 93-103

Knoefel, P. 1984, *Felice Fontana. Life and Works,* Studi su Felice Fontana, 2, Trento: Società di Studi Trentini de Scienze Storiche.

Kurzer, F. 2003, «William Haseldine Pepys FRS: A Life in Scientific Research, Learned Societies and Technical Enterprises», *Annals of Science*, 60, 137-183.

Langins, J. 2004, «Diverging Parallel Lives in Science: Unpublished Correspondence from Georges-Frédéric Parrot to Georges Cuvier», *Journal of Baltic Studies*, 35, 3, 297-300.

Lefebvre, E.; Bruijn, J.G. (eds.) 1976, *Martinus van Marum. Life and Work, Selection of Letters from Martinus van Marum Correspondence*, Vol. 6. Haarlem: Noordhoff International Publishing, 6 vols.

Levere, T.H. 1999, «The Hauch Cabinet. Chemical Apparatus and the Chemical Revolution», *Bulletin of the Scientific Instrument Society*, 60, 11-15.

_____ . 2000, «Measuring Gases and Measuring Goodness», In *Instruments and Experimentation in the History of Chemistry*, edited by L.H. Holmes and T.H. Levere, Cambridge, MA and London: The MIT Press, 105-135.

Levere, T.H.; Holmes, F.L. 2000, «Introduction: A Practical Science», In *Instruments and Experimentation in the History of Chemistry*, edited by L.H. Holmes and T.H. Levere, Cambridge, MA and London: The MIT Press, i-xvii.

Lundgren, A. 1990, «The Changing Role of Numbers in the 18th- Century Chemistry», In *The Quantifying Spirit in the Eighteenth Century*, edited by T. Frägsmyr, J.L. Heilbron and R.E. Reider, Berkeley, Los Angeles, Oxford: University of California Press, 245-266.

Maffiodo, B. 1996, *I borghesi taumaturghi. Medici, cultura scientifica e società in Piemonte fra crisi dell'antico regime ed età napoleònica*, Firenze: Leo S. Olschki editore.

Magiels, G. 2010, *From Sunlight to Insight. Jan Ingen Housz, the discovery of photosynthesis & science in the light of ecology*, Brussels: Uitge Verij Vubpress, Brussels University Press.

Malaquias, I. 2008, «Aspects of the Scientific Network and Communication of John Hyacinth de Magellan in Britain, Flanders and France», *Ambix*, 55, 3, 255-273.

Manzini, P., 1996, «Il manoscritto ritrovato», *La «Mal-laria» di Lazzaro Spallanzani e la respirabilità dell'aria nel Settecento*, Firenze: Leo S. Olschki Editore, 57-69.

Martin-Decaen, A. 1912, *Le dernier ami de J.J. Rousseau. Le Marquis René de Girardin (1735-1808)*, Paris: Perrin et Cie.

Mason, S.F. 1991, «Jean Hyacinthe De Magellan, F.R.S., and the Chemical Revolution of the Eighteenth Century», *Notes and Records of the Royal Society of London*, 45, 155-164.

McCrory, D. 2010, *Nature's Interpreter. The Life and Times of Alexander von Humboldt*, Cambridge: The Lutterworth Press.

McKie, D. 1956, «Priestley's Laboratory and Library and Other of His Effects», *Notes and Records of the Royal Society*, 12, 114-137.

Mercier, P. 1976, «Armand François Séguin (1765-1835)», *Bulletin de la Section d'Histoire des Usines Renault*, année 7, 2, 12, 218-233.

Morris, P.J.T. 2015, *The Matter Factory. A History of the Chemistry Laboratory*, London: Reaktion Books.

Müller, H.H. 2002, *Franz Carl Achard. 1753-1821. Biographie*, Berlin: Verlag Dr. Albert Bartens KG.

Müürsepp, P. 2013, «Georges Frédéric Parrot and the 'New' Enlightenment», *Acta Baltica Historiae et Philosophiae Scientiarum*, 1, 2, 15-25.

Oliver, R.W.A.; Carrier, M. 2006, *The Library of John Dalton*, Kendal: Titus Wilson & Son.

Osman, W. A. 19587, «Alessandro Volta and the Inflammable-Air Eudiometer», *Annals of Science*, 14, 4, 215-242.

Pancaldi, G, 2003, *Volta. Science and Culture in the Age of Enlightenment*, Princeton, NJ: Princeton University Press.

Partington, J.R. 1961-1970, *A History of Chemistry*, London: Macmillan; New York, NY: St. Martin's Press, 4 vols.

Paris, J.A. 1831, *The Life of Sir Humphry Davy*, London: Henry Colburn and Richard Bently.

Poncet, S.; Dahlberg, L. 2011, «The Legacy of Henri Victor Regnault in the Arts and Science», *International Journal of Arts and Sciences*, 4, 13, 377-400.

Powers, J. C. 2012, *Inventing Chemistry. Herman Boerhaave and the Reform of the Chemical Arts*, Chicago, IL and London: The Chicago University Press.

Principe, L. M. 2000, «Apparatus and Reproducibility in Alchemy», In *Instruments and Experimentation in the History of Chemistry*, edited by L.H. Holmes and T.H. Levere, Cambridge, MA and London: The MIT Press, 55-74.

Prinz, J.P. 2005, «Lavoisier Experimental Method of his Research of Human Respiration», In *Lavoisier in Perspective*, edited by M. Beretta, Munchen: Deutsches Museum, 43-52.

Proverbio, E. 2007, «Sulle ricerche pneumatiche, sulla respirazione, circolazione e composizione del sangue, sulla salubrità dell'aria e sullo studio delle arie, su nuovi strumenti meteoroligici e sui primi strumenti a registrazione continua progettati e utilizzati a Milano nella seconda metà del settecento, come applicazione delle nuove scienze di chimica dei gas allà medicina sociale: il contributo di Pietro Moscat», *Atti de la Fondazione Giorgio Ronchi*, Anno LXII, 1, 3-144.

Reif-Acherman, S. 2012, «The Contributions of Henri Victor Regnault in the Context of Organic Chemistry of the First Half of the Nineteenth Century», *Quim. Nova*, 35, 2, 438-443.

Quintana-Marí, A., 1935, «Antoni de Martí i Franquès. Memòries originals. Estudi biogràfic i documental», *Memòries de l'Acadèmia de Ciències i Arts de Barcelona*, Tercera època, 24, Barcelona: Nebots de López Robert i Cª.

_____ . 1985, «Biografia desapassionada d'Antoni de Martí i Franquès», In *Miscel·lània Antoni de Martí i Franquès amb motiu de la commemoració del 150è aniversari de la seva mort, 1832 - 1982*, Tarragona: Excm. Ajuntament de Tarragona, 47-88.

Roberts, L. 1991, «A Word and the World. The Significance of Naming the Calorimeter», *Isis*, 82, 198-222.

Rocke, A.J. 1984, *Chemical Atomism in the Nineteenth Century. From Dalton to Cannizzaro*, Columbus, OH: Ohio University Press.

_____ . 1993, *The Quiet Revolution. Hermann Kolbe and the Science of Organic Chemistry*, Berkeley, Los Angeles, CA and London: University of California Press.

_____ . 2000, «Organic Analysis in Comparative Perspective: Liebig, Dumas, and Berzelius, 1811-1837», In *Instruments and Experimentation in the History of Chemistry*, edited by L.H. Holmes and T.H. Levere, Cambridge, MA and London: The MIT Press, 274-310.

_____ . 2005, «In Search of El Dorado: John Dalton and the Origins of the Atomic Theory», *Social Research*, 72, 1, 125-158.

Roscoe, H. E.; Harden, A. 1896, *A New View of the Origin of Dalton's Atomic Theory. A Contribution to Chemical History*, London and New York: Macmillan and Co.

Schaffer, S. 1990, «Measuring Virtue, Eudiometry, Enlightenment and Pneumàtic Chemistry», In *The Medical Enlightenment of the Eighteenth Century*, edited by A. Cunnigham and R. French, Cambridge: Cambridge University Press, 281-318.

Schofield, R.E. 1966, *A Scientific Autobiography of Joseph Priestley (1733-1804)*, Cambridge, MA and London: The MIT Press,

_____ . 2004, *The Enlightened Joseph Priestley. A Study of His Life and Work from 173 to 1804*, Pennsylvania, PA: The Pennsylvania State University Press.

Seitz, F. 2005, «Henry Cavendish; The Catalyst for the Chemical Revolution», *Notes & Records. The Royal Journal of the History of Science*, 59, 175-199.

Seligardi, R., 2000, «Volta and the Synthesis of Water: Some Reasons for a Missed Discovery», In *Nuova Voltiana. Studies on Volta and his Times*, edited by F. Bevilacqua and L. Fregonese, Vol. 2, 33-48. Pavia: Università degli Studi di Pavia; Milano: Editore Ulrico Hoepli, 5 vols.

Sibum, H.O. 1995, «Reworking the Mechanical Value of Heat: Instruments of Precision and Gestures of Accuracy in Early Victorian England», *Studies in History and Philosophy of Science*, 26, 1, 73-106.

Smeaton, W.A. 1966, «The Portable Chemical Laboratories of Guyton de Morveau, Cronstedt and Göttling», *Ambix*, 13, 84-91.

_____ . 1986, «Carl Wilhelm Scheele (1742–1786)», *Endeavour*, 10, 1, 28-30.

_____ . 1992, «Carl Wilhelm Scheele (1742–1786): provincial Swedish pharmacist and world-famous chemist», *Endeavour*, 16, 3, 128-130.

_____ . 2000, «Platinum and Ground Glass: Some Innovations in Chemical Apparatus by Guyton de Morveau and Others», In *Instruments and Experimentation in the History of Chemistry*, edited by L.H. Holmes and T.H. Levere, Cambridge, MA and London: The MIT Press, 211-237.

Stefani, M. 2010, «Le vicende editoriali delle memorie sulla respirazione animale di Lazzaro Spallanzani», *Edizione nazionale delle opere di Lazzaro Spallanzani*, Parte quinta, Vol. 5, 35-54, Modena: Mucchi Editore, 6 parts.

Stewart, L. 2008, «The Laboratory, the Workshop, and the Theater of Experiment», In *Science and Spectacle in the European Enlightenment*, edited by B. Bensaude-Vincent and C. Blondel, 11-24. London and New York: Routledge, Taylor and Francis Group.

Thackray, A. 1972, *John Dalton. Critical Assessments of His Life and Science*, Cambridge, MA: Harvard University Press.

Tartar, H. V. 1914, «The Reaction between Sulfur and Calcium Hydroxide in Aqueous Solution», *Journal of the American Chemical Society*, 36, 3, 495-498.

Taub, L. 2009, «Introduction: On Scientific Instruments», *Studies in History and Philosophy of Science*, 40, 337-343.

Troost, J. L. 1866, *Un laboratoire de chimie au dix-huitième siècle. Scheele. Conférence fait à la Sorbonne dans la soirée scientifique du 19 janvier 1866*, Paris: Étienne Giraud.

Turner, G. L'E. 1973, «Descriptive Catalogue of van Marum's Scientific Instruments in Teyler's Museum», In: *Martinus van Marum. Life and Work*, edited by E. Lefebvre and J.G. Bruijn, Vol. 4, Part II, 127-396. Leyden: Noordhoff International Publishing, 6 vols.

Usselman, M. C. 2000, «Multiple Combining Proportions: The Experimental Evidence», In *Instruments and Experimentation in the History of Chemistry*, edited by L.H. Holmes and T.H. Levere, Cambridge, MA and London: The MIT Press, 243-272.

_____ . 2003, «Liebig's Alkaloid Analyses: the Uncertain Route from Elemental Content to Molecular Formulae», *Ambix*, 50, 1, 71-89.

Usselman, M. C.; Reinhart, C.; Foulser,K. 2005, «Restaging Liebig: A Study in the Replication of Experiments», *Annals of Science*, 62, 1, 1-55.

Usselman, M.C.; Leaist, D.G.; Watson, K.D. 2008, «Dalton's Disputed Nitric Oxide Experiments and the Origins of His Atomic Theory», *ChemPhysChem*, 9, 106-110.

Van Helden, A., Hankins, T.L. 1994, «Introduction: Instruments in the History of Science», *Instruments, Osiris*, 9, 1-6.

Van Slyke, L. L.; Bosworth, A. W.; Hedges, C. C. 1910, «Chemical Investigation of the Best Conditions for Making the Lime-Sulphur Wash», *New York Agricultural Experiment Station Bulletin*, 329, December.

Warner, D. 1990, «What is a Scientific Instrument, When Did It Become One, and Why?», *British Journal for the History of Science*, 23, 83-93.

Watermann, R. 1968, «Eudiometrie, 1772-1805», *Technikgeschichte*, 35, 4, 293-319.

Weindling, P.J. 1982, «A Platinum Gift to King George III. A Gesture by William Haseldine Pepys, Cutler and Instrument Maker», *Platinum Metals Review*, 26, 1, 34-37.

Wilson, G. 1851, *The Life of the Honourable Henry Cavendish*, London: Harrison and Son.

Index

A

abridged procedure, 72, 104, 108, 112, 116, 299
abridged version, 90, 104, 155
absorbents, 215, 275, 287, 289
Académie des Sciences, 42, 63, 131, 137, 172, 186, 285
accuracy, xxviii, 47, 60, 74, 85, 92, 95, 101, 105, 111, 116, 128, 152, 167, 176, 188, 205, 222, 231, 241, 253, 262, 272, 282, 288, 296, 301
Achard, 119, 132, 142, 159, 181, 199, 300, 303, 318
Adams, 74, 110
aerial acid, 48, 145
aerial medicine, 20
affinity, 147, 183
affordability, 88, 301
agricultural chemistry, 201
air measure, 27, 69, 74, 79, 112, 117, 298
algebraic approach, 275, 294
algebraization, 275, 277
alkaline sulphides, 103, 142, 145, 152, 157, 182, 188, 199, 212, 218, 243, 301
Ampichel, 105
analytical power, 245, 253
Anforni, 134
animal respiration, 3, 98, 115, 167, 171, 191, 196, 213, 278, 285, 296, 303, 304, 308
Animal respiration, 198
apprenticeship, 118, 295
artificial illumination, 259, 296, 306
artisans, xxiii, 66
authority, 65, 110, 298

B

balloon ascent, 242
Banks, 81, 114
Beccaria, 37
Beddoes, 214, 219
Bérard, 238, 247, 256
Bergman, 10, 100
Berthollet, 63, 95, 154, 155, 163, 179, 182, 184, 188, 198, 202, 210, 221, 230, 242, 246, 249, 255, 294, 301, 309
Biot, 159, 163, 242
Black, 219, 227
blast furnace, 268, 290
Boerhaave, xxvi, 234
Bonaparte, 181, 201
Bonpland, 213
Boyle, 2, 91, 125
Brugnatelli, 191
bubbles, 46, 52, 59, 80, 81, 94, 110, 176, 274, 294
Bunsen, 267, 278, 284, 286, 295, 297, 300, 306, 308

C

cabinet, xxvi, xxviii, xxx, 73, 114, 313
calcination, xxiv, 91, 97, 138, 145, 169, 170, 306, 308
calcium sulphide, 145, 158, 160, 164, 218, 221, 228, 235, 243, 253, 258, 297, 301, 302
calibration, 60, 62, 79, 293
caloric, 174, 183
calorimeter, 98, 171, 307
Campi, 39, 48
carbon dioxide, 15, 135, 137, 159, 162, 196, 223, 226, 255, 303, 307

Q

R

S